Other McGraw-Hill Communications Books of Interest

In order to receive additional information on these or any other
McGraw-Hill titles, in the United States please call 1-800-722-4766.
Or visit us at www.ee.mcgraw-hill.com. In other countries, contact
your local McGraw-Hill representative.

Practical Cellular and PCS Design

Clint Smith, P.E.

McGraw-Hill

New York San Francisco Washington, D.C. Auckland Bogotá
Caracas Lisbon London Madrid Mexico City Milan
Montreal New Delhi San Juan Singapore
Sydney Tokyo Toronto

Library of Congress Cataloging-in-Publication Data

Smith, Clint
   Practical cellular and PCS design / Clint Smith.
       p. cm.
   Includes index.
   ISBN 0-07-059287-X
   1. Wireless communication systems.   2. Personal communication
service systems.   I. Title.
TK5103.2.S6524   1997
621.3845—dc21

                                              97-33009
                                                 CIP

McGraw-Hill

A Division of The McGraw-Hill Companies

1 2 3 4 5 6 7 8 9 DOC/DOC 9 0 2 1 0 9 8 7

ISBN 0-07-059287-X

*The sponsoring editor of this book was Steve Chapman. The editing supervisor was
Ruth W. Mannino, and the production supervisor was Tina Cameron. It was set in
New Century Schoolbook per the MHT design by Joanne Morbit of McGraw-Hill's
Professional Book Group composition unit, Hightstown, New Jersey.*

Printed and bound by R.R. Donnelley & Sons Company.

McGraw-Hill books are available at special quantity discounts to use as premiums
and sales promotions, or for use in corporate training programs. For more information,
please write to the Director of Special Sales, McGraw-Hill, 11 West 19th Street,
New York, NY 10011. Or contact your local bookstore.

This book is printed on recycled, acid-free paper containing a
minimum of 50% recycled de-inked fiber.

To Sam, Mary, and Rose, whose continued support made this effort possible ... and to all my colleagues who continue to provide technical support at those critical junctures of dire need.

Contents

Preface

The wireless industry is moving extremely fast. The pace that the industry is moving at has not slowed down for more than a decade and shows no sign of relaxing. It continues to be one of, if not the most, exciting field to be involved in. From many aspects, wireless applications are just beginning to emerge. The rapid expansion of operators in many markets has placed a premium on finding good skilled radio engineers for manufacturers and operators alike.

Fortunately or unfortunately, depending on your viewpoint, this has lead to many opportunities for those skilled in radio engineering. To compensate for the acute shortage of qualified RF engineers, many programs have been implemented both in industry and in academia to help provide the skilled RF engineers needed for the industry as a whole. The number of new engineers entering the wireless industry has been a mainly positive experience. However, there remains a lack of experienced RF engineers who can act as mentors to the new engineers. One of the best methods for helping both novice and experienced engineers is to provide sound documentation for the theory and practical aspects of project management.

This objective behind writing this book was to compile the vast amount of practical information needed by a radio engineer working in the wireless industry. All too often I found myself utilizing multiple references for projects I have been involved with, because a single reference for most of the material was not available. Therefore, the book combines in one key source information pertaining to basic RF engineering issues that are important to practicing engineers and their managers engaged in the wireless industry.

This book endeavors to cover RF engineering issues with application examples, showing where an approach is applied and how it can be utilized to enhance the design and performance of your wireless system. The book can and should be utilized by cellular, PCS, and other wireless engineers, both experienced and novice, in their efforts to understand and refine radio frequency engineering concepts and issues. The examples included are for the purpose of reinforcing the topics presented in the chapters with descriptions on how one could utilize the information.

The book is structured so that the major building blocks of a radio communication system are covered, starting with a overview of systems and then progressing into the transport mechanism utilized for conveying the information content desired. The topics discussed provide a convenient source of information that is well structured and contains the necessary breadth and depth for the RF engineer engaged in initial system design, expansion of an existing system, or improvement of the performance of a site or system. Several chapters are devoted to the topic of cell sites, the RF design process, and technology-specific examples other than radio engineering issues. But more specifically, the concepts and examples presented here can be utilized for any technology platform involving radio frequencies, whether they are in HF, VHF, UHF, or microwave. The book also builds on the work presented in *Cellular System Design and Optimization*, which primarily focused on what you need to do after a system goes into commercial service.

Simply put, this book focuses on practical cellular and PCS design as it pertains to the RF engineer. Each of its 11 chapters covers a specific topic associated with RF engineering. The appendixes include a traffic table, antenna and feed-line sweeping procedures, and a trigonometric identity table that can be used for multiple purposes.

Chapter 1 introduces the general aspects of a wireless system. The chapter is basically an overview of the general network and RF engineering as it pertains to cellular and PCS systems.

Chapter 2 covers more basic elements of radio engineering and how they apply to any radio system, regardless of the frequency band of operation or the technology utilized. This chapter provides a brief overview of various technology platorms, focusing on several of the more prevalent platforms in use today.

Chapter 3 discusses link budgets—what they are, how they apply to a communication system, and how they can be formulated. A discussion of propagation models is also included with several field expedient methods presented to help facilitate the initial analysis phase. In addition, a proposed link budget report that should be part of the engineering documentation for any system is presented.

Chapter 4 covers some of the forms of modulation techniques utilized in wireless communication. Here the intention is to highlight some of the different forms of communication formats used today in wireless communications. The generation of each modulation technique is discussed, and brief examples of their potential applications are presented.

Chapter 5 focuses on the antennas and antenna systems for a wireless network, including the different types of antennas, where they are utilized, and some practical considerations as to the selection of an antenna for a particular system. Also discussed are some critical performance criteria for determining the type and manufacturer of an antenna system.

Chapter 6 covers the important topic of filters and how they are utilized in a communication system. It discusses several types of filters, focusing on their practical use. Included are some critical performance criteria that need to be factored into the filter selection process.

Chapter 7 discusses the various aspects of a radio receiver and how they apply to its performance. Some of the receiver types that are used are discussed in detail. Also included here are the performance criteria for evaluating the radio receiver itself. The practical aspects associated with the receiver are those that are focused on here.

Chapter 8 covers the topic of transmitters, their type and application. Included here are techniques used for different combining methods and applications as well as the various component stages of transmit combing.

Chapter 9 discusses various issues associated with a wireless communication site. The topics of interest include macro-, micro-, and picocell designs. Additional topics involve rerads, and intermodulation. The utilization of tower top amplifiers is also discussed. The chapter also contains a communication site checklist that should be used in the commissioning portion of the sites design.

Chapter 10 covers the system design requirements for building a wireless system. Topics involve the process that an engineer should utilize for designing a wireless system from the ground up or for expanding an existing system and include the information needed to effectively design a network and some proposed output forms.

Chapter 11 covers specific design issues associated with CDMA, IS-136, PCS1900, GSM, and AMPS. It focuses on specific design rules for each of these technologies, including channel charts, frequency planning, and other pertinent information. Extensive coverage is given to CDMA, IS-136, and GSM.

In closing, I want to thank the many colleagues who have helped and encouraged this latest effort. I trust that you will find this book extremely helpful and will make it part of the few select books that you refer to on a constant basis.

Clint Smith, P.E.

1

Introduction

1.1 Introduction

The rapid expansion of wireless technology in the marketplace has resulted in a plethora of radio systems deployed over the last few years. Many new engineers have been thrust into the wireless arena with little or no practical experience. Currently a multitude of reference materials are available for an engineer to learn about various aspects of radio engineering. However, it is more common than not to have to rely on such a multitude of reference and other sources to pull all the pieces together.

An excellent source for fundamental radio engineering knowledge is the textbooks used in engineering communication classes. However, the material presented in them focuses on the theory and not the practical aspects needed for operating a wireless system. The practical guidelines, however, are what is desired and needed by the RF engineer in order to expeditiously resolve many complex issues in a short period of time.

1.2 History

The concept of cellular radio was initially developed by AT&T at their Bell Laboratories to provide additional radio capacity for a geographic customer service area. The initial mobile systems from which cellular evolved were called mobile telephone systems (MTS). Later improvements to these systems occurred, and the systems were referred to as improved mobile telephone systems (IMTS). One of the main problems with these systems was that a mobile call could not be transferred from one radio base station to another without loss of communication. This problem was resolved by implementing the concept of reusing the allocated frequencies of the system and handing off calls between base stations. Reusing the frequencies in cellular systems enables a market to offer higher radio traffic capacity. The increased radio traffic allows more users in a geographic service area than with the MTS or IMTS systems. A list of the current cellular system types is provided below.

Advanced mobile phone system (AMPS) standard.   AMPS is the cellular standard that was developed for use in North America. This type of system operates in the 800-MHz frequency band. AMPS systems have also been deployed in South America, Asia, and Russia.

Code-division multiple access (CDMA).   CDMA is an alternative digital cellular standard developed in the United States. CDMA utilizes the IS-95 standard and is being implemented as the additional generation for cellular systems. The CDMA system coexists with the current analog system.

D-AMPS.   Digital AMPS system, also called NADC (North American digital cellular) is the digital standard for cellular systems developed for use in the United States. Rather than develop a completely new standard the AMPS standard was developed into the D-AMPS digital standard. This was done to quickly provide a means to expand the existing analog systems that were growing at a rapid pace. NADC is designed to coexist with current cellular systems and relies on both the IS-54 and IS-136 standards.

Global system for mobile (GSM) communications.   GSM is the European standard for digital cellular systems operating in the 900-MHz band. This technology was developed out of the need for increased service capacity due to the analog system's limited growth. This technology offers international roaming, high speech quality, increased security, and the ability to develop advanced systems features. The development of this technology was completed by a consortium of 80 pan-European countries working together to provide integrated cellular systems across different borders and cultures.

Nordic mobile telephone (NMT) standard.   NMT is the cellular standard that was developed by the Nordic countries of Sweden, Denmark, Finland, and Norway in 1981. This type of system was designed to operate in the 450- and 900-MHz frequency bands. These are noted as NMT 450 and NMT 900. NMT systems have also been deployed throughout Europe, Asia, and Australia.

Total access communications systems (TACS).   TACS is a cellular standard that was derived from the AMPS technology. TACS systems operate in both the 800- and the 900-MHz band. The first system of this kind was implemented in England. Later these systems were installed in Europe, China, Hong Kong, Singapore, and the Middle East. A variation of this standard was implemented in Japan (JTACS).

Personal communication systems (PCS).   PCS is a general name given to systems that have recently developed out of the need for more capacity and design flexibility than that provided by the initial cellular systems. There are several PCS systems that an operator can possibly utilize. Some of the systems are listed below for quick reference.

DCS1800.   DCS1800 is a digital standard based upon the GSM technology with the exception that this type of system operates at a higher frequency range, 1800 MHz. The DCS1800 technology is intended for use in PCN-type systems. Systems of this type have been installed in Germany and England.

PCS1900.   This is the same as DCS1800 and is a GSM system. The only difference between the PCS1900, also called DCS1900, and DCS1800 and GSM is the frequency band of operation. PCS1900 operates in the PCS frequency band for the United States, 1900 MHz.

Personal digital cellular (PDC).   PDC is a digital cellular standard developed by Japan. PDC-type systems were designed to operate in the 800-MHz band and in the 1.5-GHz band.

IS-661.   This technology platform is being developed by Omnipoint and is a spread-spectrum technology that relies on time-division duplexing (TDD).

IS-136.   The PCS standard that relies on the NADC system except it operates in the 1900-MHz band.

CDMA.   Another popular PCS platform that utilizes the same standard as that for CDMA in cellular except it too operates in the 1900-MHz band.

1.3 Systems

In the United States cellular operators' geographic area of operation is broken into metropolitan statistical areas (MSAs) and rural statistical areas (RSAs). Each MSA and RSA has two different cellular operators that offer service. The two cellular operators are referred to as A-band and B-band systems. The A-band system is referred to as the nonwireline system and the B band is referred to as the wireline system for the MSA or RSA. Because of acquisition, the nonwireline and wireline references are no longer valid and simply A and B bands are used. Each cellular operator is given 25 MHz of spectrum to utilize, 12.5 MHz to transmit and 12.5 MHz to receive.

The PCS spectrum in the United States was made available through an action process set up by the Federal Communications Commission (FCC). The PCs band was broken into A, B, C, D, E, and F blocks. The A, B, and C blocks involved a total of 30 MHz while D, E, and F blocks are allocated 10 MHz. The spectrum allocation for both cellular and PCS in the United States is shown in Fig. 1.1.

It should be noted that the geographic boundaries for PCS licenses are different from those imposed on cellular operators in the United States. Specifically PCS licenses are defined as MTAs and BTAs. The MTA has several BTAs within its geographic region. A total of 93 MTAs and 487 BTAs are defined in the United States. Therefore, a total of 186 MTA licenses are awarded for the construction of a PCS network, each with a total of 30 MHz of spectrum to utilize. In addition a total of 1948 BTA licenses will be awarded in the United States. Of the BTA licenses the C band will have 30 MHz of spectrum while the D, E, and F blocks will have only 10 MHz available.

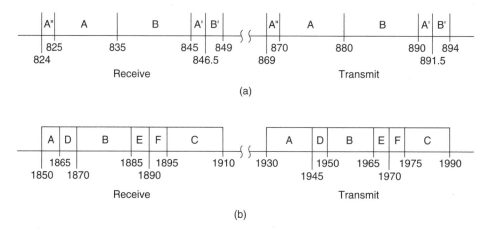

Figure 1.1 (*a*) Cellular and (*b*) personal communications spectrum (PCS) allocation in the United States.

1.4   Generic System Configuration

A generic system configuration is shown in Fig. 1.2. The configuration shown in Fig. 1.2 involves all the high-level system blocks of a cellular network. Many components make up each of the blocks shown in Fig. 1.2. Each of the individual system components of a cellular network is covered in a later chapter of this book.

Referring to Fig. 1.2 the mobile communicates to the cell site through use of radio transmissions. The radio transmissions utilize a full duplex configuration. The full duplex configuration involves having a separate transmit and receive frequency used by the mobile and cell site. The cell site transmits on the frequency the mobile unit is tuned to while the mobile unit transmits on the radio frequency the cell site receiver is tuned to.

The cell site acts as a conduit for the information transfer converting the radio energy into another medium. The cell site sends and receives information from the mobile and the mobile switching center (MSC). The MSC is connected to the cell site either by leased T1 lines or through a microwave system. The cellular system is made up of many cell sites which all interconnect back to the MSC, also referred to as MTSO (Mobile Telephone Switching Office) in some areas.

The MSC processes the call and connects the cell site radio link to the public service telephone network (PSTN). The MSC performs a variety of functions involved with call processing and is effectively the brains of the network. The MSC maintains the individual subscriber records, the current status of the subscribers, call routing, and billing information, to mention a few items.

1.5   Generic Cell Site Configuration

Figure 1.2 is an example of a generic cell site configuration. The cell site configuration shown in Fig. 1.3 is a picture of a monopole cell site. The monopole cell site has an equipment hut associated with it that houses the radio transmission equipment. The monopole, which is next to the equipment hut, supports the antennas used for the cell site at the very top of the monopole. The cable tray between the equipment hut and the monopole supports the coaxial cables which connect the antennas to the radio transmission equipment.

The radio transmission equipment used for a cellular base station, located in the equipment room, is shown in Fig. 1.4. The equipment room layout is a typical arrangement in a cell site. The cell site radio equipment consists of a base site controller (BSC), radio bay, and the amplifier (TX) bay. The cell site radio equipment is connected to the antenna interface frame (AIF) which provides the receive and transmit filtering. The AIF is then connected to the antennas on the monopole through use of the coaxial cables which are located next to the AIF bay.

The cell site is also connected to the MSC through the Telco or interconnect bay. The Telco bay provides either the T1 leased line or microwave radio link connection. The power for the cell site is secured through use of a power bay (rectifier) which converts ac electricity to dc. Batteries are used in the cell site in the event of a power disruption to ensure that the cell site continues to operate until

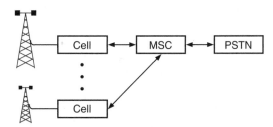

Figure 1.2 Generic wireless system. MSC = mobile switching center. PSTN = public service telephone network.

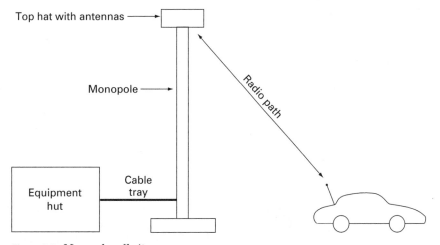

Figure 1.3 Monopole cell site.

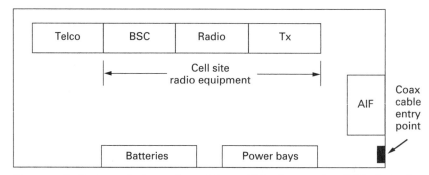

Figure 1.4 Radio transmission equipment for a cellular base station. BSC = base site controller. Tx = amplifier. AIF = antenna interface. Telco = T1, μ-wave interconnect.

power is restored or the batteries are exhausted. In some installations, generators are also used to augment the backup power system.

1.6 Handoffs

The handoff concept is one of the fundamental principles of this technology. Handoffs enable cellular to operate at lower power levels and provide high capacity. A multitude of algorithms are invoked for the generation and processing of a handoff request and eventual handoff order. The individual algorithms are dependent upon the individual vendor for the network infrastructure and the software loads utilized.

Handing off from cell to cell is fundamentally the process of transferring the mobile unit that has a call in progress on a particular voice channel to another voice channel, all without interrupting the call. Handoffs can occur between adjacent cells or sectors of the same cell site. The need for a handoff is determined by the actual quality of the RF signal received from the mobile into the cell site. As the mobile traverses the wireless network, it is handed off from one cell site to another cell site, ensuring a quality call is maintained for the duration of the conversation.

Reference

1. Smith, Clint, and Curt Gervelis, *Cellular System Design and Optimization,* McGraw-Hill, New York, 1996.

2

Radio Engineering

2.1 Introduction

This chapter covers some of the basic fundamentals for radio engineering. It is meant to give a brief overview regarding some of the various technology platforms that exist today in the wireless industry. The chapter first describes the fundamental points associated with any wireless communication system. Then a brief listing of different technology platforms is covered. The chapter then moves into a discussion of some of the major technology platforms prevalent in the wireless industry.

2.2 Radio System

The fundamental building blocks of a communication system are shown in Fig. 2.1. The simplified drawing in Fig. 2.1 represents the major components in any communication system. The major components shown in Fig. 2.1 consist of an antenna, filters, receivers, transmitter, modulation, demodulation, and propagation. Each of the major components identified in Fig. 2.1 requires the RF engineer to consider all of the various perturbations in order to achieve the optimum design for the situation.

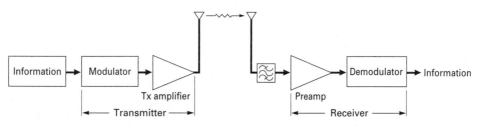

Figure 2.1 Basic radio system.

Entire books can be and are written on each section listed in Fig. 2.1. It is essential to know all the major components that actually make up the communication system that is being designed. For instance, in the design of the system if the subscriber units were simplex operated it would make little sense for a variety of reasons to design it for duplex operation. Therefore, knowing the design characteristics of each of the components is essential in building a communication system that will provide the proper transport functions for the information content.

2.2.1 Broadcast

The term broadcasting has many meanings depending on the individual engineer that it refers to. Typically broadcasting comprises the standard broadcast stations, AM from 535 to 1605 kHz, FM broadcast stations (from 88 to 108 MHz), and television broadcast stations (54 to 72, 74 to 88, 174 to 216, and 470 to 806 MHz).

A simplified drawing representing a broadcast system involves a single transmitter with multiple receivers in its coverage area (Fig. 2.2). Basically the transmitter sends the information and the receivers translate the information received into a usable format.

The separation between AM channels is only 10-kHz of spectrum. The AM station operation comes under Chapter 73 of the CFR 47 rules and regulations put forth by the FCC. An AM station is either omni or directional in its pattern. Typically an AM station will employ a directional array and the directional array consists of several towers all phased together to achieve the desired pat-

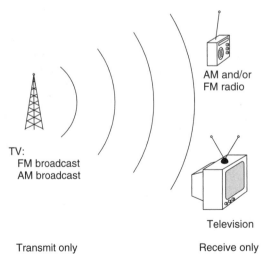

AM and/or
FM radio

TV:
FM broadcast
AM broadcast

Television

Transmit only   Receive only

Figure 2.2 Transmit-only and receive-only broadcast system.

tern. It is also common for each AM station to have both a daytime and night-time pattern.

FM broadcast stations are separated by 200 kHz of channel. The channels are assigned numbers, with channel 201 representing frequency 88.1 MHz. Frequency 107.9 MHz is channel 300, so the channel relationship is

$$[(\text{Channel no. } 201) \times 0.2 + 88.1]\,\text{MHz}$$

where channel no. is from 300 to 201

There are several different FM station classifications (A,B, and C). Each has a separate set of rules that it must adhere to. The rules for FM broadcast stations are found in Chapter 73 of CFR 47.

Television stations are allocated specific channels by the FCC and are geographically based. Depending on the channel that is granted by the FCC, different power levels are authorized to be transmitted. For example, Table 2.1 is a quick reference to maximum allowable power. It will come in handy for intermodulation and desense analysis later.

TABLE 2.1

Channel	Power
2–6	100 kW
7–13	316 kW
14–69	5000 kW

The individual TV channels and their respective frequencies are listed in Tables 2.2 and 2.3. Traditional paging systems can also be classified as a broadcast system.

2.2.2   Duplex and simplex

When most people think of radio communications they tend to think of two-way communication systems. In the two-way communication system both, or multiple, parties have the capability to converse and share information. The information exchange can be either analog, voice, or some form of data communication, or both.

There are basically two types of radio communication systems, simplex and duplex. Obviously this is a major generalization in that there are multiple variants to radio systems. However, for this example the basic two types of radio systems will be covered and are shown in Fig. 2.3. For the simplex system Fig. 2.3a shows that unit 1 sends information on f_1 and also receives information on f_1. The same can be said for unit 2. Simplex operation is a common practice for construction sites since it maximizes the available spectrum for others to utilize. Simplex operation in this case requires the other party to cease transmitting in order to get the reply message through. This is commonly referred to as "push to talk and release to listen."

TABLE 2.2   Television Channels and Frequencies

Channel	Frequency	Channel	Frequency	Channel	Frequency
2	54–60	25	542–548	48	674–680
3	60–66	26	548–554	49	680–686
4	66–72	27	554–560	50	686–692
5	76–82	28	560–566	51	692–698
6	82–88	29	560–556	52	698–704
7	174–180	30	556–572	53	704–710
8	180–186	31	572–578	54	710–716
9	186–192	32	578–584	55	716–722
10	192–198	33	584–590	56	722–728
11	198–204	34	590–596	57	728–734
12	204–210	35	596–602	58	734–740
13	210–216	36	602–608	59	740–746
14	470–476	37	608–614	60	746–752
15	476–488	38	614–620	61	752–758
16	488–494	39	620–626	62	758–764
17	494–500	40	626–632	63	764–770
18	500–506	41	632–638	64	770–776
19	506–512	42	638–644	65	776–782
20	512–518	43	644–650	66	782–788
21	518–524	44	650–656	67	788–794
22	524–530	45	656–662	68	794–800
23	530–536	46	662–668	69	800–806
24	536–542	47	668–674		

TABLE 2.3   Television Channels Allocated to Land Mobile Services

Channel	Frequency	Channel	Frequency
70	806–812	77	848–854
71	812–818	78	854–860
72	818–824	79	860–866
73	824–830	80	866–872
74	830–836	81	872–878
75	836–842	82	878–884
76	842–848	83	884–890

Figure 2.3b also shows a brief diagram of a duplex system. The duplex system can be either full duplex or half duplex. As is shown in the diagram, unit 1 transmits on f_1 and receives on f_2. The chief difference between duplex and simplex operation is the inclusion of a second channel for the communication. This of course comes at the expense of spectrum but increases the ease and overall throughput of the communication.

Mention was given to both full and half duplex operation. Full duplex operation involves being able to send and receive simultaneously throughout the entire communication linkup. An immediate example of full duplex operation is the telephone system that has become a staple in everyday life. Half duplex, as it implies, allows the user only to send or receive at any one time even though two channels are available for use.

2.3 Technology Platforms

Numerous technology platforms are available for use in wireless communication. Some of the formats are structured toward paging systems and the others toward voice/data mobile systems. The technology platform that is used has advantages and disadvantages that need to be looked at prior to their implementation. However, if the technology platform already exists in the system, it is equally important to understand the pros and cons for improving the overall system efficiency.

Tables 2.4 to 2.6 are meant to line up the different technology platforms for easy cross reference. The list of technologies in the tables is not all-inclusive but does cover a vast majority of the potential platforms.

Most technology platforms represent the digital radio category. Digital radio technology is currently being deployed in cellular with the attempt to increase the quality and capacity of today's cellular systems. The PCS operators, however, are deploying only digital formats.

In an analog cellular system the voice communication is digitized within the cell site itself for transport along the T-carrier to the MTSO. The voice representation and information transfer utilized in AMPS cellular is analog, and it is this part in the communication link that the effort is underway to convert to a digital platform; however, in most cases it is analog all the way.

The digital effort is meant to take advantages of many features and techniques that are currently not obtainable for analog cellular communication. Several competing digital techniques are being deployed in the cellular arena. The competing digital techniques fall into two primary categories, time-division and code-division multiple access (TDMA and CDMA) for cellular.

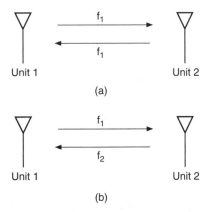

Figure 2.3 Radio communication systems. (*a*) Simplex system. (*b*) Duplex system.

TABLE 2.4

	AMPS	NADC	CDMA	GSM*
Frequency range, MHz	824–849	824–849	824–849	890–915
	869–894	869–894	869–894	935–960
Technology	FM	TDMA	CDMA	TDMA
Modulation	FM	π/4 DQPSK	QPSK	0.3 GMSK
Radio channel spacing	30 kHz	30 kHz	1.23 MHz	200 kHz
Modulation data rate	NA	48.6 kbits/s	1.2288 Mbits/s	270.833 kbits/s
CODEC	NA	VSELP	CLEP	RELP-LTP
		8 kbits/s	8/13 kbits/s	13 kbits/s
Spectrum allocation	50 MHz	50 MHz	50 MHz	50 MHz
Number of channels	832	832	10	124
Users per channel	1	3	118	8
Standard	AMPS	IS-54	IS-95	GSM

*The GSM technology, in the frequency band listed, is not utilized in the United States.

TABLE 2.5

	TACS	GSM	NMT	C450	CT2	DECT
TX frequency						
Base	935–960	935–960	935–960	461–466		
Mobile	890–915	890–915	890–915	451–456		
Technology	FM	TDMA	FM	FM	TDMA	TDMA
Modulation	FM	GMSK	FM	FM	GFSK	GMSK
Channel bandwidth, kHz	25	200	12.5	20 (b)	100	1730
				10 (m)		
Modulation data rate	NA	270.833 kbits/s	NA	NA	72 kbits/s	1152 kbits/s
CODEC	NA	RELP-LTD	NA	NA	ADPCM	ADPCM
		13 kbits/s				
Spectrum allocation	50 MHz	50 MHz	50 MHz	10 MHz		
Users per channel	1	8	1	1	1	12
Number of channels	1000	125	1999	222		
				444		

TABLE 2.6

	CDMA	DCS 1900	IS-136	PCS 2000	DCS 1800
TX frequency					
Base	869–894/	1930–1990	869–894/	1930–1990	1805–1880
	1930–1990		1930–1990		
Mobile	824–849/	1850–1910	824–849/	1850–1910	1710–1785
	1850–1910		1850–1910		
Technology	CDMA	TDMA	TDMA	TDD	TDMA
Modulation	QPSK	0.3 GMSK	π/4 DQPSK	QPSK	0.3 GMSK
Channel bandwidth, kHz	1.23 MHz	200 kHz	30 kHz	5 MHz	200 kHz
Modulation data rate	1.2288 Mbits/s	270.833 kbits/s			270.833 kbits/s
CODEC	CELP	RELP-LTP	VSELP	CELP	RELP-LTD
Spectrum allocation	25,30,10 MHz	30 MHz/10 MHz	25,30,10 MHz	10/30 MHz	150 MHz
Users per channel	64	8	3	64	8
Standard	CDMA	GSM	IS–136	IS–661	GSM

PCS, however, currently has three major competing technologies, CDMA, GSM, and IS-661 in the United States. Which technology platform is best depends on the application desired, and at present each platform has its pros and cons. Tables 2.4 to 2.6 represent some of the different technology platforms in the cellular band.

The major benefits associated with utilizing digital radios for a cellular environment involve

- Increased capacity over analog
- Reduced capital infrastructure costs
- Reduced capital per subscriber cost
- Reduced cellular fraud
- Improved user features
- Improvement in customer perceived performance
- Encryption
- High level of privacy

Figure 2.4 is a block diagram representation of the differences between analog and digital radios. Reviewing the digital radio portion of the diagram, the initial information content, usually voice, is input into the microphone of the transmission section. The speech then is processed in a vocoder which converts the audio information into a data stream utilizing a coding scheme to minimize the amount of data bits required to represent the audio. The digitized data then go to a channel coder which takes the vocoder data and encodes the information even more so it will be possible for the receiver to reconstruct the desired message. The channel-coded information is then modulated onto an RF carrier utilizing one of the several modulation formats referred to above. The modulated RF carrier is then amplified, passes through a filter, and is transmitted out an antenna.

The receiver, at some distance away from the transmitter, receives the modulated RF carrier through use of the antenna which then passes the information though a filter and into a preamp. The modulated RF carrier is then downconverted in the digital demodulator section of the receiver to an appropriate intermediate frequency. The demodulated information is then sent to a channel decoder which performs the inverse of the channel coder in the transmitter. The digital information is then sent to a vocoder for voice information reconstruction. The vocoder converts the digital format into an analog format which is passed to an audio amplifier connected to a speaker for the user at the other end of the communication path to listen to the message sent.

2.4   Spread Spectrum

Spread spectrum is used to describe a variety of different technology platforms in wireless communications. A spread-spectrum system is any communication

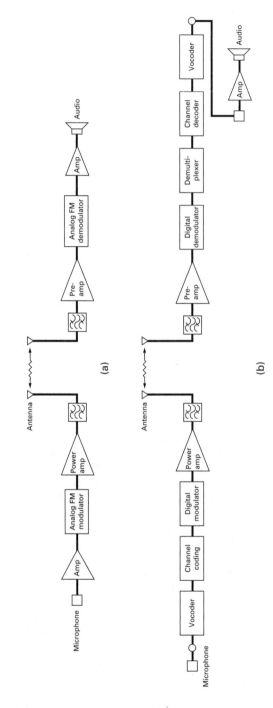

Figure 2.4 (a) Analog and (b) digital radio.

system that deliberately will occupy more channel bandwidth than the minimum required for data transfer. The rationale behind utilizing spread spectrum is to gain an improvement in the signal-to-noise ratio of the communication system itself. Three basic types of spread-spectrum formats are utilized with many perturbations. The three basic spread-spectrum types are coder-division multiple access (CDMA), time-division multiple access (TDMA), and frequency-division multiple access (FDMA). An example of an FDMA system in wireless is analog cellular.

2.5   Conventional Radio (Two-Way)

Conventional radio, the most common form of communication, normally operates in the VHF 25- to 175-MHz or UHF 400- to 500-MHz frequency ranges. Conventional radio systems usually operate in geographically defined areas and are primarily used for voice communication. Two-way radio operations can either be a point-to-point system, utilize repeaters, or be dispatch-oriented. It is also not uncommon for a two-way system to take on all these methods in order to exploit the most economic worth from the system itself. Obviously there are practical limits to how many different configurations are possible from an economic point of view. Also there are certain rules and regulations that two-way operations must adhere to. They are covered and discussed in part 90 of CFR 47.

Figure 2.5*a* is an example of a conventional radio system that is simplex in operation. Figure 2.5*b*, however, reflects a radio system that is either full or half duplex in nature. Please note that the communication can be from base to mobile or mobile to mobile in either case.

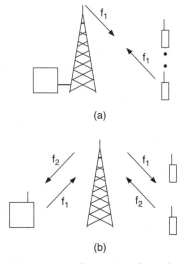

Figure 2.5 (*a*) Conventional simplex and (*b*) full or half-duplex radio systems.

An obvious issue comes to mind: If several people utilize the same radio channel, as is the case in all wireless operations, how is the discrimination process done? Specifically, in the discrimination process the subscriber unit will unsquelch its receiver only when the proper side tone is received. The side tone can be either analog or digital in nature. The side tone, often referred to as PL, serves a similar function to SAT in cellular communication.

The analog PL tones are shown in Table 2.7. The digital codes can range from 1 to 255. For the digital or analog PL method to work, each unit in the fleet needs to be programmed to the same code. The obvious drawback with the systems shown in Fig. 2.5 is that when the subscriber exits the coverage area of the initial cell site it no longer can communicate with the dispatch station.

To increase the capacity of a system and extend its coverage area several methods have been utilized. One method that is used to extend the effective coverage area of a system is to deploy multiple stations in a given area (Fig. 2.6). Each station has a defined set of radio channels in it such that the channels in the adjacent station will be different for interference reasons. For this system each subscriber unit or group of units is assigned a unique identification number and is referred to as the user's call control ID (CCID). Each subscriber unit associated with a CCID is assigned a home channel. When the subscriber unit is in the primary cell's coverage area (No. 1), it accesses the channel that it is assigned to for the individual cell. The subscriber unit will camp on the channel and await a page to unsquelch its receiver.

When the subscriber unit wishes to gain access to a radio channel, it first tries to gain access to the primary channel. However, if the channel is busy, being

TABLE 2.7   CTSS Tones

Group A		Group B		Group C	
Code*	Frequency, Hz	Code*	Frequency, Hz	Code*	Frequency, Hz
XZ	87.0	XA	71.9	YA	85
XB	77	YZ	82.5	ZZ	91
YB	88.5	ZA	94.8		
1Z	100.0	1A	103.5		
1B	107.2	2X	110.9		
2A	114.4	2B	118.8		
3Z	123	3A	127.3		
3B	131.8	4Z	136.5		
4A	141.3	4B	146.2		
5Z	151.4	5A	156.7		
5B	162.2	6Z	167.9		
6A	173.8	6B	179.9		
7Z	186.2	7A	192.8		
M1	203.5	M2	210.7		
M3	218.1	M4	225.7		
	233.1		241.8		
	250.3				

*Genave/NRC.

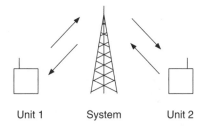

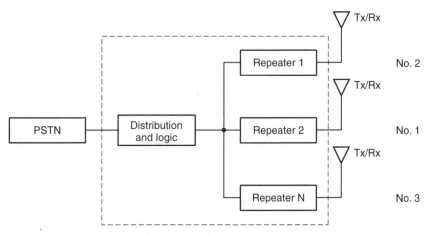

Figure 2.6 Deployment of multiple stations for more effective area coverage.

used by someone else at the time, it is directed to go to another channel in cell 1 and gain access that way. When the subscriber unit dekeys, the system tries to assign it to the primary channel again for the next burst of information. Next when the subscriber unit is, say, in cell 2's area it registers with the system and is assigned to utilize a channel other than its home channel.

The beauty of this type of system lies in the fact that the control information to the subscriber unit does not exhaust a radio channel, as trunking does. However, trunking systems allow for more capacity in the network. Trunked radios have up to 28 channels associated with each and normally operate in the UHF 800- and 900-MHz bands. Whenever a subscriber unit presses the push to talk button, the system searches the available channels and assigns the subscriber unit to it. A different radio channel can and often does get assigned to the subscriber unit for every time it depresses the push to talk button.

2.6   Paging

Paging is a technology platform that has many useful purposes and continues to provide a very cost-effective method for delivering a message to an individual. Until recently paging was a one-way, broadcast-type system where the sender was never really sure if the intended party really received the message.

The recent auctions that took place for narrowband PCS have made it possible for two-way (acknowledge paging) to become a reality. However, the majority of pagers and paging systems in operation today are traditional one-way paging.

Figure 2.7 is a simplified one-way paging system, and Figure 2.8 is a simplified two way paging system.

The subscriber unit in paging systems is the pager. Pagers are essentially FM receivers tuned to a specific frequency. Each pager is able to decode the information that is sent from the transmitter so it can recognize when its individual CAP code is sent. The CAP code is unique to each paging subscriber and is similar in nature to the MIN for cellular. Sometimes the same CAP code is assigned to a group of users who wish to receive the same information.

The pager's primary function is to alert the user of a message. This is done by either a tone, physical vibration, visual methods, or a combination of all three.

There are two general types of paging systems from the operator and user perspective, manual and automatic. In a manual system a user dials a number and converses with an operator who works with the paging company either as an employee or as a member of an adjunct service provider for the paging company. The operator then enters the information into the paging network, encoding it, and monitors the paging channel for activity. When there is no activity, in the case of a shared frequency, the operator keys the transmitter and the information is sent out into the network.

In an automatic system a telephone interconnect system answers the phone automatically. When the desired information has been delivered, the system goes on hook and terminates the call. When accessing the system the paging terminal also prompts the user for various inputs. The database for the paging system is also queried for the user's CAP code and current status; that is, have they paid their bills?

The system then encodes the information into analog or binary codes for the pager and sends it to the transmitter. The transmitter control monitors the channel, for a shared channel, keys the transmitter, and sends the information out into the network.

The flow diagram for a one-way paging system message is shown in Fig. 2.9. The diagram does not include repages that the operator often employs to ensure that the subscriber unit does in fact receive the desired message.

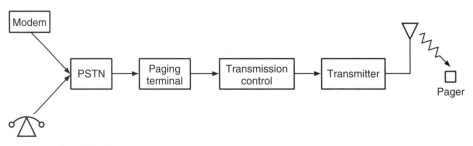

Figure 2.7 Simplified one-way paging system.

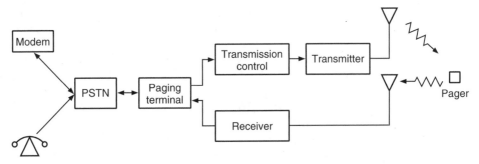

Figure 2.8   Simplified two-way paging system.

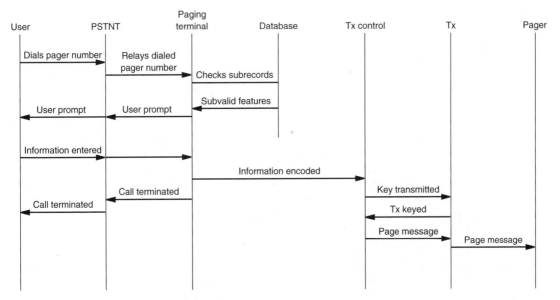

Figure 2.9   Generalized one-way page flow.

The conventional paging sequence involves the following:

1. The caller dials a number associated with a pager, say, 1-800-652-2594. The caller is actually dialing into the paging terminal for the paging network.

2. A voice prompt or tone is then heard by the caller.

3. The caller then enters the number that is desired to be called back and presses the # sign.

4. The paging terminal then converts the message into a code for the pager and sends the message to the various transmitters in the system.

5. The transmitters then send the information (code) by transmitting the information over the air.

6. The paging code is then received by all the pagers in the network, but only the pager with the proper code alerts the user of the message that was sent.

The paging systems are run by a radio common carrier company, normally referred to as the RCC. Often the RCC utilizes resellers to promote their system under third-party names. In addition to different paging systems different paging services can be offered by any RCC. The basic paging services involve tone, numeric, alphanumeric, and global alphanumeric paging and of course two-way, or rather acknowledge, back paging. The RCCs through their resellers normally offer some level of adjust services for a subscriber to utilize. Some of the adjunct services include voice mail, fax on demand, broadcast messaging, and ethnic paging.

Paging systems utilize a variety of different paging protocols. Each of the protocols has advantages and disadvantages. Several common formats for one-way paging are GOLAY, POCAG, FLEX, APOC, and ERMES. For narrowband PCS involving two-way paging the formats involve FLEX, ReFLEX, and InFLEXion.

2.7   AMPS

AMPS, typically called cellular, was first introduced in 1978 and went into commercial service in 1983. The unique feature associated with cellular at the time was its ability to reuse frequencies at short distances and at the same time hand off from one cell to another while maintaining the conversation.

Cellular is based on the concept of frequency reuse, and this is determined by a few simple parameters. Since cellular is based on frequency modulation, the frequency reuse for a system is driven by its carrier-to-interference ratio C/I. The C/I level for a system has a direct relationship to how closely spaced the cell sites and thus the reusing cell sites are.

Currently there are many different types of frequency reuse patterns. Some of the reuse patterns deployed in cellular involve N = 4,7,8,9,12, and 21, to mention a few perturbations. The separation between reusing cell sites is determined through the D/R ratio, which can be computed using the simple relationship shown below.

D	N
3.46R	4
4.6R	7
6R	12
7.55R	19

The D/R ratio is directly influenced by the frequency reuse pattern as shown when the lower the reuse rate, i.e., 12 vs. 4, the larger the separation is between the reusing cell sites (see Fig. 2.10).

The spectrum that AMPS occupies in the United States is shown in Fig. 2.11. The spectrum indicates that the system is a full duplex system and the separation between transmit and receive frequency pairs is 45 MHz. Each of

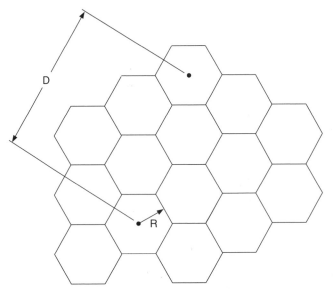

Figure 2.10 The D/R ratio. $D/R = \sqrt{\underset{3}{\wedge} N}$. N = reuse factor.

the channels in an AMPS cellular system occupies 30 kHz of spectrum. Obviously as the reuse distance between the sites increases the spectral efficiency decreases.

A typical cellular system (Fig. 2.12) can consist of omni, sector, or a combination of both types of cells within a given network. The choice between omni or sector is driven by the capacity requirements of the network so that the proper C/I ratio can be maintained, ensuring good call quality.

A typical $N = 12$ cell layout scheme is shown in Fig. 2.13, and an $N = 7$ pattern is shown in Fig. 2.14. The flow diagram in Fig. 2.15 represents the call processing flow for a mobile-to-land call, and Fig. 2.16 represents a mobile-to-mobile call.

2.8 TACS

TACS is a cellular system that is common outside of the United States. A TAC's system is very similar to that of an AMPS system with just a few variants. Specifically the variants involve the frequency of operation, as shown in Fig. 2.17.

The primary difference between AMPS and TACS is the data rate that is used by TACS. TACS uses 8 kHz while AMPS uses 10 kHz. This fundamental difference enables the TACS to have closer channel spacing than AMPS, 25 vs. 30 kHz, which means TACS appears to be more spectral-efficient. However, the lower data rate means signaling between the base station and the mobile is slower.

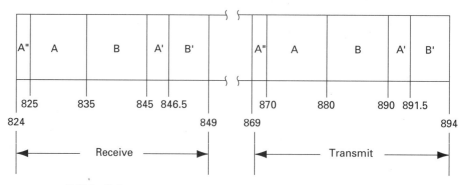

Figure 2.11 AMPS cellular spectrum.

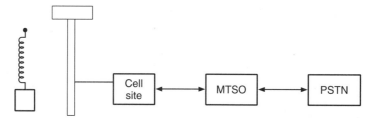

Figure 2.12 Typical cellular system.

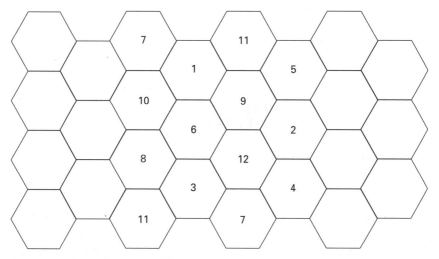

Figure 2.13 $N = 12$ frequency grid.

2.9 IS-95 Code-Division Multiple Access (CDMA)

Code-division multiple access (CDMA) is a spread-spectrum technology platform which enables multiple users to occupy the same radio channel (frequency spectrum) at the same time. CDMA has been and is being utilized for microwave point-to-point communication, for satellite communication, and also by the military. Both cellular and PCS operators are deploying CDMA

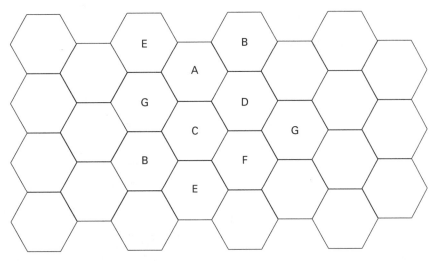

Figure 2.14 $N = 7$ frequency grid.

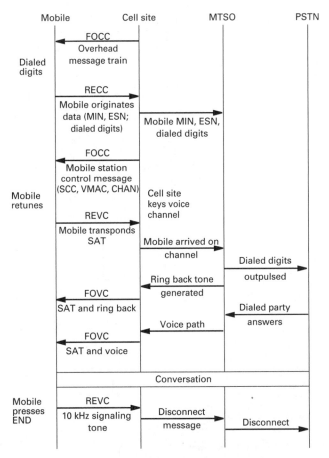

Figure 2.15 Call processing flow for a mobile-to-land call.

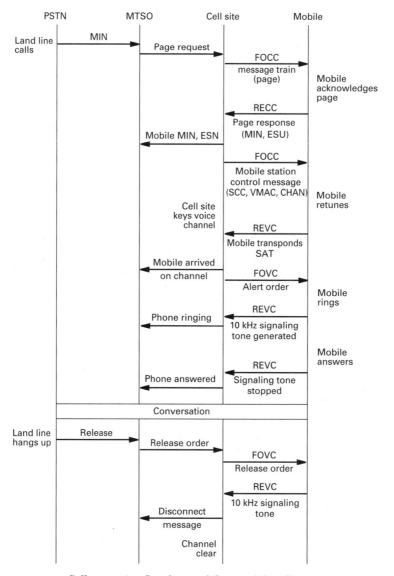

Figure 2.16 Call processing flow for a mobile-to-mobile call.

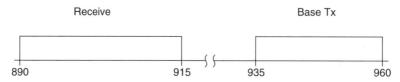

Figure 2.17 TACS cellular system spectrum.

technology using the standard IS-95. The IS-95 technology is the same for PCS as it is for cellular, with the primary difference being the frequency of operation.

With CDMA each of the subscribers, or users, utilizes a unique code to differentiate themselves from the other users. CDMA offers many unique features, including the ability to thwart interference and improved immunity to multipath effects due to its bandwidth.

The benefits associated with CDMA are:

Increased system capacity over analog and TDMA

Improved interference protection

No frequency planning required between CDMA channels

Improved handoffs due to soft handoffs

Fraud protection due to encryption and authentication

New wireless features accommodated

CDMA is based on the principle of direct sequence (DS) and is a wideband spread-spectrum technology. The CDMA channel utilized is reused in every cell of the system and is differentiated by the pseudo-random number (PN) code that it utilizes.

CDMA spreads the energy of the RF carrier as a direct function of the chip rate that the system operates at. The CDMA system utilizing the Qualcom technology has a chip rate of 1.228 MHz. The chip rate is the rate at which the initial data stream (original information) is encoded and then modulated. The chip rate is the data rate output of the PN generator of the CDMA system. A chip is simply a portion of the initial data or message which is encoded through use of a XOR process.

The receiving system also must despread the signal utilizing the exact same PN code sent through an XOR gate that the transmitter utilized in order to properly decode the initial signal. If the PN generator utilized by the receiver is different or is not in synchronization with the transmitter's PN generator, then the information being transmitted will never be properly received and will be unintelligible. Figure 2.18 represents a series of data that are encoded, transmitted, and then decoded back to the original data stream for the receiver to utilize.

The chip rate also has a direct effect on the spreading of the CDMA signal. Figure 2.19 shows a brief summary of the effects on spreading the original signal that the chip rate chosen has. The heart of CDMA lies in the point that the spreading of the initial information distributes the initial energy over a wide bandwidth. At the receiver the signal is despread through reversing the initial spreading process where the original signal is reconstructed for utilization. When the CDMA signal experiences interference in the band, the despreading process despreads the initial signal for use but at the same time spreads the interference so it minimizes its negative impact on the received information.

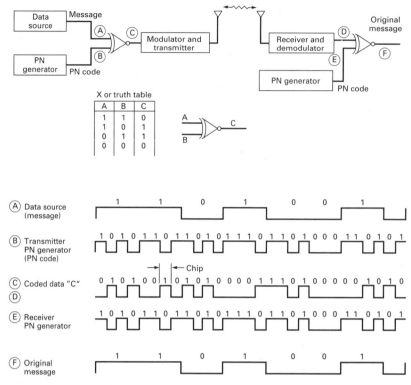

Figure 2.18 CDMA system.

The number of PN chips per data bit is referred to as the processing gain (Fig. 2.20) and is best represented by Eq. (2.1). Another way of referencing processing gain is the amount of jammer (interference) power that is reduced going through the despreading process. Processor gain is the improvement in the signal-to-noise ratio of a spread-spectrum system.

$$G_P = B_S/B_D \qquad (2.1)$$

2.9.1 Handoff

The handoff process for CDMA can take on several variants. The variants for handoffs in CDMA are soft handoff, softer handoff, and hard handoff. Each of the handoff scenarios is a result of the particular system configuration and of where the subscriber unit is in the network.

The handoff process begins when a mobile detects a pilot signal that is significantly stronger than any of the forward traffic channels assigned to it. When the mobile detects the stronger pilot channel, the following sequence should take place. The subscriber unit sends a pilot strength measurement message to the base station instructing it to initiate the handoff process. The cell site then sends a handoff direction message to the mobile unit directing

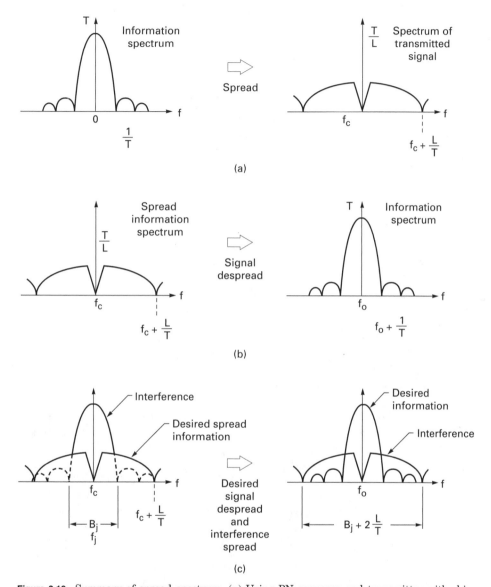

Figure 2.19 Summary of spread spectrum. (*a*) Using PN sequence and transmitter with chip (*PN*) duration of T/L. (*b*) Using correlation and a synchronized replica of the *pn* sequence at the receiver. (*c*) When interference is present. L/T = chip duration; f_j = jamming frequency; Bj = jammer's bandwidth.

it to perform the handoff. Upon the execution of the handoff direction message the mobile unit sends a handoff completion message on the new reverse traffic channel.

In CDMA a soft handoff involves an intercell handoff and is a make-before-break connection. The connection between the subscriber unit and the cell site is maintained by several cell sites during the process. Soft handoff can occur

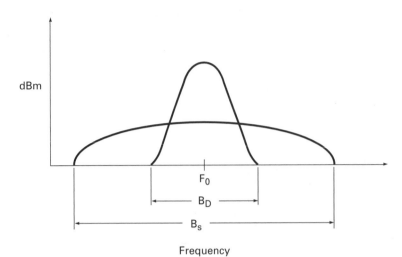

dBm

F_0

B_D

B_S

Frequency

Figure 2.20 Processor gain: $G_P = B_S/B_D$. B_D = bandwidth of initial signal. B_S = bandwidth of initial signal spread.

only when the old and new cell sites are operating on the same CDMA frequency channel.

The advantage of the soft handoff is path diversity for the forward and reverse traffic channels. Diversity on the reverse traffic channel results in less power being required by the mobile unit, reducing the overall interference, which increases the traffic-handling capacity.

The CDMA softer handoff is an intracell handoff between sectors of a cell site and is a make-before-break type. The softer handoff occurs only at the serving cell site.

The hard handoff process is meant to enable a subscriber unit to hand from a CDMA call to an analog call. The process is functionally a break-before-make and is implemented in areas where there is no longer CDMA service for the subscriber to utilize while on a current call. The continuity of the radio link is not maintained during the hard handoff.

A hard handoff can also occur between two distinct CDMA channels which are operating on different frequencies.

2.9.2 Forward CDMA channel

The forward CDMA channel (Fig. 2.21a) consists of the pilot channel, one sync channel, up to seven paging channels, and potentially 64 traffic channels. The cell site transmits the pilot and sync channels for the mobile to use when acquiring and synchronizing with the CDMA system. When this occurs the mobile is in the mobile station initiation state. The paging channel also transmitted by the cell site is used by the subscriber unit to monitor and receive messages that might be sent to it during the mobile station idle state or system access state.

The pilot channel is continuously transmitted by the cell site. Each cell site utilizes a time offset for the pilot channel to uniquely identify the forward CDMA channel to the mobile unit. There are a possible 512 different time offset values for the cell site to utilize. If multiple CDMA channels are assigned to a cell site the cell will still utilize only one time offset value. The time offset is utilized during the handoff process.

The sync channel is a forward channel that is used during the system acquisition phase. Once the mobile acquires the system it will not normally reuse the sync channel until it powers on again. The sync channel provides the mobile with the timing and system configuration information. The sync channel utilizes the same spreading code (time offset) as the pilot channel for the same cell site. The sync channel frame is the same length as the pilot PN sequence. The information sent on the sync channel is the paging channel rate and the time of the base stations pilot PN sequence with respect to the system time.

The cell site utilizes the paging channel to send overhead information and subscriber-specific information. The cell site will transmit at the minimum one paging channel for each supported CDMA channel that has a sync channel.

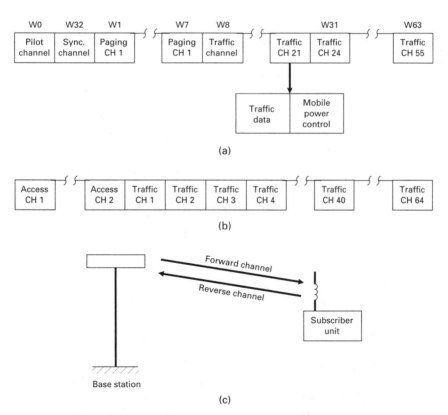

Figure 2.21 CDMA channels. (*a*) Forward channel. (*b*) Reverse channel. (*c*) Forward and reverse channel movement. *W* = Walsh codes.

Once the mobile has obtained the paging information from the sync channel the mobile will adjust its timing and begin monitoring the paging channel. Each mobile, however, monitors only a single paging channel. The paging channel conveys four basic types of information. The first set of information conveyed by the paging channel is the overhead information. The overhead information conveys the systems configuration by sending the system and access parameter messages, the neighbor lists, and CDMA channel list messages. Paging is another message type sent where a mobile unit is paged by the cell site for a land-to-mobile or mobile-to-mobile call. The channel assignments messages allow the base stations to assign a mobile to the traffic channel, alter the paging channel assignment, or redirect the mobile to utilize the analog FM system.

The forward traffic channel is used for the transmission of primary or signaling traffic to a specific subscriber unit during the duration of the call. The forward traffic channel also transmits the power control information on a subchannel continuously as part of the closed-loop system. The forward traffic channel will also support the transmission of information at 9600, 4800, or 1200 bits/s utilizing a variable rate which is selected on a frame-by-frame basis, but the modulation symbol rate remains constant.

2.9.3   Reverse CDMA channel

The cell site contiguously monitors the reverse access channel to receive any message that the subscriber unit might send to the cell site during the system access state. The reverse CDMA channel (Fig. 2.21b) consists of an access channel and the traffic channel. The access channel provides communication from the mobile to the cell site when the subscriber unit is not utilizing a traffic channel. One access channel is paired with a paging channel and each access channel has its own PN code. The mobile responds to the cell site's messages sent on the paging channel by utilizing the access channel.

The forward and reverse control channels utilize a similar control structure which can vary from 9600, 4800, 2400, or 1200 bits/s, which enables the cell or mobile to alter the channel rate dynamically to adjust for the speaker. When there is a pause in the speech the channel rate decreases to reduce the amount of energy received by the CDMA system increasing the overall system capacity.

There are four basic types of control messages on the traffic channel. The four messages involve messages that will control the call itself, handoff messages, power control, security, and authentication. CDMA power control is fundamentally different from that utilized for AMPS or IS-54. The primary difference is that the proper control of total power coming into the cell site, if limited properly, will increase the traffic-handling capability of that cell site. As more energy is received by the cell site, its traffic-handling capabilities will be reduced unless it is able to reduce the power coming into it.

The forward traffic power control is composed of two distinct parts. The first part, the cell site, will estimate the forward link's transmission loss utilizing the mobile subscribers' received power during the access process. Based on the estimated forward link path loss the cell site will adjust the initial digital gain for each of the traffic channels. The second part of the power control involves the cell site making periodic adjustments to the digital gain, which is done in concert with the subscriber unit.

The reverse traffic channel signals arriving at the cell site vary significantly and require a different algorithm from that of the forward traffic power control. The reverse channel also has two distinct elements used for making power adjustments. The first element is the open-loop estimate of the transmit power which is performed solely by the subscriber unit without any feedback from the cell site itself. The second element is the closed-loop correction for these errors in the estimation of the transmit power. The power control subchannel is continuously transmitted on the forward traffic channel every 1.25 ms instructing the mobile to either power up or power down, affecting its mean power output level. There are a total of 16 different power control positions.

2.10   NADC (IS-54 and IS-136)

The NADC network currently is using the IS-54 standard, but most operators are converting their networks to the IS-136 standard. PCS operators who have chosen the NADC system are deploying IS-136. Both IS-54 and IS-136 utilize time-division multiple access (TDMA) technology. TDMA is another form of spread-spectrum technology allowing multiple users to occupy the same frequency spectrum. TDMA technology allows multiple users to occupy the same channel through the use of time division.

The primary difference between IS-54 and IS-136 is that IS-136 is the evolution of the IS-54 specification. Utilization of IS-136 enables an operator to offer short messaging services and full use of digital control channels.

IS-54/136 utilizes the same channel bandwidth as does analog cellular, 30 kHz per physical radio channel. However, TDMA enables three and possibly six users to operate on the same physical radio channel at the same time. The TDMA channel presents a total of six time slots in the forward and reverse direction. TDMA at present utilizes two time slots per subscriber with the potential to go to half-rate vocoders which require the use of only one time slot per subscriber.

IS-136 TDMA has many advantages in its deployment in a cellular system:

Increased system capacity up to three times over analog

Improved protection for adjacent channel interference

Authentication

Voice privacy

Reduced infrastructure capital to deploy

Frequency plan integration over CDMA

Short message paging

Encryption

Digital control channel

Integrating TDMA into an existing cellular system can be done more easily than that for the deployment of CDMA. The use of TDMA in a network requires the use of a guardband to protect the analog system from the TDMA system. However, the guardband required consists of only a single channel on either side of the spectrum block allocated for TDMA use. Depending on the actual location of the TDMA channels in the operator's spectrum it is possible to require only one or no guardband channel.

The TDMA (IS-54 and IS-136) has the unique advantage of affording the implementation of digital technology into a network without elaborate engineering requirements. The implementation advantages mentioned for TDMA also facilitate the rapid deployment of this technology into an existing network.

The implementation of TDMA is further augmented by requiring only one channel per frequency group as part of the initial system offering. The advantage with requiring only one channel per sector in the initial deployment is the minimization of capacity reduction for the existing analog network. Another advantage with deploying one TDMA channel per sector is elimination of the need to preload the subscriber base with dual-mode (TDMA) handsets.

The TDMA system signaling is shown in Fig. 2.22. The format of the TDMA signal involves a total of six potential conversations taking place over the same bandwidth that one 30-kHz voice conversation utilizes for analog cellular. Presently full-rate vocoders are utilized by operators. The full-rate vocoder utilizes two time slots in both the forward and reverse links. The use of the full-rate vocoder put the limit at three to the amount of TDMA users that can occupy one physical channel. In Fig. 2.22 a subscriber utilizing a full-rate vocoder would occupy two time slots, A1 and A2, while a half-rate vocoder user would use only A1 or A2.

The time slots are paired in TDMA for a full-rate vocoder system. Conversation A would utilize time slots 1 and 4, conversation B would use slots 2 and 5, and conversation C would utilize time slots 3 and 6. The modulation scheme utilized by the NADC TDMA system is a $\pi/4$ DQPSK format. The C/I levels used for frequency management associated with TDMA are the same for analog, 17 dB C/I. However, many operators are now designing for a C/I of 20 dB.

Access to the TDMA system is achieved through either the primary control channel (utilized for analog communication) or the secondary dedicated control channel. During the initial acquisition phase the mobile reads the overhead control message from the primary control channel and determines if the system is digital capable. If the system is digital capable a decision will be made whether to utilize the primary or secondary dedicated control channel. The secondary dedicated control channels are assigned as FCC channels 696 to 716 for the A-band system and channels 717 through 737 for the B-band system. The use of the secondary dedicated control channels

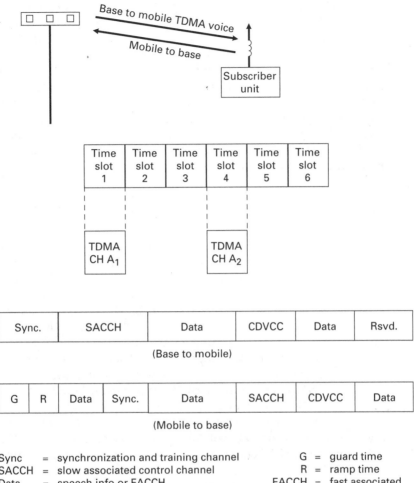

Sync   = synchronization and training channel
SACCH   = slow associated control channel
Data   = speech info or FACCH
CDVCC   = coded digital verification color code
Rsvd.   = reserved

G   = guard time
R   = ramp time
FACCH = fast associated
        control channel

Figure 2.22 TDMA channels.

enables a variety of enhanced features to be provided by the system operator to the subscribers.

One of the unique features associated with TDMA is the ability for a mobile-assisted handoff (MAHO). The MAHO process enables the mobile to constantly report back to the cell site indicating its present condition in the network. The cell site or MSC also is collecting data on the mobile through the reverse link measurements but the forward link, base to mobile, is being evaluated by the mobile itself, therefore providing critical information about the status of the call.

For the MAHO process the mobile measures the received signal strength level, (RSSI) received from the cell site. The mobile also performs a bit error rate test (BER) and a frame error rate test (FER) as another performance metric.

The mobile also measures the signals from a maximum of 12 potential digital handoff candidates utilizing either a dedicated control channel or a beacon channel. The channels utilized by the mobile for the MAHO process are provided by the serving cell site for the call. The dedicated control channel is either the primary or secondary control channel and the measurements are performed on the forward link. The mobile can also utilize a beacon channel for the performance measurement. The beacon channel is either a TDMA voice channel or an analog channel, both of which are transmitting continuously with no dynamic power control on the forward link. The beacon channel is utilized when the setup or control channel for the cell site has an omni configuration and not a dedicated setup channel per sector.

2.11   GSM

Group special mobile (GSM) is a technology platform that has gained worldwide acceptance and deployment. There are several systems, all of which are referred to as GSM. However, each of the different systems occupies a different frequency band. In addition to the frequency band differences the number of channels available for each service are different.

The three technology platforms that utilize GSM as the system standard are GSM, DCS1800 and DCS1900, or PCS1900. Each of the systems mentioned has the same features and functionality as the other with the exception that it operates at a different frequency band. The frequency band that the system operates in is dependent upon the country where the system is deployed. For instance, GSM operates in the 900-MHz range, DCS1800 in the 1800-MHz range, and DCS1900 or PCS1900 in the 1900-MHz range.

In the United States, GSM is referred to as PCS1900. The particular frequencies that the operator of this technology platform can utilize are dependent upon the PCS frequency block they won in the auction process put forth by the FCC.

The GSM system consists of a switching system, a base station system, an operations support system, and the subscriber units. Each of the functions listed above has several subcomponents that make up each major element of the network. A sample GSM system which can be applied to any of the three GSM systems is shown in Fig. 2.23.

The mobile switching center (MSC) performs the basic switching for the GSM network including call routing to and from mobile subscribers. The MSC is the heart of the GSM network and connects the GSM network with the PSTN and SS7 networks. The MSC can also be connected to a gateway or GMSC for interconnecting to other GSM networks. Several nodes are associated with the MSC, the home location register (HLR), the visitor location register (VLR), the equipment identify register (EIR), the authentication center (AUC), and the interworking unit (IWU).

The HLR is the main database used for managing the subscriber units that are resident to it. A typical GSM system may utilize more than one HLR depending on the size of the system itself. The general information stored in

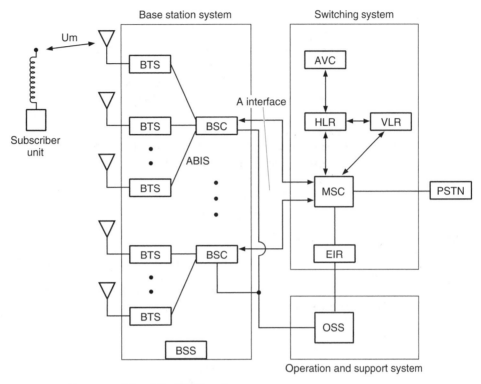

Figure 2.23   Group special mobile (GMS) system.

the HLR includes specific subscriber information that defines the services the subscriber is allowed to have. In addition, the HLR is used to track the subscriber in order to enable proper call routing to take place. In the case where the subscriber unit is not utilizing its home switch, the HLR notes which VLR the subscriber is a temporary resident to.

The VLR is a database similar to the HLR except it is a dynamic database which contains information about subscriber units utilizing a different MSC than where it is home to. It is common for several VLRs to exist in a system, since there are often several MSCs.

The VLR receives the subscriber information from the subscriber's HLR. The information received by the VLR from the HLR includes the subscriber services it is allowed to use. In addition, the VLR maintains the location information on where the subscriber is for proper call routing to take place.The EIR is used in the GSM network to identify stolen or not approved subscriber units. The EIR is used by the MSC to check the validity of the international mobile subscriber identity (IMSI). The EIR is not required to be operational in a GSM system. However, the EIR helps prevent stolen or not approved subscriber units from operating on the GSM network that uses the EIR.

The authentication center (AUC) is connected to the HLR. Its purpose is to provide the HLR with the subscriber authentication parameters and the ciphering keys used by the subscriber unit. The AUC is an integral part of the GSM system.

The interworking unit is a part of the MSC that provides all the necessary functions to allow correct interworking between the GSM network and other nonvoice services. Examples of nonvoice services the IWU is responsible for include asynchronous and synchronous data, fax services, and packet switched data for examples.

The base station system consists of the base station controller (BSC) and the base transceiver station (BTS). The BSC can either be colocated with a MSC or remote to the MSC depending on the infrastructure vendor chosen. The BSC connects the BTSs to the MSC. The number of BTSs that a BSC can have are again dependent upon the infrastructure vendor.

The BSC is the mobility manager for the GSM system. The BSC is responsible for controlling the BTSs and handovers. Power control is another task completed by the BSC.

The BTSs are the individual cell sites that contain radio elements. The BTS is designed to provide the radio link, both forward and reverse, between the subscriber unit and the base station itself. A BTS may be omni or directional and can contain several GSM radios per cell or sector.

2.12   CDPD

Cellular Digital Packet Data (CDPD) is a packetized data service utilizing its own air interface standard. The CDPD systems utilized by cellular operators are functionally a separate data communication service which physically shares the cell site and cellular spectrum.

CDPD has many applications but is most applicable for short bursty-type data applications and not for large file transfers. CDPD applications of short messages are E-mail, telemetry applications, credit card validation, and global positioning, to mention a few potentials.

CDPD does not establish a direct connection between host and server locations. Instead it relies on the OSI model for packet switching data communications which routes the packet data throughout the network. The CDPD network comprises various layers. Layer 1 is the physical layer, layer 2 is the data link itself, and layer 3 is the network portion of the architecture. CDPD utilizes an open architecture and has incorporated authentication and encryption technology into its air-link standard. The CDPD system consists of several major components. A block diagram of a CDPD system is shown in Fig. 2.24.

The mobile end system (MES) is a portable wireless computing device that moves around the CDPD network communicating to the MDBS. The MES is typically a laptop computer or other personal data device which has a cellular modem.

The mobile database station (MDBS) resides in the cell site itself and can utilize some of the same infrastructure that the cellular system does for transmitting and receiving packet data. The MDBS acts as the interface between the MES and the MDIS. One MDBS can control several physical radio chan-

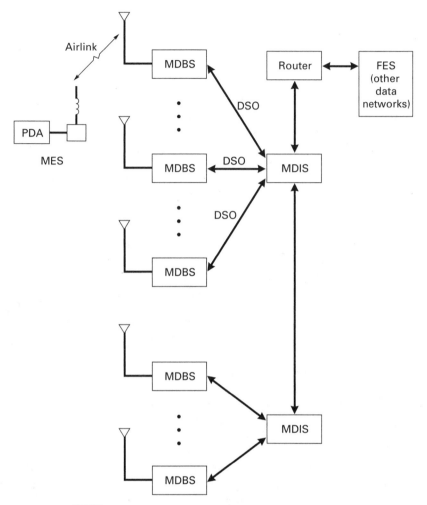

Figure 2.24 CDPD system.

nels depending on the site configuration and loading requirements. The MDBS communicates to the MDIS via a 56 kbits/s data link. Often the data link between the MDBS and MDIS utilizes the same facilities as that for the cellular system but occupies a dedicated time slot.

The mobile data intermediate system (MDIS) performs all the routing functions for CDPD. The MDIS performs the routing tasks utilizing the knowledge of where the MES is physically located within the network itself. Several MDISs can be networked together to expand a CDPD network.

The MDIS also is connected to a router or gateway which connects the MDIS to a fixed-end system (FES). The FES is a communication system that handles layer 4 transport functions and other higher layers.

The CDPD system utilizes a gaussian minimum-shift keying (GMSK) method of modulation and is able to transfer packetized data at a rate of 19.2 kbits/s over the 30-kHz-wide cellular channel. The frequency assignments for CDPD can take on two distinct forms. The first form of frequency assignment is a method of dedicating specific cellular radio channels to be utilized by the CDPD network for delivering the data service. The other method of frequency assignment for CDPD is to utilize channel hopping where the CDPD's mobile database station (MDBS) utilizes unused channels for delivering its packets of data. Both methods of frequency assignments have advantages and disadvantages.

Utilizing a dedicated channel assignment for CDPD has the advantage of not having the CDPD system interfere with the cellular system with which it shares the spectrum. By enabling the CDPD system to operate on its own set of dedicated channels there is no real interaction between the packet data network and the cellular voice network. However, the dedicated channel method reduces the overall capacity of the network, and depending on the system loading conditions this might not be a viable alternative.

If the method of channel hopping is utilized for CDPD, and this is part of the CDPD specification, the MDBS for that cell or sector will utilize idle channels for the transmission and reception of data packets. In the event the channel that is being used for packet data is assigned by the cellular system for a voice communication call, the CDPD MDBS detects the channel assignment and instructs the mobile end system (MES) to retune to another channel before it interferes with the cellular channel. The MDBS utilizes a scanning receiver or sniffer, which scans all the channels it is programmed to scan to determine which channels are idle and/or in use.

The disadvantage of the channel hopping method involves the potential interference problem to the cellular system. Coexisting on the same channels with the cellular system can create mobile-to-base station interference. The mobile-to-base station interference occurs because of the different handoff boundaries for CDPD and cellular for the same physical channel. The difference in handoff boundaries occurs largely because CDPD utilizes a BER for handoff determination and the cellular system utilizes RSSI at either the cell site, analog, or MAHO for digital.

References

1. Smith, Clint, and Curt Gervelis, *Cellular System Design and Optimization,* McGraw-Hill, New York, 1996.
2. *Reference Data for Radio Engineers,* 6th ed., Sams, 1983.
3. Rappaport, *Wireless Communications,* IEEE, 1996.
4. Hoffman, Wayne, *A New Breed of RF Components,* Glenayre, 1996.
5. Boucher, Neil, *The Paging Technology Handbook,* Quantum Publishing, 1992.
6. American Radio Relay League, *The ARRL 1986 Handbook,* 63 ed., The American Radio Relay League, Newington, CT, 1986.
7. DeRose, *The Wireless Data Handbook,* Quantum Publishing, Inc., Mendocino, CA, 1994.
8. Dixon, *Spread Spectrum Systems,* 2d ed., Wiley, New York, 1984.
9. Kaufman, M., and A. H. Seidman, *Handbook of Electronics Calculations,* 2d ed., McGraw-Hill, New York, 1988.
10. Lee, W. C. Y., *Mobile Cellular Telecommunications Systems,* 2d ed., McGraw-Hill, New York, 1996.
11. Qualcom, *An Overview of the Application of Code Division Multiple Access (CDMA) to Digital Cellular Systems and Personal Cellular Networks,* Qualcom, San Diego, CA, May 21, 1992.

12. Sklonik, M. I., *Introduction to Radar Systems,* 2d ed., McGraw-Hill, New York, 1980.
13. Fink, Donald, and Donald Christiansen, *Electronics Engineers Handbook,* 3d ed., McGraw-Hill, New York, 1989.
14. Rappaport, *Wireless Communications Principles and Practices,* IEEE, 1996.
15. Steele, *Mobile Radio Communications,* IEEE, 1992.
16. Jakes, W. C., *Microwave Mobile Communications,* IEEE, New York, 1974.
17. Carlson, A. B., *Communications Systems,* 2d ed., McGraw-Hill, New York, 1975.
18. Johnson, R. C., and H. Jasik, *Antenna Engineering Handbook,* 2d ed., McGraw-Hill, New York, 1984.
19. Lathi, *Modern Digital and Analog Communication Systems,* CBS College Printing, New York, 1983.
20. MacDonald, "The Cellular Concept," *Bell Systems Technical Journal,* vol. 58, no. 1, 1979.
21. Mouly, Pautet, *The GSM System for Mobile Communications,* Mouly Pautet, 1992.

3

Link Budget Topics

What exactly is a link budget? The link budget is a power budget that is one of the fundamental elements of radio system design. The link budget is the part of the system design where all the issues associated with propagation are included. Put simply, the link budget can be either forward- or reverse-oriented and must account for all the gains and losses that the radio wave will experience as it goes from the transmitter to the receiver.

The link budget is the primary method that a RF engineer must determine in order to ascertain if a valid communication link can and does exist between the sender and the recipient of the information content. The link budget, however, incorporates many elements of the communication path. Unless the actual path loss is measured empirically, the RF engineer has to estimate or rather predict just how well the RF path itself will perform. The many elements involved in the communication path incorporate assumptions made regarding various path impairments.

Figure 3.1 shows what part of the radio communication path the link budget tries to account for. There are two portions to the link budget, up link and down link (Fig. 3.2). The up-link path is the path from the subscriber unit to the base station. The down-link path is the path from the base station to the subscriber unit. Both the up-link path and the down-link path are reciprocal,

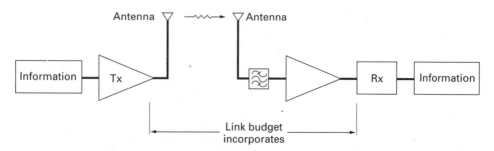

Figure 3.1 Link budget components.

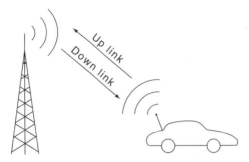

Figure 3.2   Up and down link.

provided they are close enough in frequency. However, the actual paths should be the same with the exception of a few key elements that are hardware-related. The actual path loss associated with the path the radio wave traverses from antenna to antenna is the same, however, whether it is up-link or down-link directed.

This chapter goes over the basic radio principle of RF propagation and the various issues that must be factored in for a proper design to take place. It will serve as a fundamental building block for determining the system requirements involving any communication system whether it is a one-way or two-way system. The radio principles covered in this chapter are not all-encompassing since the field of radio engineering is very vast. However, the intention here is to cover the most relevant issues that are needed as a basic foundation for a wireless engineer to know.

3.1   Propagation Models

The radio wavelength for cellular and especially for PCS is rather small in size, and as a result of this it has some unique propagation characteristics which have been modeled by numerous technical people over the years. The relative difference in physical wavelength is shown in Eq. (3.1).

$$c = f\lambda \tag{3.1}$$

where f = frequency, Hz
$c = 3 \times 10^8$ m/s
λ = wavelength, m

For cellular models,

$$f = 880 \text{ MHz}$$

$$\lambda = \frac{c}{f} = \frac{3 \times 10^8 \,\text{m/s}}{880 \times 10^6 \,\text{Hz}} = 0.34 \text{ m} \approx 13 \text{ in}$$

For PCS,

$$f = 1960 \text{ MHz}$$

$$\lambda = \frac{c}{f} = \frac{3 \times 10^8 \text{ m/s}}{1960 \times 10^6 \text{ Hz}} = 0.153 \text{ m} \approx 6 \text{ in}$$

Some of the more popular propagation models used are Hata, Carey, Elgi, Longley-Rice, Bullington, Lee, and Cost231. Each of these models has advantages and disadvantages. Specifically some baseline assumptions used with any propagation model need to be understood before they are used. Most cellular operators use a version of the Hata model for conducting propagation characterization. The Carey model, however, is used for submitting information to the FCC with regard to cell site filing information. The personal communication services (PCS) operators, however, are utilizing either Hata or Cost231 as their primary method for determining path loss.

Regardless of the frequency band of operation the model used for predicting coverage needs to factor into it a large number of variables which directly impact the actual RF coverage prediction of the site. The positive attributes affecting coverage are the receiver sensitivity, transmit power, antenna gain, and antenna height above average terrain. The negative factors affecting coverage involve line loss, terrain loss, tree loss, building loss, electrical noise, natural noise, antenna pattern distortion, and antenna inefficiency, to mention a few.

With the proliferation of cell sites the need to theoretically predict the actual path loss experienced in the communication link is becoming more and more critical. To date no overall theoretical model has been established that explains all the variations encountered in the real world. However, as the cellular and PCS communication systems continue to increase, growing reliance is placed on the propagation prediction tools. The reliance on the propagation tool is intertwined in the daily operation of the wireless communication system. The propagation model employed by the cellular and PCS operator has a direct impact on the capital build program of the company for determining the budgetary requirements and cell locations for the next few fiscal years.

Therefore, it is essential that the model utilized for the propagation prediction tool utilized be understood. The model should be understood in terms of what it can actually predict and what it cannot predict. Over the years numerous articles have been written with respect to propagation modeling in the cellular communications environment. With the introduction of PCS there has been increased focus on refining the propagation models to assist in planning the networks. However, no one model predicts every variation that will take place in the environment. To overcome this obstacle some operators have resorted to utilizing a combination of models depending on the environmental conditions relevant to the situation.

Presently most available cellular propagation tools utilize a variation of the Hata model,[21] while the PCS and some cellular operators are utilizing the

Cost231 model.[24] However, there are several variants to the Cost231 model as there are with the Hata model. Which model should be used, its baseline assumptions, and its relative performance are briefly covered here.

In addition to which model would be the best to utilize, other perturbations to the model need to be considered. One of the most basic considerations is the determination of the morphology that the model will be applied to. Morphologies are normally defined in four categories, dense urban, urban, suburban, and rural. The selection of which morphology to utilize at times is more of an art than a direct science, and this often leads to gross assumptions being made for a geographic area.

The morphologies are generally defined using the rough set of criteria:

Dense urban. This is normally the dense business district for a metropolitian area. The buildings for the area generally are 10 to 20 and above stories consisting of skyscrapers and high-rise apartments.

Urban. This type of morphology usually consists of building structures that are from 5 to 10 stories in height.

Suburban. This morphology is a mix of residential and business, with the buildings ranging from 1 to 5 stories but mainly consisting of 1- and 2-story structures.

Rural. This morphology, as the name implies, is generally open areas with structures not exceeding 2 stories and with that being sparsely populated.

From the above morphologies it may seem obvious that how you classify an area is rather ambiguous, since the geographic size of the area is left to the engineer to define.

As mentioned before, several propagation models are currently utilized throughout the industry. Each of the models has pros and cons. It is through understanding the advantages and disadvantages of each of the models that a better engineering design can actually take place in a network.

The specific models discussed here are free space, Hata, Carey, Cost231, and Quick.

3.1.1 Free space

Free space path loss is usually the reference point for all the path loss models employed. Each propagation model points out that it more accurately predicts the attenuation experienced by the signal over that of free space. The equation that is used for determining free space path loss is based on the relation $1/R^2$, or 20 dB/decade path loss.

The free space path loss equation (3.2) has a constant value that is used for the air interface loss, a distance and frequency adjustment.[24] Using some basic values the different path loss values can be determined for comparison with later models discussed.

$$L_f = 32.4 + 20 \log_{10} R + 20 \log_{10} f_c \tag{3.2}$$

where $R = \text{km}$
$\quad f_c = \text{MHz}$
$\quad L_f = \text{dB}$

The baseline assumptions in Table 3.1 are distances in kilometers and frequencies of 880 and 1960 MHz.

Looking at Table 3.1, it would be very nice if the frequency band utilized by cellular and PCS operators behaved with this path loss. In addition the increase in path loss from the cellular to PCS band is in the neighborhood of 7 dB. One temptation is to just add a 7-dB correction factor to the propagation model used for the cellular band and apply it to the PCS band. However, in real life frequency propagation does not behave like free space loss and therefore requires another equation.

TABLE 3.1   Path Loss Measured in Decibels

Distance, km	Frequencies	
	880 MHz	1960 MHz
1.0	91.29	98.25
2.0	97.31	104.27
3.0	100.83	107.79
4.0	103.33	110.29
5.0	105.27	112.23

3.1.2 Hata

The most prolific path loss model employed in cellular presently is the empirical model developed by Hata or some variant of it. The Hata model [Eq. (3.3)] is an empirical model derived from the technical report made by Okumura, so the results could be used in a computational model.[21] The Okumura report[44] is a series of charts that are instrumental in radio communication modeling.

$$L_H = 69.55 + 26.16 \cdot \log_{10} f_c - 13.87 \log_{10} h_b - a(h_m)$$
$$+ (44.9 - 6.55 \log_{10} h_b) \cdot \log_{10} R \tag{3.3}$$

Utilizing the same values used for the free space calculation a similar chart is derived. It should be noted that some additional conditions are applied when using the Hata model. The values utilized are dependent upon the range over which the equation is valid. If the equation is used with parameters outside the values the equation is defined for, the results will be subject to error.

The range for which the Hata model is valid follows: f_c = 150–1500 MHz; h_b = 30–200 m; h_m = 1 – 10 m; R = 1 – 20 km. Therefore, the Hata model should not be employed when trying to predict path loss less than 1 km from the cell site or if the site is less than 30 m in height. This is an interesting point to note since cellular sites are placed at times less than 1 km apart and often below the 30-m height.

In the Hata model the value $a(h_m)$ is used to correct for the mobile antenna height. The interesting point is that if you assume a height of 1.5 m for the mobile, that value nulls out of the equation.

Glancing back at the Hata model valid range, a curious point should come to light: The model that is referenced is valid to 1500 MHz, which is below the frequency range for PCS. The Hata model presented here is not valid for the PCS band but is valid for a host of other frequency bands ranging from 150 to 1500 MHz. This implies that the Hata model, as shown, should not be used for evaluating PCS propagation performance.

However, the Hata model is very applicable to the cellular band and assuming the following conditions for the Hata model the comparison shown in Table 3.2 can be made for different morphologies: f_c = 880 MHz; h_b = 30 m; h_m = 1.5 m

A critical point to mention here is that the Hata model employs three correction factors based on the environmental conditions that path loss prediction is evaluated over. The three environmental conditions are urban, suburban, and open. For Table 3.2 I have assumed an open environment to best compare the two equations.

The environmental correction values are easily calculated but vary for different values of mobile height. For the values listed below a mobile height of 1.5 m was assumed.

Urban	0 dB
Suburban	−9.88 dB
Open	−28.41 dB

3.1.3 Carey

The Carey model has many applications, but one of the key applications is its use with the FCC filing process. The Carey curves were initially specified for 450 to 460 MHz but are effective from 450 to 1000 MHz. A few assumptions, or rather design constraints, are placed on the use of this model, as is the case with any other propagation model.

TABLE 3.2   Path Loss Measured in Decibels

Distance, km	Free space model	Hata model Rural	Hata model Urban
1.0	91.29	97.75	126.16
2.0	97.31	108.36	136.77
3.0	100.83	114.56	142.97
4.0	103.33	118.96	147.37
5.0	105.27	122.37	150.79

The Carey model is defined in Eqs. (3.4).

$$L_1 = 110.7 - 19.1 \cdot \log_{10} h_b + 55 \log_{10} R \qquad 8 \le R \le 48 \text{ km}$$
$$L_2 = 91.8 - 18 \cdot \log_{10} h_b + 66 \cdot \log_{10} R \qquad 48 \le R \le 96 \text{ km}$$

(3.4)

where h_m = mobile height = 1.8 m

h_b = 30 to 1500 m

3.1.4 Cost231 Walfisch/Ikegami

The Cost231 Walfisch/Ikegami propagation model[24] is used for estimating the path loss in an urban environment for cellular communication. The Cost231 model is a combination of empirical and deterministic modeling for estimating the path loss in an urban environment over the frequency range of 800 to 2000 MHz. The Cost231 model is used primarily in Europe for GSM modeling and in some propagation models is used for cellular in the United States.

The Cost231 model [Eqs. (3.5)] is composed of three basic components.

1. Free space loss

2. Roof to street diffraction loss and scatter loss

3. Multiscreen loss

$$L_c = \begin{cases} L_f + L_{RTS} + L_{ms} \\ L_f \text{ where } L_{RTS} + L_{ms} \le 0 \end{cases}$$

(3.5a)

L_f = free space loss

L_{RTS} = roof top to street diffraction and scatter loss

L_{ms} = multiscreen loss

$$L_f = 32.4 + 20 \log_{10} R + 20 \log_{10} f_c$$

(3.5b)

where R = km

f_c = MHz

$$L_{RTS} = -16.9 - 10 \log_{10} W + 10 \log f_c + 20 \log \Delta h_m + L_0$$

(3.5c)

where w = street width, m

$\Delta h_m = h_r - h_m$

$$L_0 = \begin{cases} -10 + 0.354\phi & 0 \le \phi \le 35 \\ 2.5 + 0.075\,(\phi - 35) & 35 \le \phi \le 55 \\ 4.0 - 0.114\,(\phi - 55) & 55 \le \phi \le 90 \end{cases}$$

(3.5d)

where ϕ = incident angle relative to street

$$L_{ms} = L_{bsh} + k_a + k_d \log_{10} R + k_f \log_{10} f_c - 9 \log b \qquad (3.5e)$$

where b = distance between buildings along radio path

$$L_{bsh} = \begin{cases} -18 \log_{10}(1 + \Delta h_b) & h_b > h_r \\ \phi & h_b < h_r \end{cases} \qquad (3.5f)$$

$$k_a = \begin{cases} 54 & h_b > h_r \\ 54 - 0.8 h_b & d \geq 500 \text{ m}, h_b \leq h_r \\ 54 - 1.6 \Delta h_b \cdot R & d < 500 \text{ m}, h_b \leq h_r \end{cases} \qquad (3.5g)$$

NOTE: Both L_{bsh} and k_a increase the path loss with lower base station antenna heights.

$$k_d = \begin{cases} 18 & h_b > h_r \\ 18 - 15 \dfrac{\Delta h_b}{\Delta h_r} & h_b \leq h_r \end{cases} \qquad (3.5h)$$

$$k_f = \begin{cases} 4 + 0.7 \cdot \left(\dfrac{f_c}{925} - 1\right) & \text{for midsized city and suburban area with moderate tree density} \\[2mm] 4 + 1.5 \cdot \left(\dfrac{f_c}{925} - 1\right) & \text{metropolitan center} \end{cases} \qquad (3.5i)$$

In Eqs. (3.5), which comprise the Cost231 model, the following items bound the equations' useful range: f_c = 800 to 2000 MHz; h_b = 4 to 50 m; h_m = 1 to 3 m; R = 0.02 to 5 km. It is important, as always, to know what the valid ranges are for the model. Table 3.3 shows that when the range of the site is less than 1 km the Cost231 model makes a better choice than the Hata model.

Some additional default values apply to the Cost231 model when specific values are not known. (The default values recommended are listed below.) The default values can and will significantly alter the path loss values arrived at: b = 20 to 50 m; $w = b/2$; $h_r = 3 \cdot$ (No. of floors) + roof; Roof = 3 m for pitched roof and 0 m for flat roof; $\theta = 90°$. Figure 3.3 is a graphical illustration of the various Cost231 parameters. Assuming the following values for the Cost231 and the same values used for the previous equations for free space and the Hata path loss models, Table 3.3 is presented for the cellular band: f_c = 880 MHz; h_m = 1.5 m; h_b = 30 m; Roof = 0 m; h_r = 30 m; $\theta = 90°$; b = 30 m; w = 15.

Figure 3.4 is a plot of the propagation models referenced here for several environments. The plot shows the similarity and differences between models utilizing the same, or near same, parameters.

The assumptions made for the models employed in the graph utilize the following baseline conditions: f_c = 880 MHz; h_m = 1.5 m; h = 30 m; Roof = 0 m; h_r = 30 m; b = 30 m; w = 15.

TABLE 3.3   Path Loss for the Cellular Band, Measured in
Decibels

Distance, km	Free space	Model	
		Hata (urban)	Cost231
1.0	91.29	126.16	139.45
2.0	97.31	136.77	150.89
3.0	100.83	142.97	157.58
4.0	103.33	147.37	162.33
5.0	105.27	150.79	166.01

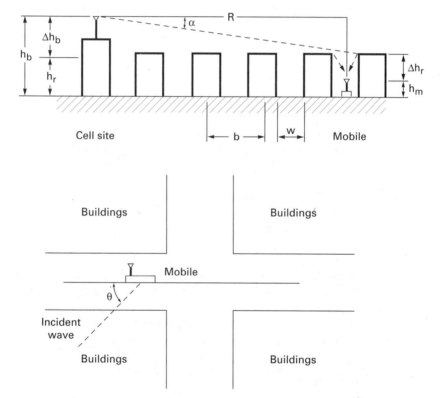

Figure 3.3   Cost231 parameter diagram.

3.1.5   Cost231 Hata

The Cost231 Hata model is shown next. This model has been tailored for the PCS 1900 MHz environment and is used by many PCS operators in establishing their system design. Equation (3.6) is utilized for Cost231 Hata:

$$PL = 46.3 + 33.9 \log_{10}(f_c) - 13.82 \log h_b + (44.9 - 6.55 \log h_b)\log_{10}d + c \quad (3.6)$$

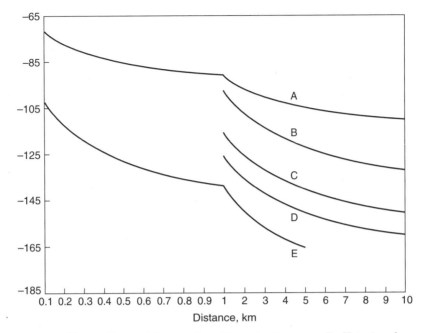

Figure 3.4 Propagation model comparison. Curve A = free space; B= Hata (rural; C = Hata (suburban); D = Hata (urban); E = Cost231 (urban).

where c = + 3 dense urban
           0 = urban
         −12 suburban
         −27 rural

The Cost231 Hata model is very similar to the Hata model with the exception of frequency and correction factors added based on the morphology that the model is applied to.

A quick look at the differences in propagation for Cost231 Hata and free space is shown in Table 3.4.

TABLE 3.4   Path Limiting Factors Associated with Operating in the 1900-MHz Band

		Cost231 Hata	
Distance, km	Free space, dB	Rural, dB	Urban, dB
1	97.98	111.03	137.03
2	103.99	121.63	147.63
3	107.51	127.84	153.84
4	110.01	132.24	158.24
5	111.95	135.65	161.656

Table 3.4 shows the severe path limiting factors associated with operating in the 1900-MHz band. Figure 3.5 is a propagation plot showing the path loss as

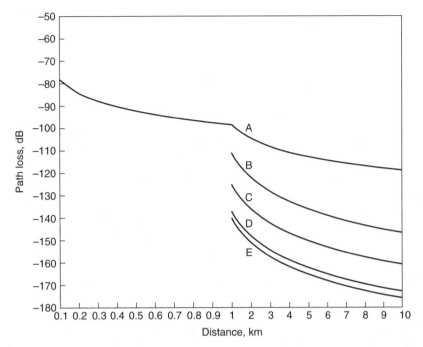

Figure 3.5 Propagation models for the 1900-MHz band. Curve A = free space; B = Cost231 Hata (rural); C = Cost231 Hata (suburban); D = Cost231 Hata (urban); E = Cost231 Hata (dense urban).

a function of distance and morphology. The assumptions for the plot are the same as those referenced for Fig. 3.4, with the exception that the frequency of operation is changed to reflect the PCS band.

A quick comparison between the two different bands, cellular and PCS, follows in Fig. 3.6. Obviously, the assumptions made for entering values into the equations play a major role in defining the outcome of the path loss value.

3.1.6 Quick

The Quick model is a down-and-dirty type of estimate that can be used for a field-expedient method of estimating the general propagation expectations for the area. The model is rather simplistic and straightforward. Its advantage is its quickness for use in roughly estimating the situation at hand; the disadvantage is it lacks the refinement of the other models.

The Quick method should be used when conducting some generalized approaches to a cell design and a rough answer is needed.

$$880 \text{ MHz} \qquad PL = 121 + 36 \log (\text{km})$$

$$1900 \text{ MHz} \qquad PL = 130 + 40 \log (\text{km})$$

The Quick method as compared to two Hata propagation models is shown in Fig. 3.7. The Quick method gives a reasonable approximation for propagation

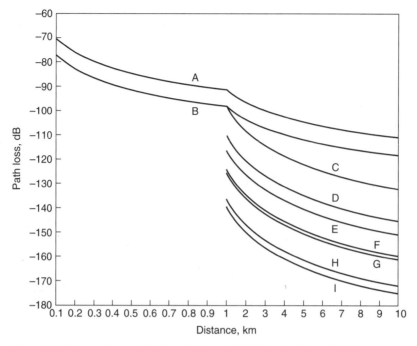

Figure 3.6 Propagation models for cellular and PCS bands. Curve A = free space (880); B = free space (1900); C = Hata (rural); D = Cost231 Hata (rural); E = Hata (suburban); F = Cost231 Hata (suburban); G = Hata (urban); H = Cost231 Hata (urban); I = Cost231 Hata (dense urban).

prediction over a variety of morphologies and can be used when details regarding the particular environment may not be readily available.

Regardless of which model is used for your analysis, the propagation model or models employed by your organization must be chosen with extreme care and undergo a continuous vigil to ensure they are truly a benefit to the company as a whole. The propagation model employed by the engineering department not only determines the capital build program but also is a direct factor in the performance of the network.

The capital build program obviously is directly affected by the propagation model chosen and particularly the underlying assumptions that accompany the use of the particular model. The propagation model is normally used as part of the 1-, 2-, and 5-year growth study performed by a cellular and/or PCS operator. The propagation model is used to determine how many sites are needed to provide a particular coverage requirement for the network. In addition the coverage requirement is coupled into the traffic loading requirements, which rely on the propagation model chosen to determine the traffic distribution, offloading, from existing site to new sites as part of the capacity relief program. The propagation model helps determine where the sites should be placed to achieve an optimal position in the network. If the propagation model used is not effective in helping place sites correctly, the probability of incorrectly justifying and deploying a site into the network is high.

The performance of the network is also affected by the propagation model chosen, since it is used for interference prediction plots. If the propagation model is inaccurate by say 6 dB, then you could be designing for a 23 or 11 dB C/I. Based on your traffic loading conditions, designing for a very high C/I level could have a negative impact on your financial picture, while on the other hand designing for a low C/I would have the obvious impact of degrading the quality of service to the very people who pay your salary. The propagation model is also used in a multitude of other system performance aspects that include handoff tailoring, power level adjustments, and antenna placements, to mention a few.

Reiterating the point that while no model can account for all the perturbations experienced in the real world, it is essential that you utilize one or several propagation models for determining the path loss of your network.

3.2 Terrain (Environmental Attenuation)

Focusing back on one of the more popular propagation models, the Hata model uses an alpha of about 3.5 for cellular. The alpha associated with PCS frequencies when taken by itself is in the order of 4.0 to 4.5. There have been

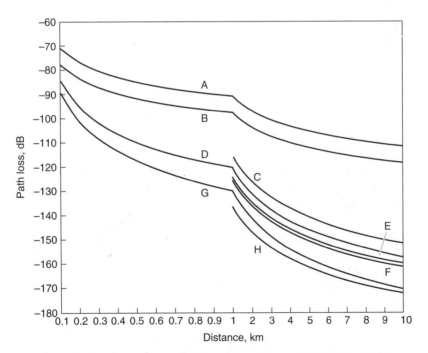

Figure 3.7 Quick propagation model. Curve A = free space (800); B = free space (1900); C = Hata (suburban, 880); D = Quick 800; E = Cost231 Hata (suburban); F = Hata (urban, 880); G = Quick 1900; H = Cost231 Hata (urban).

many debates regarding the propagation tools dB/decade slope used. However, it is the other variables associated with the propagation characteristics that have the largest impact on how the model truly predicts the propagation characteristics of the potential cell site.

The difference between alphas of 3.5 and 3.8 is only 3 dB/decade. The 3 dB can easily be aborted in environmental attenuation issues. The environmental issues associated with propagation characteristics require the highest amount of focus. Table 3.5 is a listing of some generalized attenuation characteristics which are used in propagation models. Glancing at Table 3.5, the difference between the urban and suburban environments is more than 3 dB. Obviously each area has its own unique propagation characteristic, and generalizing the characteristics of an area is a best guess fit. Table 3.5 is meant to be used as a general guide for determining environmental effects on path loss. However, it is obvious that the 3 or even 5 dB/decade is not the leading indicator for modeling propagation for a particular area.

TABLE 3.5 Environmental Effects on Path Loss, db

Foliage	
Sparse	6
Light	10
Medium	15
Dense	20
Very dense	25
Buildings	
Water/open	0
Rural	5
Suburban	8
Urban	22
Dense urban	27
Vehicle	10–14

3.3 Diffraction

Diffraction of the RF signal also has a very important role in predicting and attenuating the signal. How to calculate the actual attenuation from diffraction is shown in an excellent paper by Bullington.[45]

Several types of diffraction methods are modeled in RF, smooth and knife-edge. Each diffraction method yields a different value. If you choose wisely for the terrain issues at hand, the calculation method presented will accurately predict the attenuation experienced. Equations (3.7)[41] can be used to determine the amount of path loss that will be experienced as a result of a single obstacle in the path.

$$v = h\sqrt{\frac{2(d_1 + d_2)}{\lambda d_1 \cdot d_2}}$$

$$G_D(\text{dB}) = 20\log_{10}|F_v|$$

(3.7)

$$\therefore G_a(\text{dB}) = 0 \qquad\qquad\qquad\qquad\qquad v \le -1$$

$$= 20\log_{10}(0.5 - 0.62v) \qquad\qquad -1 \le v \le 0$$

$$20\log_{10}(0.5^{-0.95v}) \qquad\qquad\quad 0 \le v \le 1$$

$$20\log_{10}(0.4) - \sqrt{0.1184 - (0.38 - 0.1v)^2} \quad 1 \le v \le 2.4$$

$$20\log_{10}\left(\frac{0.225}{v}\right) \qquad\qquad\qquad v > 24$$

Figure 3.8 shows that there is a direct relationship between the distance from the obstruction and the height of the transmitter and receiver along with the frequency of operation. The diffraction loss can be calculated by using either a nomogram or Eqs. (3.7). Either method produces the same result.

Diffraction loss in both the cellular and PCS bands can be illustrated by Eqs. (3.8).

For 880 MHz

$$d_1 = 1\text{ km} \qquad d_2 = 1.5 \qquad h = 200\text{ m} \qquad \lambda_{880} = 0.34\text{ m}$$

$$v_{880} = 200\sqrt{\frac{2(2500)}{(0.34)(1.5\times 10^6)}} = 19.8$$

$$G_d = 20\log_{10}\left(\frac{0.225}{v}\right) = 20\log_{10}\left(\frac{0.225}{19.8}\right) = -38\text{ dB} \qquad (3.8a)$$

or a diffraction loss of 38 dB.

For 1900 MHz

$$v_{1900} = 200\sqrt{\frac{2(2500)}{(0.157)(1.5\times 10^6)}} = 29.14$$

$$G_d = 20\log_{10}\left(\frac{0.225}{29.14}\right) = -42.2\text{ dB} \qquad (3.8b)$$

or a diffraction loss of 42.2 dB.

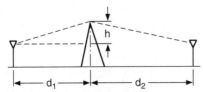

Figure 3.8 Relationship between distance from an obstruction and height of the transmitter and receiver. $d_1 \le d_2$.

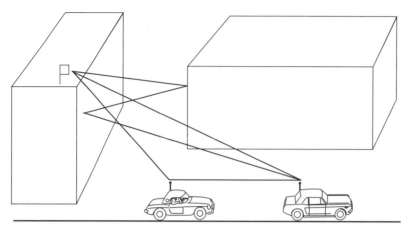

Figure 3.9 Multipath signals in an urban environment.

Knowing how to calculate diffraction is useful for trying to determine the loss of signal expected in a major valley. Another method is calculating the signal loss expected by using a mountain for containing the signal through placement of the antenna. Another valuable application for determining diffraction is to reduce the tilt angle of a cell site and estimate the positive and negative impact of improving the signal over a ridge or in a valley.

3.4   Multipath and Delay Spread

The communication quality between a mobile subscriber unit and the cell site depends on a variety of factors affecting the path over which the radio signal travels. Several types of signal impairments take place over the radio signal path. The four basic impairments experienced in a communication path involve path loss, shadowing, multipath, and Doppler shift.

The impact on the communication link from path loss is covered earlier in this chapter. Path loss is a direct result of the distance between the transmitter and receiver in the communication path. Shadowing, also called slow fading, is caused largely by partial blockage or environmental absorption such as trees. This is also covered in an earlier section of this chapter.

In any communication system multipath propagation presents one of the most challenging problems for designing a mobile communication system. Multipath propagation is the predominant form of transmission path in an urban environment since more than one reflection in the transmission path is normal (Fig. 3.9). The issues of multipath problems to a communication system show up a delay spread and Rayleigh fading.

In an analog communication system multipath, the same information taking multiple paths causes fading. The fading itself can sound like a flutter at low speeds. The mutipath is a result of Rayleigh or fast fading when the receiving antenna moves through constructive and destructive wavefronts. The receivers susceptibility to fading is a function of the frequency of operation and the receiver bandwidth. The higher the frequency the short the distance is between wave crests. The wider the bandwidth the less susceptible the receiver is to fading.

Delay spread, time dispersion, is to the digital radio system what multipath is to analog. However, with a digital system the delay in the signal's arrival is more important than that of the signal's received level. Dispersion occurs when multiple signals arrive at different times to the receiver and the difference in time between the signals arriving is in the order of a bit period. The multiple signals arriving at the receiver within a bit period cause a distortion in the representation of the bit desired or even the wrong bit's being decoded because the delayed reflected signal is stronger than the direct signal itself. When the time delay spread is no longer negligible with respect to the modulation bandwidth the received information will be distorted because the different paths are transferred with the multipath incoming waves. The delay spread is more pronounced with higher data rates since the effect can cause symbols to overlap, producing intersymbol interference. Figure 3.10 is an example of intersymbol interference (ISI). In Fig. 3.10 two multipath signals are received but there is enough delay between the symbols received to cause signal ambiguity, resulting in ISI.

The interesting point about delay spread is that the signal level of the incident, or desired, signal does not play as important a role as does the time delay between received signals. The dispersion problem for a digital communication system could be so severe for a cell site that you could have line of sight (LOS) communication and still not successfully demodulate the signal properly. There are a few solutions that one can utilize for correcting the dispersion problem.

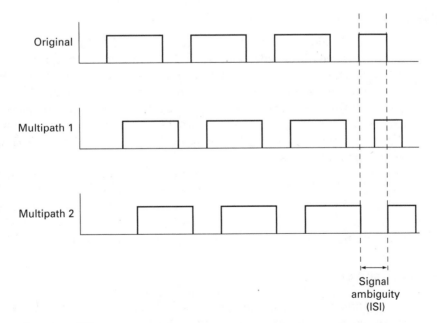

Figure 3.10 ISI.

One method of resolving the delay spread, or dispersion, problem is to employ adaptive equalizers in the demodulation portion of the receiver, utilized for TDMA systems. The equalizer provides a method for detecting the delayed signals and locking onto the strongest signal received. The equalizer operates by utilizing a training sequence which is sent at the start of the data communication burst, and then the equalizer adjusts itself to provide the maximum response on the channel, negating the effects of the radio channel itself. The use of equalizers is an integral part of TDMA cellular communication.

A CDMA system, however, does not utilize an adaptive equalizer to try to mitigate the effect of delay spread. Instead the CDMA system utilizes a rake receiver which has the ability to discriminate, or pick out, the desired signal from the other. The rake receiver is an integral part of a CDMA communication system.

One additional method that can be utilized in minimizing dispersion problems is through reducing the effective radiated power of the offending cell site itself. However, reducing the ERP of a site will also have a negative impact on the effective coverage area and may require more cell sites to provide the same level of coverage. Another method for reducing dispersion problems is to alter the antenna orientation of the cell site, therefore changing the dispersion path which is negatively affecting the radio communication.

When there is any motion between the transmitter and receiver, a frequency (Doppler) shift is experienced between the transmitted and received frequencies. The relative change in the Doppler shift can be serious enough to produce a random frequency modulation on the original signal itself. The Doppler shift also can have a positive or negative impact on the multipath propagation.

3.5   Cable Loss

One of the major items that comes into play for RF design is the proper selection of the feedline cable that is used to connect the radios to the antenna. Based on the situation, it is not uncommon to have different specifications for transmit and receive in terms of cable attenuation allowed.

Table 3.6 reflects the differences in cable losses as a function of distance, cable size, and frequency of operation. Please note that the table here assumes that there are no mismatches or faulty installations that can cause further impediments to the signal's attenuation.

Table 3.7 is a listing similar to Table 3.6 except that this applies to leaky feeder cable which has direct applications for in-building and tunnel installations, to mention a few. Table 3.8 lists coupling losses at 20 ft. The coupling losses are those that would be expected at 20 ft from the cable itself, provided you installed the cable according to the proper offsets required.

3.6   Cell Radius and ERP

The actual effective radiated power (ERP) for the cell site or transmitter used for the communication site, coupled with its antenna height, will determine its

TABLE 3.6   Attenuation per 100 Feet of Cable

MHz	LDF 50 Ω cable diameter*			
	$\frac{1}{2}$ in, dB	$\frac{7}{8}$ in, dB	$1\frac{1}{4}$ in, dB	$1\frac{5}{8}$ in, dB
150	0.845	0.458	0.34	0.280
450	1.51	0.834	0.617	0.515
824	2.10	1.17	0.866	0.731
894	2.20	1.23	0.907	0.767
1700	3.15	1.79	1.32	1.13
2000	3.45	1.97	1.45	1.25

*LDF= foam dialectric cable.

TABLE 3.7   Leaky Feeder
Attenuation Cable Path Loss

MHz	Cable diameter, dB		
	$\frac{7}{8}$ in	$1\frac{1}{4}$ in	$1\frac{5}{8}$ in
150	0.55	0.39	0.33
450	1.2	0.8	0.61
900	2.1	1.4	1.0
1700	3.5	2.3	1.7

*Free space with proper offsets for
mountains.

TABLE 3.8   Coupling Losses*

MHz	Cable diameter, dB		
	$\frac{7}{8}$ in	$1\frac{1}{4}$ in	$1\frac{5}{8}$ in
150	71	73	77
450	75	74	82
900	80	81	85
1700	81	81	86

*At 20 ft ± 10 dB.

transmit radius. The ERP setting should be balanced with the receive path to ensure that there is not a disparity between talk-out (transmit) and talk-back (receive) paths for a cell site.

The ERP for the site is set with reference to a dipole antenna. The method for calculating the ERP for a site is as follows using the data provided below.

Transmitter output	44	dBm (25 W)
Combining losses	−1	dB
Feedline loss	−3	dB
Antenna gain	10	dBd
	50	dBm (100 W)

If you are using dBi (isotropic gain) or want to convert dBi to dBd, all that is involved is a simple conversion:

$$dBd = 2.14 \text{ dB} + dBi$$

The ERP for the site has a dramatic impact on the cell site radius. An increase in ERP could reduce the build program requirements. The change to the build program is best explained through use of an example. Altering the ERP by 3 dB can alter the geographic area the cell site serves, assuming flat earth. In reality if there is a major obstruction, increasing the ERP may have little or no real effect depending upon the actual terrain conditions. However, referring back to the propagation model which has a slope of 35 dB/decade, this value is then used to determine the effective cell radius (ECR). Example 3.1 evaluates a 3-dB increase and decrease in ERP and its effect on the ECR.

Example 3.1

$$\text{Area} = \pi R^2$$

3-dB increase in ERP:

$$+ 3 \text{ dB} = 35 \log(R_{new})$$

$$\therefore R_{new} = 1.218 \qquad R_{old} = 1.0$$

$$\frac{\text{Area (new)}}{\text{Area (old)}} = \frac{\pi R_{new}^2}{\pi R_{old}^2} = \frac{R_{new}^2}{R_{old}^2} = R_{new}^2 = 1.484$$

$$\text{Area (new + 3 dB)} = 1.484 \text{ area}_{old}$$

3-dB decrease in ERP:

$$- 3 \text{ dB} = 35 \log (R_{new})$$

$$\therefore R_{new} = 0.820 \qquad R_{old} = 1.0$$

$$\frac{\text{Area (new)}}{\text{Area (old)}} = \frac{\pi R_{new}^2}{\pi R_{old}^2} = \frac{R_{new}^2}{\pi R_{old}^2} = R_{new}^2 = 0.674$$

$$\text{Area (new} - 3 \text{ dB)} = 0.674 \text{ area}_{old}$$

The actual effect the ERP has on the build program is best shown in Example 3.2.

Example 3.2 Total geographic area to cover = 50 km^2.

$$\text{Area (old)} = \pi R_{old}^2 = 3.14 \text{ km}^2 \qquad R_{old} = 1 \text{ km}$$

$$\text{Area (+3 dB)} = \pi R_{new}^2 = 4.66 \text{ km}^2 \qquad R_{new+3 \text{ dB}} = 1.484 \text{ km}$$

$$\text{Area (-3 dB)} = \pi R_{new}^2 = 2.16 \text{ km}^2 \qquad R_{new-3 \text{ dB}} = 0.674 \text{ km}$$

Approximate number of cells needed to cover 50 km^2:

$$\text{Number of cell sites} = \frac{\text{geographical area}}{\text{cell site area}}$$

$$\text{Number of cell sites (old)} = \frac{50 \text{ km}^2}{3.14 \text{ km}^2} \approx 16$$

$$\text{Number of cell sites (new} + 3 \text{ dB)} = \frac{50 \text{ km}^2}{4.66 \text{ km}^2} \approx 11$$

$$\text{Number of cell sites (new} - 3 \text{ dB)} = \frac{50 \text{ km}^2}{2.16 \text{ km}^2} \approx 23$$

The simple example shown points out that a mere 3-dB increase in ERP across all the cells in the network results in a decrease in the total amount of cell sites required by 31 percent. Conversely a 3-dB reduction in ERP has an effect of increasing the cell site build program by 43 percent. Increasing or decreasing the ERP by itself should not be done without a careful analysis of the link budget for the communication system, and of course the impact to the D/R ratio.

3.7 Link Budget Items

The maximum path loss, or limiting path, for any communication system used determines the effective range of the system. Table 3.9 involves a simple calculation of a link budget for the determination of which path is the limiting case to design from. In this table the receiver sensitivity value has the thermal noise, bandwidth, and noise figures factored into the final value presented. The up-link path, defined as mobile to base, is the limiting path case. From Table 3.9 the talk-back path is 6 dB less than the talk-out path. The limiting path loss is then used to determine the range for the site using the propagation model for the network.

TABLE 3.9   Link Budget

	Down link	Up link
Transmit	50 dBm	36 dBm
Antenna gain	3 dBd	12 dBd
Cable loss	2 dB	3 dB
Receiver sensitivity	−116 dBm	−116 dBm
C/N ratio	17 dB	17 dB
Maximum path loss	150 dB	144 dB

When calculating the actual link budget, the following items are recommended to be included in the calculation. The following examples should

Link Budget—Up Link

Frequency range, MHz

Mobile parameters
  Tx PA output (max)
  Cable loss
  Antenna gain
                                        _____
                                                    (Subscriber ERP max)

Environmental margins
  Fading margin
  Environmental attenuation
  Cell overlap
                                        _____

Base station parameters
  Rx antenna gain
  Rx jumper loss
  Rx tower top amp gain (net)
  Rx cable loss
  Rx lightning arrester loss
  Rx duplexer loss
  Rx diversity gain
  Rx coding gain
  Rx sensitivity
                                        _____
                                                    Up-link budget, dB

Figure 3.11   Link budget.

include more items than what may be utilized in the physical system being installed. However, items that can impact the link budget are included for reference. It is also highly possible that other devices can be added in the path to either enhance or potentially degrade the performance of the network.

Figure 3.11 is an example of a link budget information sheet that has more details attached to it. It is important to obtain the link budget sheet items at the onset of any design. The information obtained from Fig. 3.11 is then used to populate the link budget form in Fig. 3.12. The link budget form here is for the up-link path only. If the limiting path happened to be the down-link path, the appropriate adjustment to the form could be made reflecting the down-link path. In either case, up-link or down-link, the limiting path should not be exceeded. One final note on the path is that certain wireless technologies utilize different modulation formats on both the up-link and down-link paths. If this is the case, some of the reciprocity may not be applicable.

3.8   Link Budget Report

The following is a suggested outline for presenting the link budget assumptions that are made for determining the system coverage requirements. The

Cell type L	
Vendor Model No. PA Tx max Rx sensitivity Rx noise figure Modulation	

Subscriber	Type	Type	Type
max Tx min Tx Rx sensitivity			

Antenna			
Type	Vertical (3 dB)	Horizontal (3 dB)	Gain

System	Area 1	Area 2	Area 3

Cell ERP
  max
  min			

Cell height
  max
  min
  Nominal			

Cable loss
  max
  Nominal			

Cell configuration omni sector (3, 4, 6)			

Figure 3.12 Link budget for up-link path.

format presented is a boiler plate and therefore can and should be modified to reflect the company's objectives.

The format described below is what I believe to be a logical format that is easy to follow and also to modify as the system requirement changes owing to system coverage maturity and marketing-driven issues.

1. Cover sheet

2. Introduction

3. Revision

4. Table of contents

5. System overview

6. Coverage objectives

7. Quality of coverage

8. Intersystem coverage

9. Technology and infrastructure (ERP, Rx sensitivity, antennas, height, auxiliary equipment)

10. Propagation model (model, terrain, clutter, outdoor, indoor, fade margin)

11. Calculation

3.8.1   Cover sheet

This is the cover sheet for the report. It should include the following items as a minimum.

1. System it is meant for, for example, New York metro

2. Date of issuance (for example, 12/7/96)

3. Revision number

4. Who or what group issued the report

5. Confidentiality statement (this should be on every page of the document, usually at the foot)

3.8.2   Introduction

This is the description of what the link budget objectives are meant to resolve. The issue here should cover what market, general types of equipment, and who this document is intended for. Included with the introduction is the time frame this report is meant to cover. Specifically, if this is a new system, the time frame for the validity of this report could be 1 year. However, if this is an existing system or one that is practically built out, the time frame may also be 1 year but should really be 2 years as a minimum.

3.8.3 Revision

This documents which version of the report this particular version is. The sign-off section that is included is meant to ensure that the version that is under scrutiny is the current one and has undergone a design review. The format is shown in Fig. 3.13.

Date	Originator	Reviewed by	Comments	Review No.

Figure 3.13 Revision page format.

3.8.4 Table of contents

This section pulls the document together into a simple reference point so that anyone can quickly reference the pertinent section without having to read the whole document.

The suggested format for the Table of Contents is shown in Fig. 3.14.

Page

Introduction ....

Revision..........

Coverage objectives ....

Figure 3.14 Table of contents format.

3.8.5 System overview

This section describes what areas the system incorporates. A map showing the physical boundaries will also be necessary for this section. The key elements that need to be included are technology used currently and what changes are envisioned in the future.

3.8.6 Coverage objectives

This section describes what the coverage objectives are for the system. The following are suggested points that need to be covered.

1. What is the current coverage of the system?

2. What are the coverage requirements?

3. What areas need coverage?

This information should be derived from the marketing, operations, RF, and system performance engineering departments.

This section should include a map of the geographic area that encompasses the system. It should include an indication of what type of coverage objective is desired and its approximate differentiation on a map. The map could be of either the existing system or an area that is currently being considered for building a system. An example of a possible map is shown in Fig. 3.15.

The map shows the desire to have different types of coverage in different sections of the network. The objectives are defined in this case as being in-building and in-car for simplicity of the illustration. However, the amount of different coverage levels and perturbations is vast and should be driven by the company's objective. I would, however, suggest keeping the number of levels to less than four.

If multiple phases are associated with the build program, this should also be reflected in the coverage objective section. In the event of multiple phases, a

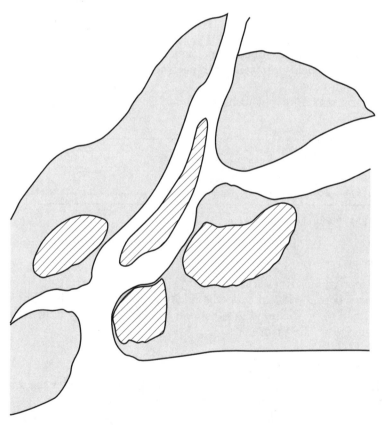

Figure 3.15 Representation of proposed system. Light-gray areas = in car; hatching = in building.

TABLE 3.10   Coverage Quality at Cell Edge, %

	In-building	In-car	On-street
Dense urban	90	95	99
Urban	85	90	99
Suburban	75	85	95
Rural	50	80	90

map showing the overall plan differentiating the different phases should be included also. This map indicating the different phases will be used as the foundation for deployment of resources that will be tied into the overall system design document that this report will play a part in.

If the system currently exists, a coverage map should also be included with this section. The coverage map should be of a scale that is sufficient to convey where the current system coverage problems are.

3.8.7   Quality of coverage

This section describes the coverage quality requirements for the system. The coverage quality is a series of parameters that will be used in helping to define better the link budget requirements for the system and geographic areas within the network.

The coverage quality (CQ) is meant to define the different morphologic requirements that will be used in determining how much of an area will need to be satisfied by the coverage requirements. Specifically Table 3.10 shows the different coverage requirements at the cell edge which reflects that 100 percent of coverage in all situations is not a design criterion. The CQ could also be differentiated to include not only the cell edge coverage requirements but also the overall coverage requirements for the cell itself depending on the morphology the cell is referenced to.

3.8.8   Intersystem coverage

This section of the report includes coverage requirements needed to provide contiguous coverage into another market. Specifically this would be applicable for an area of the system that say interfaces to another BTA, MTA, CGSA, or RSA where the ability to handle roaming traffic is desired.

The coverage objective here should define what the overlap should be in terms of dB and where the coverage objective should be. A map indicating the desired geographic areas would be directly applicable here. Any comments with regard to the other systems' build program and coverage objectives should be listed here also.

3.8.9   Technology and infrastructure

This section of the link budget report defines what the technology and infrastructure guidelines will be for the RF design. Specifically it defines the type

Link Budget—Down Link

Frequency range, MHz   Tx

Base station parameters
  Tx PA output power
  Tx combiner loss
  Tx duplexer loss
  Tx lightning arrester loss
  Tx cable loss
  Tx jumper loss
  Tx tower top amp gain
  Tx antenna gain
                                    _____
                                                (Cell ERP)

Environmental margins
  Tx diversity gain
  Fading margin
  Environmental attenuation
  Cell overlap
                                    _____

Mobile parameters
  Antenna gain
  Rx diversity gain
  Antenna cable loss
  Coding gain
  Rx sensitivity
                                    _____
                                                Down-link budget, dB

Figure 3.16 Link budget outline.

of radio equipment that will be utilized, the various receiver sensitivities, antenna types, possible structure, and the different auxiliary equipment that will be available to the RF engineer for the design phase.

This section of the document will play a critical role in defining the fundamental infrastructure assumptions that will permeate the entire design process. The technology and infrastructure section coupled with the propagation model assumptions will define how many cells and of what type will be required to meet the coverage objective defined earlier in the report.

Figure 3.12, the link budget form, is recommended for defining the technology and infrastructure that will be utilized by the RF engineering group. It is meant as a general guide and should be populated for each type of infrastructure equipment that is available. For example, if four types of cell sites are available, macro, micro, mini, and pico, each should have the information included in this document for reference. Either the chart can be filled out or the specification sheet provided by the manufacturer can be included. The chart also references the types of standard antennas that will be used in the design, the subscriber equipment types, and some of the basic system design constraints related to a propagation link budget. Additional items that might pertain to this section are the inclusion of a tower top amplifier, smart antennas, or any other potential performance enhancements relative to improving the link budget.

3.8.10   Propagation model

This section should include the propagation model or models utilized for the link budget, also a listing of the various margins that are included in the propagation model utilized for the analysis.

The propagation model selected should also include the frequency of operation and any generalized model assumptions utilized. For example, if Cost231 is chosen for the propagation model, the specific frequency that is chosen should be specified in addition to whether the default parameters for say street width are utilized and what it specifically is.

Also the various terrain loss data should be included here in addition to the C/I or Eb/No requirements defined for nominal system operation. The loss factors including the fade margin protection, building loss, and clutter factors, to mention but a few, need to be referenced clearly since their inclusion or exclusion will have a dramatic effect upon the link budget calculations.

3.8.11   Calculation

This section is where the various values defined throughout the report are correlated and placed in a series of tables. The format in Fig. 3.11 can be utilized for consistency and a separate link budget sheet can accompany the different coverage requirement classifications and equipment types.

When the design is carried to an individual cell site, a separate link budget sheet should be made up for each cell site. The link budget sheet format is shown in Figs. 3.11 and 3.16.

References

1. Smith, Clint, and Curt Gervelis, *Cellular System Design and Optimization,* McGraw-Hill, New York, 1996.
2. Ikegami Fumio, Tsutomu Takeuchi, and Susumu Yoshida, "Theoretical Prediction of Mean Field Strength for Urban Mobile Radio," *IEEE Trans. on Antennas and Propagation,* vol. 39, no. 3, March 1991.
3. Amitay Noach, "Modeling and Computer Simulation of Wave Propagation in Lineal Line of Site Microcells," *IEEE Trans. on Vehicle Technology,* vol. 41, no. 4, November 1992.
4. Tan, S. Y., and H. S. Tan, "UTD Propagation Model in an Urban Street Scene for Microcellular Communications," *IEEE Trans. on Electromagnetic Compatibility,* vol. 35, no. 4, November 1993.
5. Erceg, Vink, Saeed Ghassemzadeh, Maxwell Taylor, Dong Li, and Donald Schilling, "Urban/Suburban Out-of-Sight Propagation Modeling," *IEEE Communications Mag.,* June 1992.
6. Steele, Raymond, John Williams, Derek Chandler, Shirin Dehghan, and Aidan Collard, "Teletraffic Performance of GSM900/DCS1800 in Street Microcells," *IEEE Communications Mag.,* March 1995.

7. Sousa, Elvino, Vladan Jovanovic, and Christian Daigneault, "Delay Spread Measurements for Digital Cellular Channel in Toronto," *IEEE Trans. on Vehicle Technology,* vol. 43, no. 4, November 1994.

8. Keenan, J. M., and A. J. Motley, "Radio Coverage in Buildings," *BR Telecom Technical J.,* vol. 8, no. 1, January 1990.

9. Mahbobi, Kamran, "Radio Wave Propagation in Urban Microcellular Environment," *IEEE Trans. on Vehicle Technology,* 1992.

10. Gudmundson, Mikael, *Cell Planning in Manhattan Environments,* IEEE, 1992.

11. Erecg, Vinko, A. J. Rustako, and R. S. Roman, "Diffraction around Corners and Its Effects on the Microcell Coverage Area in Urban and Suburban Environments at 900 MHZ, 2 GHz, and 6 GHz," *IEEE Trans. on Vehicle Technology,* vol. 43, no. 3, August 1994.

12. Xia, Howard, Henry Bertoni, Leonard Maciel, Andre Linsay-Stewart, and Robert Rowe, "Microcellular Propagation Characteristics for Personal Communication in Urban and Suburban Environments," *IEEE Trans. on Vehicle Technology,* vol. 43, no. 3, August 1994.

13. American Radio Relay League, *The ARRL 1986 Handbook,* 63d ed., The American Radio Relay League, Newington, CT, 1986.

14. American Radio Relay League, *The ARRL Antenna Handbook,* 14th ed., The American Radio Relay League, Newington, CT, 1984.

15. AT&T, *Engineering and Operations in the Bell System,* 2d ed., AT&T Bell Laboratories, Murray Hill, NJ, 1983.

16. Brewster, *Telecommunications Technology,* Wiley, New York, 1986.

17. Carr, J. J., *Practical Antenna Handbook,* McGraw-Hill, New York, 1989.

18. Code of Federal Regulations, CFR 47 Parts 1, 17, 22, 24, and 90.

19. DeRose, *The Wireless Data Handbook,* Quantum Publishing, Inc., Mendocino, CA, 1994.

20. Dixon, *Spread Spectrum Systems,* 2d ed., Wiley, New York, 1984.

21. M. Hata, "Empirical Formula for Propagation Loss in Land Mobile Radio Services," *IEEE Trans. on Vehicle Technology,* vol. VT-29, no. 3, pp. 317–325, 1980.

22. Kaufman, M., and A. H. Seidman, *Handbook of Electronics Calculations,* 2d ed., McGraw-Hill, New York, 1988.

23. Lee, W. C. Y., *Mobile Cellular Telecommunications Systems,* 2d ed., McGraw-Hill, New York, 1996.

24. Low, K., "Comparison of Urban Propagation Models with CW Measurements," 42d IEEE Vehicle Technology Conference, Denver, May 1992, pp. 317–325.

25. Qualcom, "An Overview of the Application of Code Division Multiple Access (CDMA) to Digitial Cellular Systems and Personal Cellular Networks," Qualcom, San Diego, CA, May 21, 1992.

26. Sklonik, M. I., *Introduction to Radar Systems,* 2d ed., McGraw-Hill, New York, 1980.

27. Simo, "IS-95 Based SS-CDMA: Operational ISSUES," seminar, January 1995.

28. White, Duff, *Electromagnetic Interference and Compatibility,* Interference Control Technologies, Gainesville, GA, 1972.

29. Yarborough, *Electrical Engineering Reference Manual,* 5th ed., Professional Publications, Belmont, CA, 1990.

30. Qualcom, *CDMA Handbook,* Qualcom, San Diego, CA, 1996.

31. Fink, Donald and Donald Christiansen, *Electronics Engineers Handbook,* 3d ed., McGraw-Hill, New York, 1989.

32. Schwartz, Bennett, Stein, *Communication Systems and Technologies,* IEEE, New York, 1996.

33. Rappaport, *Wireless Communications Principles and Practices,* IEEE, New York, 1996.

34. Webb, Hanzo, *Modern Amplitude Modulations,* IEEE, New York, 1994.

35. Steele, *Mobile Radio Communications,* IEEE, New York, 1992.

36. Jakes, W. C., *Microwave Mobile Communications,* IEEE, New York, 1974.

37. Carlson, A. B., *Communications Systems,* 2d ed., McGraw-Hill, New York, 1975.

38. Johnson, R. C., and H. Jasik, *Antenna Engineering Handbook,* 2d ed., McGraw-Hill, New York, 1984.

39. Lathi, *Modern Digital and Analog Communication Systems,* CBS College Printing, New York, 1983.

40. MacDonald, "The Cellular Concept," *Bell Systems Technical J.,* vol. 58, no. 1, 1979.

41. Rappaport, *Cellular Radio and Personal Communications,* IEEE, New York, 1995.

42. *Reference Data for Radio Engineers,* 6th ed., Sams, 1983.

43. Mouly, Pautet, *The GSM System for Mobile Communications,* Mouly Pautet, 1992.

Chapter

4

Modulation

4.1 Introduction

To convey voice and data information from one location to another without physically connecting them it is necessary to send the information by another method. Many methods exist for conveying information to and from locations that are not physically connected. Some involve talking or using flags, drums, and lights, to mention a few. Each method has its advantages and disadvantages. However, the problem common to all the methods for communicating involves the physical distance the sender and the receiver have to be from each other.

In order to increase the distance and increase the information transfer rate between sender and receiver an electromagnetic wave is employed. The use of electromagnetic waves is fundamental to radio communication. However, to utilize an electromagnetic wave it is necessary to modulate the carrier wave at the transmitting source and then demodulate it at the receiver. The modulation and then demodulation of the carrier wave forms the principle of a radio communication system.

The generalized radio system is shown in Fig. 4.1. The choice of modulation and demodulation utilized for the radio communication system is directly dependent upon the information content to be sent, the available spectrum to

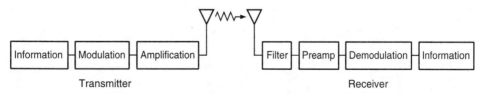

Figure 4.1 Basic radio system.

convey the information, and the cost. The fundamental goal of modulating any signal is to obtain the maximum spectrum efficiency, or rather information density per hertz.

Many types of modulation and demodulation formats are utilized for the conveyance of information. However, all the communication formats rely on one, two, or all three of the fundamental modulation types, amplitude modulation (AM), frequency modulation (FM), and phase modulation (PM). Figure 4.2 and Eq. (4.1) highlights the differences between the modulation techniques in terms of their impact on the electromagnetic wave itself.

$$E(t) = A \sin (2\pi f_c t + \phi) \tag{4.1}$$

where A = amplitude
     f_c = carrier frequency
     ϕ = phase
     t = time
     E = instantaneous electric field strength

  amplitude modulation (AM) modifies A

  frequency modulation (FM) modifies f_c

  phase modulation (PM) modifies ϕ

Amplitude modulation (AM) has many unique qualities; however, this form of communication is not utilized in cellular communication, primarily because it is susceptible to noise.

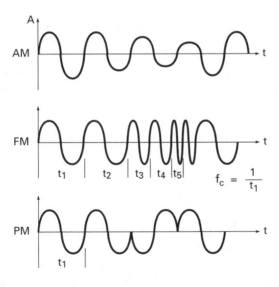

Figure 4.2 Modulation techniques.

Frequency modulation (FM) is utilized for AMPS and TACS analog communication. FM is utilized for analog cellular communication since it is more robust to interference. Cellular utilizes a channel bandwidth of 30 kHz for FM modulation. GSM also utilizes a form of FM referred to as GMSK.

Phase modulation (PM) is used for TDMA and CDMA communication systems. There are many variations to phase modulation. Specifically many digital modulation techniques rely on modifying the RF carrier phase and amplitude.

Textbooks and technical articles abound in the industry that focus purely on each of the modulation schemes discussed next. The discussion that follows is a general overview and not a detailed dissertation regarding each possible method of modulation and demodulation. The reference material listed at the end of this chapter should be used as a basis for further learning regarding modulation and demodulation schemes.

4.2 Information Bandwidth

The information and the channel bandwidth play a major role in determining which modulation scheme to utilize for the communication system. Often the channel bandwidth is defined and then the appropriate modulation technique must be applied.

The channel theoretical capacity is defined by the Shannon-Hartley equation (4.2).

$$C = B \cdot \log_2 (1 + S/N) \tag{4.2}$$

where S/N = signal-to-noise ratio
       B = bandwidth
       C = capacity

For example,

$$B = 30 \text{ kHz} \qquad S/N = 18 \text{ dB} = 1$$

$$C_{18} = (30 \times 10^3) \log_2(1 + 63)$$

$$= 3 \times 10^3 \cdot \frac{\log_{10}(64)}{\log_{10}(2)} = 18{,}000 \ or \ 18 \text{ kbits/s}$$

If

$$S/N = 13 \text{ dB}$$

$$C_{13} = (30 \times 10^3) \log_2(1 + 19.9)$$

$$C_{13} = 131{,}672 \text{ bits/s or} \approx 131.5 \text{ kbits/s}$$

an increase in 7.3 times for a relaxation of 5 dB.

The above example shows that for a relaxation of 5 dB in the S/N an increase in 7 occurs for the information content of the channel. However, the bandwidth of the channel exceeds that of the information bandwidth for the system requiring only 13 dB of S/N. The second system therefore requires some level of coding.

Table 4.1 highlights the differences in some of the digital modulation formats of PSK and MSK. The table shows that as the modulation levels increase so does the S/N requirement to achieve the same BER. Table 4.2 is a brief chart of the differences in spectral efficiency for the more popular wireless communication platforms. It should be noted that the spectral efficiency noted here does not take into account frequency reuse or the services that can be supported on each.

TABLE 4.1   Comparison of Digital Modulation Schemes[6]

Modulation	Spectral efficiency, bits/s/Hz	Required sin, dB
BPSK	1	11.1
QPSK	2	14
PSK (16-level)	4	26
MSK (2-level)	1	10.6
MSK (4-level)	2	13.8

BER = 1×10^6.

TABLE 4.2   Digital Modulation Schemes

System	Modulation technique	Channel bandwidth, kHz	Data rate, kbits/s	Spectral efficiency, bits/s/Hz
NADC	QPSK	30	48.6	1.62
GSM	GMSK (0.3)	200	270.8	1.35
CDMA	QPSK OQPSK	1230	1230	1.0
JDC	$\pi/4$ QPSK	25	42	1.68
CT-2	GMSK	100	72	0.72
DECT	GMSK (0.5)	1728	1572	0.67

4.3   Amplitude Modulation (AM)

Amplitude modulation is one of the methods that is used for modulating and demodulating signals. Many forms of amplitude modulation are used in wireless communications. Some involve simple AM, ASK, SSB, DSB, VSB, and QAM, to mention a few.

The fundamental principle of AM modulation is represented in Eq. (4.3).

$$E_{(t)} = A \sin (2\pi f_c t + \phi) \tag{4.3}$$

Here, the only real variation that takes place is in altering the value represented by the symbol A.

$$m(t) \longrightarrow \bigotimes \longrightarrow A_c[1 + m(t)] \sin(w_c t + \phi) \qquad (4.4)$$

$$A_c \sin(w_c t + \phi)$$

where $w_c = 2\pi f_c$

Expression (4.4) shows the basic method for modulating the AM signal while Eqs. (4.5) and (4.6) show the two basic methods for demodulating an AM signal, coherent and noncoherent. The noncoherent method [Eq. (4.5)] is the envelope detector and the coherent method [Eq. (4.6)] involves the product detector circuit shown.

$$A_m \sin(2\pi f_c t + \phi_m) \longrightarrow \mathrm{A} \longrightarrow \mathrm{B} = \frac{1}{\pi} A_m \qquad (4.5)$$

$$x_c(\sin w_{ct} + \phi_m) \longrightarrow \bigotimes \longrightarrow y = \frac{(x_c \cdot A_c)}{2} \cos(\phi_m + \phi_0) \qquad (4.6)$$

$$A_c \sin(w_c t + \phi_0)$$

4.3.1   AM

The modulation of voice or tones by varying the carrier amplitude is shown in Fig. 4.3. The method of modulation is a direct conversion using a multiplier, or rather mixer. The modulation is achieved through varying the amplitude of the information signal, which in turn causes the modulation of the carrier. [See Eq. (4.7).]

(1) Information $m(t) \longrightarrow \bigotimes \longrightarrow$ Output S_{Am}

Carrier
$A_c \sin(2\pi f_c t + \phi)$

(2) $\qquad\qquad (4.7)$

$$R_T = 2\,Bm$$

$$S_{Am} = A_c[1 + m(t)](\sin 2\pi f_c t + \phi)$$

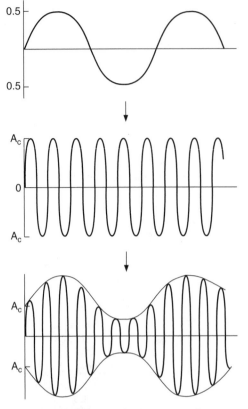

Figure 4.3 Modulation of voice or tones by varying the carrier amplitude.

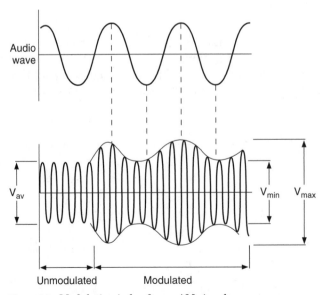

Figure 4.4 Modulation index for an AM signal.

Figure 4.4 is an example of how the modulation index is achieved for an AM signal.

Figure 4.5 shows the effects on the modulated signal based on the percentage of modulation. Note that modulation over 100 percent leads to distortion of the information content that should be delivered to the receiving party error-free [see Eq. (4.8)].

$$\text{Percent modulation} = \frac{V_{max} - V_{min}}{2V_{av}} \cdot 100 \tag{4.8}$$

If

$$V_{max} = A_C[1 + m(t)]$$

$$V_{min} = A_C[1 - m(t)]$$

$$m(t) = 0.5\, A_C$$

$$\text{Percent modulation} = \frac{A_C(1.5) - A_C(0.5)}{2\, A_C} = 50 \text{ percent}$$

Equation 4.9 shows the product detector in operation.

$$X_a = A_m \sin\left(2\pi f_c t + \phi_m\right) \cdot A_C \sin\left(2\pi f_c t + \phi_c\right) \tag{4.9}$$

$$= \frac{A_m A_c}{2} \cos[(2\pi f_c t + \phi_m) - (2\pi f_c t \phi_c)] - \cos(2\pi f_c t + \phi_m + 2\pi f_c t + \phi_c)$$

$$= \frac{A_m A_c}{2} [\cos(\phi_m - \phi_c) - \underbrace{\cos(4\pi f_c t + \phi_m + \phi_c}_{\text{Removed due to low-pass filter}}]$$

$$X_b = \frac{A_m A_c}{2} \cos(\phi_m - \phi_c)$$

$$X_b = \frac{A_m A_c}{2} \qquad \text{if } \phi_m = \phi_c$$

4.3.2   ASK

Amplitude shift keying (ASK) is one method of sending digital information via AM modulation. The principle of ASK is to alter the amplitude of the carrier

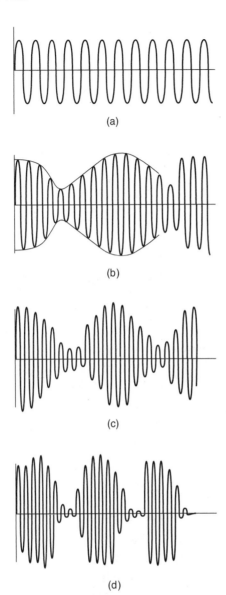

(a)

(b)

(c)

(d)

Figure 4.5 Effects on the modulated signal by percentage of modulation: (*a*) 0 percent; (*b*) 50 percent; (*c*) 100 percent; (*d*) over 100 percent. In *d*, the modulation envelope is clipped. In addition, there is a percentage of time during which no transmission takes place, thus changing the original wavelength of the modulating signal.

by a predetermined amount depending on whether the data is a one or zero. ASK can either have the carrier completely suppressed for a one or zero or simply alter the amplitude.

Figure 4.6 is an example of ASK modulation of a signal. The calculation is shown as Eq. (4.10).

$$X_t = \sin(2\pi f_c t + \phi) \tag{4.10}$$

m(t) ————⊗———— $A_c[1 + m(t)] (\sin 2\pi f_c t + \phi)$

$A_c \sin (\tau\pi f_c t + \phi)$

where $m(t) = 1A_c$ for 1

$\qquad\qquad = 0.5A_c$ for 0

The example shows how a data stream can be represented by an ASK signal. Now to reverse the process, Fig. 4.7 shows the demodulation process of the ASK signal. The demodulation process assumes no conversion losses or distortions in the modulation and demodulation process [see Eq. (4.11)].

$$X_m = A_c[1 + m(t)\sin(u_c + \phi_m)] \tag{4.11}$$

X_m ————⊗————[≈≈]———→ $\dfrac{A_c[1 + m(t)]}{2} = X_D$

$A_c \sin (w_c t + \phi_d)$

$$\therefore m(t) = 1 \qquad X_O 1 = A_c$$

$$m(t) = 0 \qquad X_D 0 = \dfrac{A_c}{2}$$

4.3.3 Sideband

Sideband modulation consists of double sideband (DSB), single sideband (SSB) of which SSB comprises either upper sideband (USB) or lower sideband (LSB). Lastly vestige sideband (VSB) is another form of sideband modulation. Each of the modulation formats has its advantages for the situation at hand. The method of modulating the signal is for either DSB, LSB, USB, or VSB is shown in Fig. 4.7. The chief advantage that any of the SSB modulation formats has is the suppression of the carrier wave, leaving more energy to be applied to the information content of the signal.

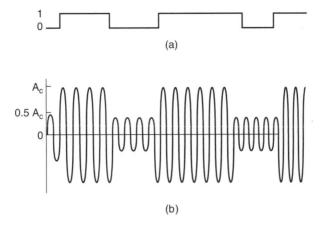

(a)

(b)

Figure 4.6 ASK modulation. (a) Data stream. (b) Analog, $A_c[1 + m(t)] \sin(w_c + \theta)$.

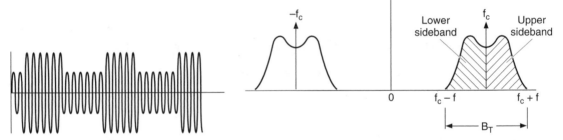

Figure 4.7 ASK demodulation. $X_m = A_c[1 + m(t)] \sin(u_c + \phi_m)$.

Figure 4.8 Single sideband modulation. $B_T = 2f$.

Figure 4.8 shows the output of the modulator. The output of the mixer is the same for DSB, LSB, USB, and VSB. The difference between the modulation formats lies in the next stage of filtering prior to the actual transmission of the signal.

The output of the modulator shown in Fig. 4.9 is best represented by the expression in Eq. (4.12). The expression is the product of the signals mixing, but this time a Fourier transform was performed on the mathematical expression, converting the expression from time to the frequency domain.

$$S_{AM}(f) = \frac{1}{2} A_c[\delta(f - f_c) + m(f - f_c) + \delta(f + f_c) + m(f + f_c)] \qquad (4.12)$$

where δ = unit impulse function
m = message signal spectrum
f = information frequency
f_c = carrier frequency

What follows is a representation of how a USB signal is arrived at in Fig. 4.10. The expression for a LSB signal is shown in Fig. 4.11.

The expression for a VSB signal is shown in Fig. 4.12. Careful analysis of the VSB signal as compared to, say, the USB signal shows a remarkable similarity. Specifically VSB is a compromise between the DSB and SSB. The bandwidth for a VSB signal is generally about 25 percent more than that of a USB or LSB signal. The reason for the increased bandwidth over a LSB or USB is

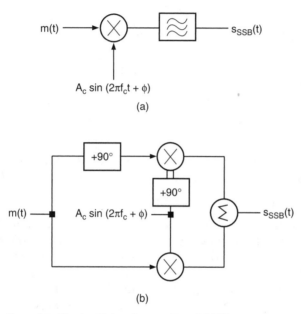

Figure 4.9 Single sideband generation. (*a*) Filter. (*b*) Balanced modulation.

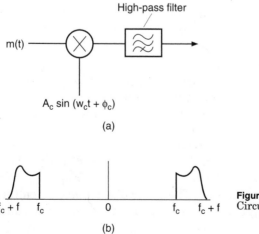

Figure 4.10 Upper sideband. (*a*) Circuit. (*b*) Spectrum.

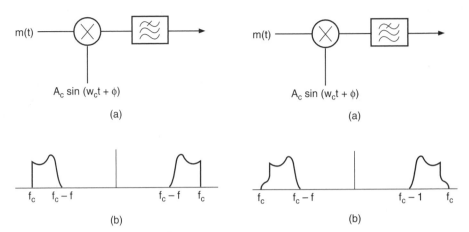

Figure 4.11   Lower sideband. (*a*) Circuit. (*b*) Spectrum.

Figure 4.12   Vestigial sideband. (*a*) Circuit. (*b*) Spectrum.

the fact that either the USB or the LSB that is unused is not altogether removed, just severely attenuated (the reason for the name vestigial).

4.4   Frequency Modulation (FM)

Frequency modulation is also referred to as angle modulation and is obtained by varying the carrier frequency linearly as a function of the information content without changing the carrier amplitude. The chief advantage with FM modulation is its immunity to noise as compared to AM modulation methods.

Frequency modulation can be used to send voice or data information. The most common form of FM communication that is experienced in everyday life is the FM radio station. FM is also utilized in many wireless communications systems for conveying voice information.

FM also can also send data, and many formats that utilize FM modulation achieve this. Some FM modulation formats that are used in wireless communication involve FSK, MSK, and GMSK.

4.4.1   FM analog modulation

The generation of an FM signal conveying voice information can be achieved through several methods. The method chosen has a direct impact on the modulation technique utilized for transferring the information onto the carrier wave itself. There are primarily two methods of transferring voice information onto the carrier, direct and indirect methods.

The direct method is shown in Fig. 4.13 and the indirect method in Fig. 4.14. The method each utilizes is to deviate the carrier frequency as a function of the amplitude of the information. The method is achieved by altering the oscillator frequency by changing the reactance to the oscillator circuit used for generation of the carrier frequency through use of a variable-control oscillator, VCO or VFO. The FM signal can be represented by Eqs. (4.13).

$$E_{(t)} = A_c \sin(2\pi f_c t + \theta)$$

$$FM_{\text{modulation}}\ \theta = \beta \sin 2\pi f_m t$$

(4.13)

where β = modulation index
f_m = modulation frequency

$$\beta = \text{modulation index} = \frac{k_f A_m}{w} = \frac{\Delta f}{w}$$

where A_m = peak value for modulating signal
Δf = peak frequency deviation
w = max bandwidth of modulating signal
K_f = frequency deviation constant, Hz/W

$$\therefore E_{(t)f_m} = A_c \sin(2\pi f_c t + \beta \sin 2\pi f_m t)$$

4.4.2   Preemphasis and deemphasis

Preemphasis and deemphasis are techniques utilized in FM modulation to improve the S/N of the signal. Preemphasis is used in the modulation process while deemphasis is used in the demodulation process.

The preemphasis is done before the actual modulation takes place and is done at the audio level. In radio communication higher frequencies in the audio band are less likely to have noise and other distortions. One way to take advantage of this is to amplify the higher-frequency signals in the audio

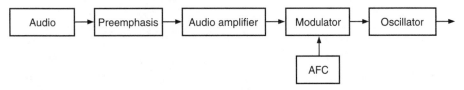

Figure 4.13   Direct FM.

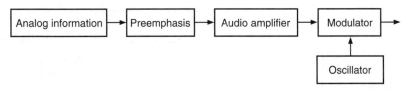

Figure 4.14   Indirect FM.

passband. The preemphasis circuit, also referred to as a compander, is shown in Fig. 4.15. The bode plot that accompanies the preemphasis circuit is a high-pass filter which attenuates lower-frequency signals in favor of the higher ones.

The deemphasis process is achieved at the receiver and performs the exact opposite function to that of the preemphasis process. Specifically the deemphasis is where the lower-frequency signals are amplified more than the higher-frequency audio signals. The deemphasis takes place after the demodulation takes place, since the deemphasis process is at the audio level.

Figure 4.16 is a diagram of the deemphasis process. Note that the bode plot representing the circuits function is really a low-pass filter. The deemphasis circuit is referred to as an expander.

4.4.3 FSK

Frequency shift keying (FSK) is another form of frequency modulation. FSK is used for sending data from one point to another by changing the frequency at which the carrier or information content operates. FSK can be used to send simple binary information by sending information as one of two different fre-

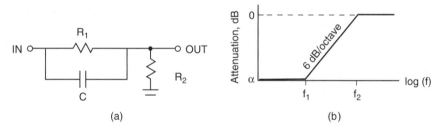

Figure 4.15 (a) Compander circuit. (b) Preemphasis.

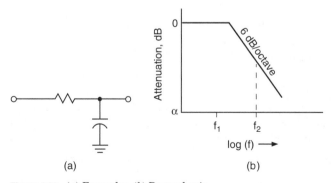

Figure 4.16 (a) Expander. (b) Deemphasis.

quencies. FSK also can be sent by more than two different frequencies, say, four discrete frequencies, representing a four-level FSK system.

FSK can be generated by several methods. One method is to simply transmit on a different frequency for, say, data bit 1 and another for data bit 0. The best method for use of FSK is to modulate the carrier frequency by the frequency used for generation of a 1 or 0. [See Fig. 4.17 and Eq. (4.14).]

$$E(t) = A_c \sin(2\pi f_c t + \phi) \tag{4.14}$$

$$\phi_1 = \beta \sin(2\pi f_1 t) \qquad \text{data bit 0}$$

$$\phi_2 = \beta \sin(2\pi f_2 t) \qquad \text{data bit 1}$$

The modulation of FSK can be done by either a direct or indirect modulation method discussed for FM analog transmission.

4.4.4 MSK

Minimum shift keying (MSK) is an FSK signal that has a modulation index of 0.5, and therefore the frequency shift is twice the data rate. MSK has the unique distinction of being referred to as many different forms of modulation. For example, MSK is also referred to as fast FSK, FFSK, and continuous phase frequency shift keying (CPFSK), and a variant of offset quadrature phase shift keying (OQPSK).

The block diagram of an MSK modulator and demodulator is shown in Figs. 4.18 and 4.19. See Eqs. (4.15).

$$E_1(t) = \cos\left(2\pi f_c t + \frac{\pi t T}{2}\right)$$

$$E_2(t) = \cos\left(2\pi f_c t - \frac{\pi t T}{2}\right)$$

$$X(t) = \cos\left(\frac{\pi t}{2T}\right) \cos(2\pi f_c t) \tag{4.15}$$

$$Y(t) = \sin\left(\frac{\pi t}{2T}\right) \sin(2\pi f_c t)$$

$$\therefore S(t) = \cos\left(2\pi f_c t \pm \frac{\pi t}{2T}\right)$$

The MSK signal as a general rule will occupy 1.2 times the bit rate.

4.4.5 GMSK

Gaussian minimum shift keying is another modulation technique that is a derivative of the MSK method for modulation. GMSK is the same as MSK except that it uses a gaussian low-pass filter prior to modulating the signal.

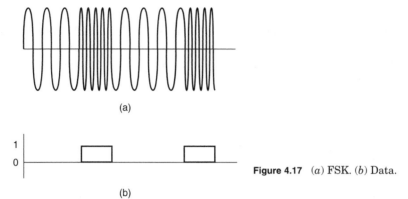

(a)

(b)

Figure 4.17 (*a*) FSK. (*b*) Data.

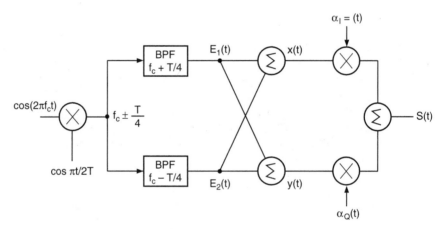

Figure 4.18   MSK modulation.

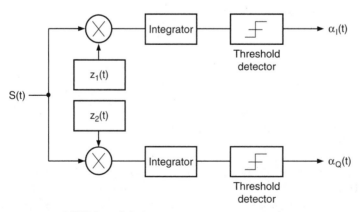

Figure 4.19   MSK demodulation.

The objective with using the filter is to improve the bandwidth efficiency for the signal.

4.5 Phase Modulation (PM)

Phase modulation is another form of angle modulation. Phase modulation can take on many different forms; however, the primary issue with phase modulation is that the phase of the carrier is altered based on the information content to be sent. The phase change with the carrier is an ideal method for transporting digital information, and phase modulation is therefore used in many digital communication formats. Some phase modulation formats involve PSK, BPSK, QPSK, and OPQSK.

Phase modulation can be represented by Eq. (4.16).

$$E_{(t)} = A_c \sin (2\pi f_c t + \theta) \tag{4.16}$$

There is no difference between frequency and phase modulation provided the phase change in the signal is varying smoothly. It is only when there are quick transitions between phases with digital signals that PM is chosen as a method for improving the spectral efficiency for sending the data.

When the phase of the signal is altered as a function of the data stream, this method of modulating the signal is referred to as phase shift keying (PSK).

4.5.1 BPSK

Binary phase shift keying is a method of modulating the carrier phase by two discrete angles. The discrete angles are normally referenced at 0 and 180°; however, other variants can and do exist. The BPSK signal (see Fig. 4.20a) is represented by Eq. (4.16).

4.5.2 QPSK

Quadrature phase shift keying (QPSK) is one form of digital modulation which has a total of four unique phase states to represent data. The four phase states

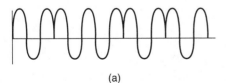

(a)

(b)

Figure 4.20 (a) BPSK signal. (b) Data diagram.

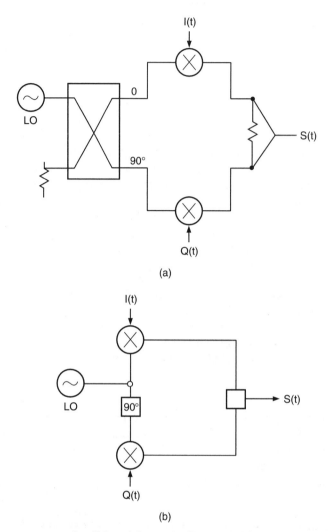

(a)

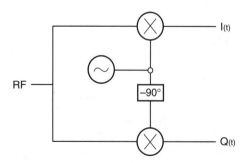

(b)

Figure 4.21 *I* and *Q* modulation. (*a*) Circuit. (*b*) Block diagram.

Figure 4.22 *I* and *Q* demodulation.

are arrived at through different I and Q values, utilizing four phase states; i.e., quadrature allows each phase state to represent two data bits.

The modulation and demodulation methods for QPSK are shown in Figs. 4.21 and 4.22. QPSK involves two data bits per I and Q function, and they are mapped on the IQ chart. The coordinate system for QPSK is best realized if you think in terms of an XY coordinate chart where X is now represented by the I, or in-phase, and y is the quadrature portion (I,Q). The distinct IQ location (phase state) shown represents a symbol made up of two distinct bits. The advantage of utilizing QPSK is the bandwidth efficiency. Since two data bits are now represented by a single symbol, less spectrum is needed to transport the information. The symbol rate = bit rate/(no. of bits/symbol).

4.5.3 DQPSK

Differential quadrature phase shift keying (DQPSK) is a modulation technique similar to that of QPSK. However, the primary difference between DQPSK and QPSK is that DQPSK does not require a reference from which to judge the transition. Instead DQPSK's data pattern is referenced to the previous DQPSK's phase state.

DQPSK has four potential phase states with the data symbols defined relative to the previous phase state as shown in Table 4.3.

$\pi/4$ differential quadrature phase shift keying ($\pi/4$ DQSK) modulation is similar to that of DQSK. However, the difference between $\pi/4$DQPSK and DQPSK is that the $\pi/4$DQPSK phase transitions are rotated 45° from that of DQPSK. Like DQPSK, $\pi/4$DQPSK has four transition states defined relative to the previous phase state shown in Table 4.4.

4.6 Digital Waveform Display Methods

Several methods can be utilized to view digitally modulated signals. Each of the methods has its positive and negative aspects, but the method chosen

TABLE 4.3

Symbol	DQPSK phase transition, °
00	0
01	90
10	−90
11	180

TABLE 4.4

Symbol	$\pi/4$ DQPSK phase transition, °
00	45
01	135
10	−45
11	−135

needs to match the objective at hand. The four primary methods for viewing digitally modulated signals are spectrum display, vector diagram, constellation diagram, and the eye diagram.

4.6.1   Spectrum display

The spectrum display is probably the most common method for viewing any RF modulation scheme. However, for viewing digital modulation, the spectrum display has limited value for analysis. The primary benefit of utilizing a spectrum display is to view the entire spectrum. The spectrum display (Fig. 4.23) can be used to view out-of-band emissions and the sidebands of the digitally modulated signal.

4.6.2   I&Q diagrams

One method that is utilized to represent phase and amplitude modulation is through *I&Q* diagrams. The *IQ* diagram shown in Fig. 4.24 utilizes vector notation for representing the actual *I* and *Q* values. Figure 4.25 represents some of the modulation schemes and their corresponding *I&Q* diagrams.

4.6.3   Vector diagrams

The vector diagram basically plots the *I* components as a function of *Q*. The primary purpose of utilizing a vector diagram for analysis of a digitally modulated signal is to view the transitions between the various states in a quadrature modulated signal. The transition status can be used to determine the overall modulation quality of the signal being viewed. If there is little error with the signal the locations on the vector diagram (Fig. 4.26) which represent symbol points are easily definable and the variation transition trajectories intersect closely at each of these points.

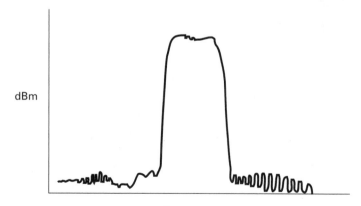

Figure 4.23   Spectrum display.

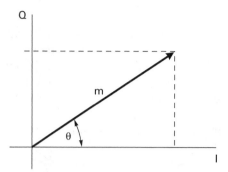

Figure 4.24 *IQ* modulation. $I = m \cos \theta$, $Q = m \sin \theta$, $m = I^2 + Q^2$, $\theta = \arctan (Q/I)$.

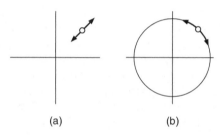

(a)   (b)

Figure 4.25 *IQ* diagrams. (*a*) AM. (*b*) FM.

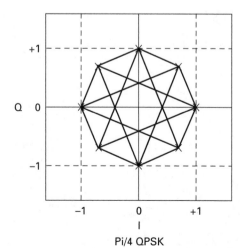

Pi/4 QPSK

Figure 4.26 Vector diagram.

4.6.4   Constellation diagrams

Another method of viewing digital modulation is the constellation diagram (Fig. 4.27), which is utilized to show the relationship between different amplitude and phase states of the modulated signal. It displays the error vector at the symbol sample time. The error vector is the difference between the theoretical symbol location and the actual symbol location on the constellation diagram.

Figure 4.28 presents additional constellation diagrams for different modulation techniques.

4.6.5   Eye diagram

The eye diagram shown in Fig. 4.29 is yet another method for the analysis of a digital communication system. For quadrature modulation the eye diagram plots both the I and Q components as a function of time. Utilizing the eye diagram you can determine the phase and amplitude errors of the system. As the number of errors in phase and amplitude increase, the eye pattern closes.

The eye diagram is better shown in Fig. 4.30, below where certain characteristics for the diagram are defined. The height of the eye is a direct indication of how well the signal is behaving in the presence of noise in the network. The width of the eye is a determination of the signal's immunity to clock phase errors. If the width decreases, it is an indication that the sampling that the system is using is not sufficient or rather optimal. And the slope s is an indication of the timing jitter. The steeper the slope the less immune the signal will be to changes in the clock sampling time.

Figure 4.31 shows the formation of an eye diagram and its relationship to the signals it is trying to represent.

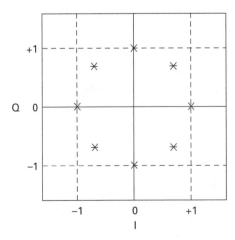

Figure 4.27 Constellation diagram.

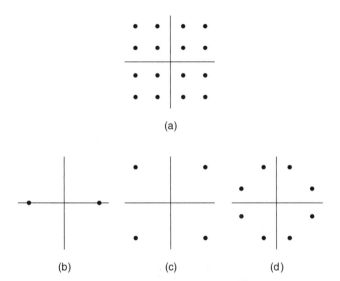

(a)

(b)   (c)   (d)

Figure 4.28   Constellation diagrams. (*a*) 16 QAM; *m* = 16.
(*b*) BPSK; *m* = 2. (*c*) QPSK; *m* = 4. (*d*) 8 PSK; *m* = 8.

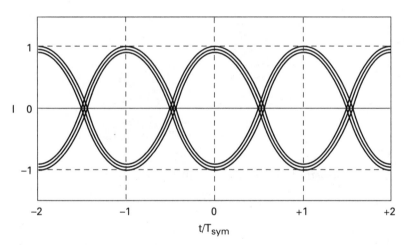

t/T_{sym}

Figure 4.29   Eye diagram.

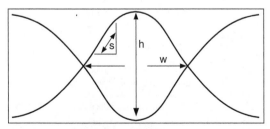

Figure 4.30   Eye diagram. *h*=vertical opening.
w=width of eye. *S*=slope of rising edge.

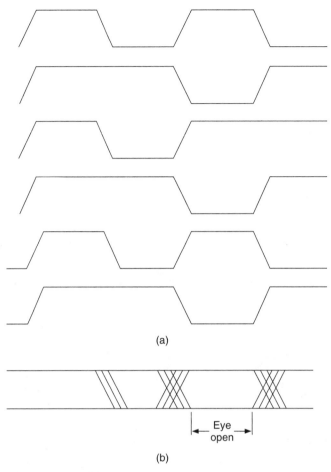

(a)

(b)

Figure 4.31 (*a*) Formation of eye diagram. (*b*) Signals represented by eye diagram.

References

1. Smith, Clint, and Curt Gervelis, *Cellular System Design and Optimization,* McGraw-Hill, New York, 1996.
2. Carlson, A. Bruce, *Communication Systems,* 2d ed., p. 356, McGraw-Hill, New York.
3. Rappaport, Theodore, *Wireless Communications Principles and Practices,* Prentice-Hall, Englewood Cliffs, NJ, 1996.
4. Lee, C. Y., *Mobile Cellular Telecommunications,* 2d ed., McGraw-Hill, New York, 1996.
5. Steele, Raymond, *Mobile Radio Communications,* Pentech, 1992.
6. Munda, Pritpal Singh, T. L. Singal, and Rakesh Kepur, *The Choice of a Digital Modulation Scheme in a Mobile Radio System,* pp. 1–4, IEEE, VTC, 1993.
7. Rappaport, Theodore, *Cellular Radio & Personal Communications Selected Readings,* IEEE, 1995.
8. Pasupathy, Subbrayan, "Minimum Shift Keying: A Spectrally Efficient Modulation," *IEEE Communications Mag.,* July 1979, pp. 14–22.
9. Howald, Robert, "Understanding the Fundamentals of Timing Recovery," *Microwaves & RF,* March 1996, pp. 89–97.
10. American Radio Relay League, *The ARRL 1986 Handbook,* 63d ed., The American Radio Relay League, Newington, CT, 1986.

11. AT&T, *Engineering and Operations in the Bell System,* 2d ed., AT&T Bell Laboratories, Murray Hill, NJ, 1983.
12. Brewster, *Telecommunications Technology,* Wiley, New York, 1986.
13. DeRose, *The Wireless Data Handbook,* Quantum Publishing, Mendocino, CA, 1994.
14. Dixon, *Spread Spectrum Systems,* 2d ed., Wiley, New York, 1984.
15. Kaufman, M., and A. H. Seidman, *Handbook of Electronics Calculations,* 2d ed., McGraw-Hill, New York, 1988.
16. Qualcom, *An Overview of the Application of Code Division Multiple Access (CDMA) to Digital Cellular Systems and Personal Cellular Networks,* Qualcom, San Diego, CA, May 21, 1992.
17. Sklonik, M. I., *Introduction to Radar Systems,* 2d ed., McGraw-Hill, New York, 1980.
18. Stimson, *Introduction to Airborne Radar,* Hughes Aircraft Company, El Segundo, CA, 1983.
19. White, Duff, *Electromagnetic Interference and Compatibility,* Interference Control Technologies, Gainesville, GA, 1972.
20. Qualcom, *CDMA Handbook,* Qualcom, San Diego, CA, 1996.
21. Fink, Donald, and Donald Christiansen, *Electronics Engineers Handbook,* 3d ed., McGraw-Hill, New York, 1989.
22. Schwartz, Bennett, and Stein, *Communication Systems and Technologies,* IEEE, New York, 1996.
23. Fink, Beaty, *Standard Handbook for Electrical Engineers,* 13th ed., McGraw-Hill, New York, 1995.
24. Webb, Hanzo, *Modern Amplitude Modulations,* IEEE, 1994.
25. Jakes, W. C., *Microwave Mobile Communications,* IEEE, New York, 1974.
26. Lathi, *Modern Digital and Analog Communication Systems,* CBS College Printing, New York, 1983.
27. *Reference Data for Radio Engineers,* 6th ed., Sams, 1983.
28. *The GSM System for Mobile Communications,* 1992.

5

Antennas

5.1 Introduction

The antenna system for any radio communication platform is one of the most critical and least understood parts of the system. The antenna system is the interface between the radio system and the external environment. It can consist of a single antenna at the base station and one at the mobile or receiving station. Primarily the antenna is used by the base station site and mobile for establishing and maintaining the communication link.

Many types of antennas are available, all of which perform specific functions depending on the application at hand. The type of antenna used by a system operator can be a collinear, log-periodic, folded dipole, or yagi, to mention a few. Coupled with the type of antenna is the notion of an active or passive antenna. The active antenna usually has some level of electronics associated with it to enhance its performance. The passive antenna is more of the classical type where no electronics are associated with its use and it simply consists entirely of passive elements.

Along with the type of antenna is the relative pattern of the antenna, indicating in what direction the energy emitted or received from it will be directed. Two primary classifications of antennas, associated with directivity for a system, are omni and directional. Omni antennas are used to obtain a 360° radiation pattern. Directional antennas are used when a more refined pattern is desired. The directional pattern is usually needed to facilitate system growth through frequency reuse or to shape the system contour.

The choice of which antenna to use will directly impact the performance of either the cell or the overall network. The radio engineer is primarily concerned in the design phase with the base station antenna, since this is the fixed location and there is some degree of control over the performance criteria that the engineer can exert on the location.

The correct antenna for the design can overcome coverage problems or other issues that are to be prevented or resolved. The antenna chosen for the application must take into account a multitude of design issues. Some issues that must be taken into account in the design phase involve the antenna's gain, its antenna pattern, the interface or matching to the transmitter, the receiver utilized for the site, the bandwidth and frequency range over which the signals to be sent will be applicable, its power-handling capabilities, and its IMD performance. Ultimately the antenna used for a network needs to match the system design objectives.

5.2   Types

Many types of antennas are available for use on the commercial market, without the need to invent more. The physical size of the antenna is directly related to the frequency of operation, or the frequency range the antenna is designed to operate with. Almost every antenna is based on one of four basic antenna elements, the dipole, monopole, loop, and patch.

5.2.1   Isotropic

The isotropic antenna is used as a reference for all antennas. It radiates energy uniformly in all directions. The dipole antenna discussed next has a power gain of 1.64, or rather 2.14 dB above that of an isotropic antenna. Equation (5.1) shows an isotropic antenna.

$$\phi \text{ dBd} = \text{dBi} + 2.14 \text{ db} \qquad (5.1)$$

5.2.2   Dipole

The dipole, the smallest self-resonating antenna, employs a conductor that is half the desired wavelength at which the system is supposed to operate. An example of a dipole is shown in Fig. 5.1. The dipole normally is fed in the center and has a characteristic impedance of 73 Ω. It is normally referenced to an isotropic antenna and has 2.14 dB of gain over an isotropic antenna.

5.2.3   Monopole

The monopole is commonly referred to as the "whip" antenna and is also referred to as the Marconi antenna. The monopole antenna is effectively a dipole antenna that is half the physical length of a dipole. The difference is the monopole obtains the same electrical length as that of a dipole through an image created by the inclusion of the ground plane used for the antenna. The length of the monopole is usually a quarter (0.25) wavelength.

The fact that the monopole is half the physical length also reduces the characteristic impedance to about one-half of that of the dipole, or rather 36 Ω. The monopole design also increases the power density to about twice that of the dipole, since the physical energy is now being sent through half of the physical size of the dipole.

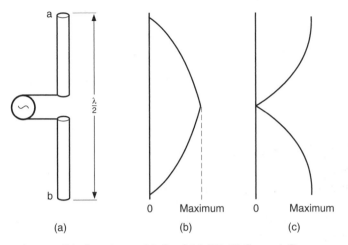

Figure 5.1 Dipole antenna. (*a*) $G = 2.14$ dBi. (*b*) Current (I).
(*c*) Electric field (E).

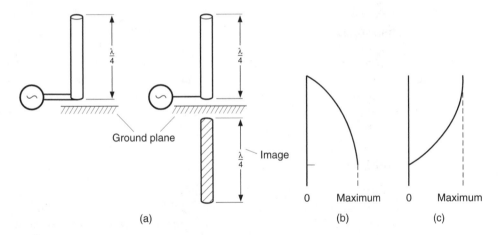

Figure 5.2 Monopole antenna. (*a*) $G = 4$ dBi. (*c*) Current. (*c*) Electric field.

An example of a monopole is shown in Fig. 5.2. The "rubber duck" antenna is an example of a monopole antenna.

5.2.4 Loop

The loop antenna shown in Fig. 5.3 illustrates the configuration for a rectangular or circular loop antenna. It is a closed-loop antenna. The physical size of the antenna is determined by the characteristic wavelength for the desired frequency. The loop antenna is fed by the transmitter or feeds the receiver as illustrated in Fig. 5.3 by simply breaking the loop anywhere along the antenna's path. This type of antenna has practical applications in the use of pagers and other PCS devices. However, the pager loop antenna length is typically a

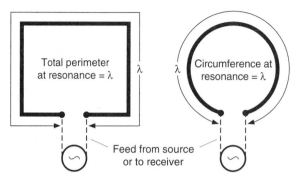

Figure 5.3 Rectangular and circular loop antennas.

fraction of the wavelength. The physical size of a loop antenna at one wavelength is 34 cm at 900 MHz.

5.2.5   Microstrip (patch)

The microstrip (patch) antenna is shown in Fig. 5.4. The microstrip antenna offers many advantages in size and cost that make it conducive for wireless applications. It consists of a conducting surface which is separated from the ground plane by a dielectric material. The patch antenna can be rectangular or circular in dimension. An additional advantage is its ability to be fabricated on a printed circuit board. The patch antenna is normally fed through a hole in the ground plane which then connects to the conductor or at the corner of the conductor.

5.3   Base Station Antennas

A multitude of antennas can be used at a base station. However, this section is restricted to collinear, log-periodic, folded dipole, yagi, and microstrip antennas. Of the antenna classifications mentioned two are more common for use in cellular and PCS communication systems for base stations. The two types of antennas used for base stations are collinear and log-periodic antennas.

5.3.1   Collinear

Collinear antennas can be either omni or directional. This type of antenna, which operates with a series of dipole elements that operate in phase, is referred to as a broadside radiator. The maximum radiation for the collinear antenna takes place along the dipole array axis, and the array consists of a number of parallel elements in one plane.

Figure 5.5 shows a four-element collinear array.

5.3.2   Log-periodic

The other general type of antenna used is a log-periodic dipole array (LPDA). The LPDA is a directional antenna whose gains, standing-wave ratio (SWR), and other key figures of merit remain constant over the operating band. The LPDA is used where a large bandwidth is needed, and the typical gain is 10 dBi.

The LPDA has a structural geometry such that its impedance and radiation characteristics repeat periodically as the log of the frequency. The actual antenna consists of several dipole elements which have different lengths and different relative spacing. Figure 5.6 shows a picture of a LPDA antenna. At the frequency of interest the short elements take on a capacitance property and receive little power by acting as a parasitic director. The longer elements, however, take on an inductive characteristic and are more dominant, acting as parasitic reflectors. The gain of the antenna indicates how it redirects or rather shapes the radiation pattern in a particular direction referenced to an ideal isotropic antenna.

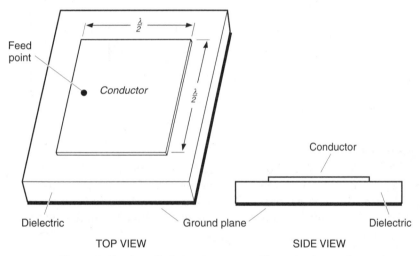

Figure 5.4 Top and side view of microstrip antenna. For a circular conductor the radius is 0.6 λ.

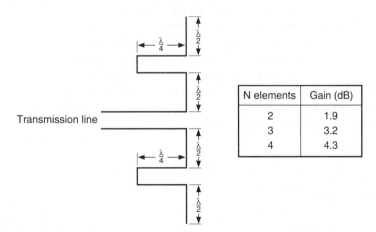

N elements	Gain (dB)
2	1.9
3	3.2
4	4.3

Figure 5.5 Four-element collinear array.

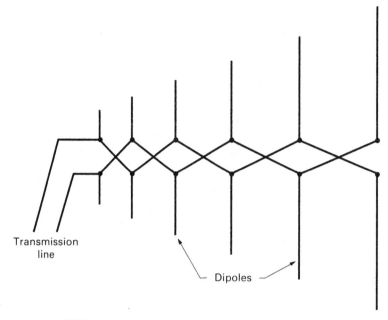

Figure 5.6   LPDA antenna.

5.3.3 Folded dipole

The folded dipole shown in Fig. 5.7 is formed by joining two dipoles at their ends and driving them at the center. The folded dipole has a better, or rather flatter, impedance versus frequency characteristic than a normal dipole. The impedance improvement over a wider frequency range is why the folded dipole is used.

5.3.4 Yagi

The Yagi antenna is commonly used for a variety of applications. It utilizes an end-fired method and therefore does not occupy much space in the vertical, or rather elevation, space. It operates off the principle of feeding a dipole and having the reflector spaced a determined distance away from the dipole, which is the driving element. A reflection of the energy takes place and is directed in the desired direction. To help increase the gain and directivity of the Yagi array directors are also added in the path but are on the side of the dipole where the energy is to be directed (Fig. 5.8).

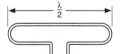

Figure 5.7   Folded dipole.

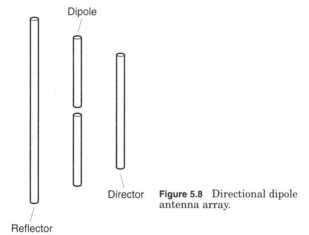

Dipole

Director   **Figure 5.8**   Directional dipole
antenna array.

Reflector

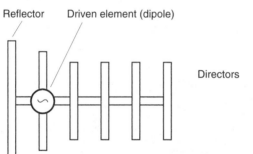

Reflector   Driven element (dipole)

Directors

Figure 5.9   Yagi antenna.

The reflector is larger in physical and electrical length than the dipole itself, and the directors are of the same size or smaller than the dipole. The spacing between the reflector and the dipole and the directors with the dipole is a function of the frequency of operation (Fig. 5.9).

5.3.5 Broadside array

A broadside antenna array shown in Fig. 5.10 has the primary direction of radiation, the main beam, perpendicular to the axis of the array. Examples of a broadside array are the collinear and folded dipole antennas.

5.3.6 End-fired array

The end-fired array is shown in Fig. 5.11. Its main beam, or radiation direction, coincides with the direction of the array itself. Examples of end-fired arrays are the LPDA and Yagi antennas.

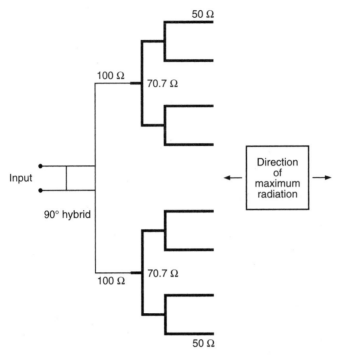

Figure 5.10 Broadside feed network.

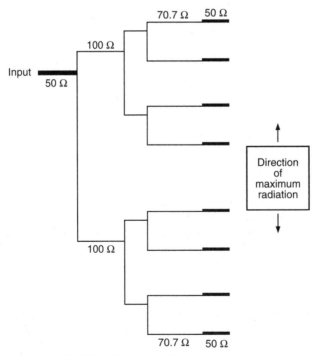

Figure 5.11 End-fired feed network.

5.3.7 Directional and omni antennas

Figures 5.12 to 5.15 are some of the available directional and omni antennas that are used for designing a network. The antenna types presented here are not all-inclusive but represent a large portion of the types that can be and are used in wireless communication for base stations.

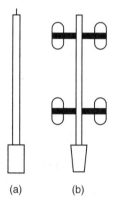

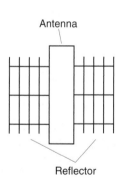

Figure 5.12 Omni antenna. (*a*) Collinear array. (*b*) Folded dipole array. (*c*) Omni (one indoor application).

(a)   (b)   (c)

Antenna

Antenna

Reflector

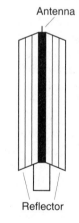

Reflector

Figure 5.13 Directional antennas.

Figure 5.14 Panel antenna.

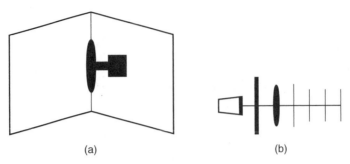

(a)   (b)

Figure 5.15   (*a*) Corner reflector and (*b*) Yagi antennas.

5.4   Performance Criteria

The performance of an antenna is not restricted to its gain characteristics and physical attributes, i.e., maintenance. Many parameters must be taken into account when looking at an antenna's performance. The parameters that define the performance of an antenna can be referred to as the figures of merit (FOM) that apply to any antenna that is selected for a communication system. Many parameters and figures of merit characterize the performance of an antenna system. The following is a partial list of the figures of merit for an antenna that should be quantified by the antenna manufacturer. The trade-offs that need to be made with an antenna chosen involve all the FOM issues discussed below.

1. *Antenna pattern.*   The graphical representation of the elevation and azimuth antenna patterns for both an omnidirectional and directional antenna is shown in Figs. 5.16 to 5.19. The antenna pattern is normally given in the form shown; however, it is very important to note the reference scale that is used since different scales can lead to different conclusions about how the antenna's pattern really will perform in the system.

The antenna pattern chosen should match the coverage requirements for the base station. For example, if the desire is to utilize a directional antenna for a particular sector of a cell site, 120°, then choosing an antenna pattern that covers 360° in azimuth would be incorrect. Care must also be taken in looking for electrical downtilt that may or may not be referenced in the literature.

2. *Main lobe.*   This is the radiation lobe containing the direction of maximum radiated power and is represented in Fig. 5.20. The main lobe is referenced to the polarization for the antenna and simply reflects the directivity of the antenna. In this case the polarization is vertical and therefore the main lobe representation is the elevation antenna pattern.

3. *Side lobe.*   This is the radiation's lobe in any direction other than the main lobe. The side lobe for an antenna is shown in Fig. 5.21. The side lobes are important to consider, since they can and do create potential problems with generating interference. Ideally there would be no side lobes for the antenna pattern. For downtilting the side lobes are important to note, since they can create secondary sources of interference.

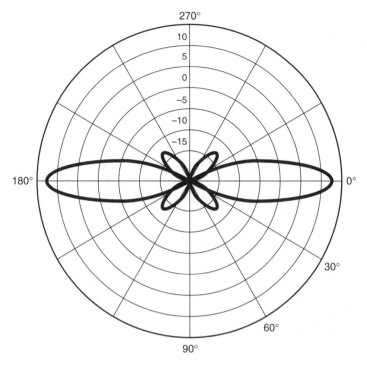

Figure 5.16 Elevation antenna pattern (omnidirectional).

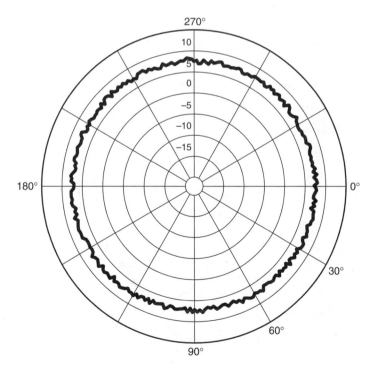

Figure 5.17 Azimuth antenna pattern (omnidirectional).

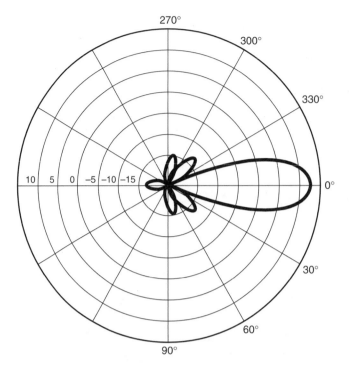

Figure 5.18 Elevation directional antenna.

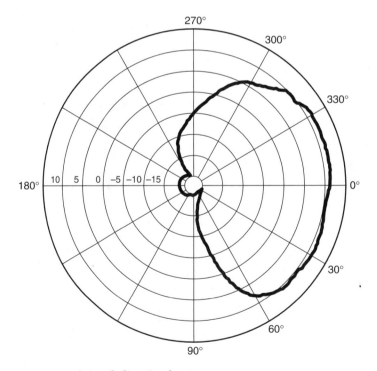

Figure 5.19 Azimuth directional antenna.

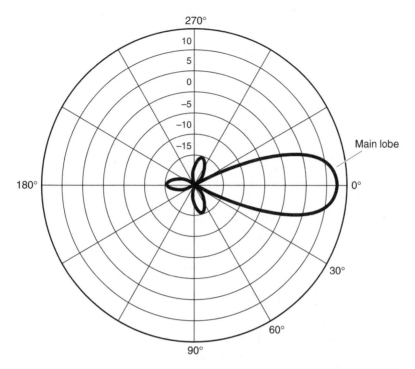

Figure 5.20 Main lobe antenna pattern.

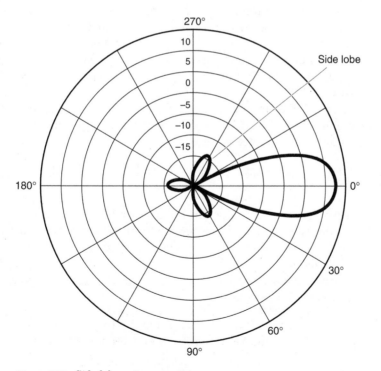

Figure 5.21 Side lobe antenna pattern.

4. *Input impedance.*   This is the impedance presented by the antenna at its terminals and is usually complex. The input impedance for an antenna should maximize the power transfer from the transmitter to the antenna for radiation. Nominal antenna input impedance is 50 Ω and matches that of the coaxial cable used to connect the transmitter to the antenna. A mismatch in input impedance obviously impacts the energy transfer and decreases its overall power output.

5. *Radiation efficiency.*   This is the ratio of total power radiated by an antenna to the net power accepted by an antenna from the transmitter, as shown in Eq. (5.2). The antenna would be 100 percent efficient if the power lost in the antenna were 0. This number indicates how much energy is lost in the antenna itself, assuming an ideal match with the feedline and the input impedance. Using the above equation, if the antenna absorbed 50 percent of the available power, it would have only 50 percent of the power for radiating, and thus the effective gain of the antenna would be reduced.

$$e = \text{power radiated/(power radiated + power lost)} \qquad (5.2)$$

6. *Beamwidth.*   This is the angular separation between two directions in which radiation interest is identical. The $\frac{1}{2}$ power point for the beamwidth is usually the angular separation where there is a 3-dB reduction off the main lobe (Figs. 5.22 and 5.23). The wider the beamwidth the lower the gain of the antenna is normally. A simple rule of thumb is for every doubling of the number of elements associated with an antenna a gain of 3 dB is realized. However, this gain comes at the expense of beamwidth. The beamwidth reduction for a 3-dB increase in gain is about one-half the initial beamwidth, so if an antenna has a 12° beamwidth and has an increase in gain of 3 dB then its beamwidth is now 6°.

7. *Directivity.*   This is the ratio of radiation intensity in a given direction to that of the radiation intensity averaged over all the other directions. Equation (5.3) shows antenna directivity:

$$G(D) = \text{maximum power radiation intensity/}$$
$$\text{average radiation intensity} \qquad (5.3)$$

The directivity of an antenna can be improved through use of reflectors or its physical location to a leg of a tower. Often the directivity of the antenna pattern can be altered through use of the tower structure itself.

8. *Gain.*   This is a very important figure of merit. The gain is the ratio of the radiation intensity in a given direction to that of an isotropically radiated signal. Equations (5.4) are for antenna gain:

$$G = \text{maximum radiation intensity from antennas/}$$
$$\text{maximum radiation from an isotopic antenna} \qquad (5.4)$$

$$G = e * G(D)$$

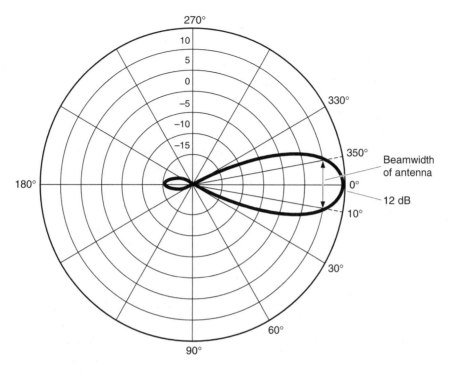

Figure 5.22 Elevation beamwidth. Beamwidth = 20°.

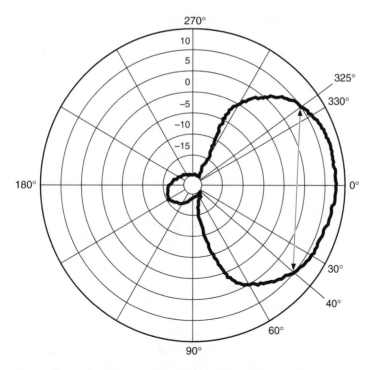

Figure 5.23 Azimuth beamwidth. Beamwidth = 75°.

If the antenna were without loss, $e = 1$, then $G = G(D)$. Referring to Figs. 5.22 and 5.23, the gain for the antenna in the elevation plan is about 12 dB. However, from the pattern it is unclear whether this gain is referenced to a dipole or isotropic antenna.

9. *Antenna polarization.*   The antenna polarization is defined as the polarization fields radiated by the antenna. The antenna's polarization is defined by the E field vector. Cellular and PCS systems utilize vertical polarization.

10. *Bandwidth.*   The bandwidth defines the operating range of the frequencies for the antenna. The SWR is usually how this is represented besides the frequency range over which it is constant. A typical bandwidth is the 1:1.5 SWR for the band of interest. Antennas are now being manufactured that exceed this, having a SWR value of 1:1.2 at the band edges.

The antenna's bandwidth must be selected with extreme care to account not only for current but also for future configuration options with the same cell site. For example, an antenna that is selected for use as the receive antenna at a cell site should also operate with the same performance in the transmit band, and vice versa. The rationale behind this dual-purpose use is to enable the antenna to be used for receive to be switched internally in the cell for use as a transmit antenna in the event that there is a problem with the transmit branch of the cell site.

11. *Front-to-back ratio.*   This is a ratio that is with respect to how much energy is directed in the exact opposite direction to the main lobe of the antenna. The front-to-back ratio is a loosely defined term. The IEEE Standard 145-1983 references the front-to-back ratio as the ratio of maximum directivity of an antenna to its directivity in a specified rearward direction. A front-to-back ratio is applicable only to a directional antenna, since obviously with an omnidirectional antenna there is no rearward direction.

Many manufacturers reference high front-to-back ratios, but care must be taken in knowing just how the number was computed. In addition, if installation is, say, on a building and the antenna will be mounted on a wall, then the front-to-back ratio is not as important a FOM. However, if the antenna is mounted so there are no obstructions between it and the reusing cell, then the front-to-back ratio can be an important FOM. Specifically in the latter case the front-to-back ratio should be at least the C/I level required for operation in the system.

An example of the front-to-back ratio is shown in Fig. 5.24. Here, the main lobe gain minus the negative lobe gain is $12 - (-15)$, giving a front-to-back ratio of 27 dB.

12. *Power dissipation.*   The total power the antenna can accept at its input terminals is its power dissipation. This is important to note, since receive antennas may not need to handle much power but the transmit antenna might have to handle 1500 W of peak power. The antenna chosen should be able to handle the maximum envisioned power load without damaging the antenna.

13. *Intermodulation.*   The amount of intermodulation the antenna will introduce to the network in the presence of strong signals should be checked against how the test was run. For instance, some manufacturers reference the

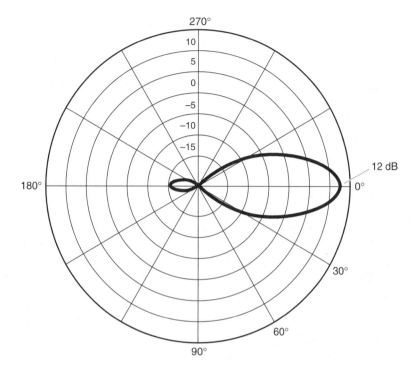

Figure 5.24 Front-to-back ratio.

IMD to 2 tones while some reference it to 3 or multiple tones. The point here is that the overall signal level that the IMD is generated at needs to be known in addition to their power levels that caused the IMD level.

14. *Construction.* The construction attributes associated with its physical dimensions, mounting requirements, materials used, wind loading, connectors, and color constitute this figure of merit. For instance, one of the items that needs to be factored into the construction FOM is the use of materials, whether the elements are soldered together or bolted. Another item is the type of metals that are used and what their expected life is in the environment that the antenna will be deployed in. For instance, if you install antennas near the ocean or an air-conditioning unit that uses salt water for cooling, it will be imperative that the material chosen will not corrode in the presence of salt water. Additionally, improper selection of metals for an antenna or its installation can and does lead to intermodulation problems.

15. *Cost.* How much the antenna costs is a critical figure of merit. No matter how well an antenna will perform in the system the cost associated with the antenna needs to be factored into the decision. For example, if the antenna chosen met or exceeded the design requirements for the system but cost twice as much as another antenna which met the requirements, the choice here would seem obvious: Pick the antenna that meets the requirement at the lowest cost.

Another example of cost implications involves selecting a new antenna type to be deployed in the network. The spares and stocking issues need to be factored into the antenna selection process. If the RF department designs every site's antenna requirements too uniquely, it is possible to have a plethora of antenna types deployed in the network, leading to a multitude of additional stocking issues for replacements. Therefore, it is important to select a specific number of antennas that should meet most if not all the design requirements for a system and utilize only those antennas.

5.5   Diversity

Diversity is used in wireless communication as a method for comparing signal fading in the environment. Diversity gain is based on the gain over fading that would have taken place in the event that a diversity technique was not used. In the case of a two-branch diversity system if the received signal into both antennas is not of an equal signal strength, then there cannot be any diversity gain. This is an interesting point considering most link budget calculations incorporate diversity gain as a positive attribute. The only way diversity gain can be incorporated into a link budget is if a fade margin is included in the link budget and the diversity scheme chosen attempts to improve (reduce) the fade margin that is included there.

5.5.1   Spatial

Spatial diversity is achieved by physically separating the antennas by a defined physical separation, either horizontal or vertical, for the sole purpose of minimizing or removing the negative effects associated with fading in a radio environment. The objective is to have enough separation to increase the chances that each antenna used in the system will not experience the same fade characteristics at the same instant in time.

However, diversity is used primarily at the base station and not at the subscriber unit, ordinarily. The reason for diversity playing a role only in the base station side of the system is the fact that most subscriber units have only one antenna and therefore cannot possibly utilize any advantage or improvement with diversity.

Horizontal.   This is done following the equation for horizontal diversity [Eq. (5.5)]. Multiple papers have been written about the reasons why horizontal separation is the desired mode of operation for a mobile radio system.

$$n = \frac{h}{D} \tag{5.5}$$

$$n = \frac{h}{D} = 11$$

where h = height (ft)
D = distance between antennas (ft)

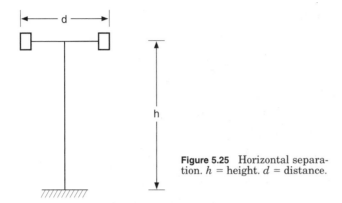

Figure 5.25 Horizontal separation. h = height. d = distance.

The objective is to have the antennas spaced from 10 to 20 wavelengths for the frequency of operation (Fig. 5.25).

Vertical.   Vertical diversity is another diversity scheme which, when applied to the mobile communication field, requires the antennas to be separated by a defined vertical separation. Vertical separation can be used as a method for diversity reception for mobile communications. However, the drawback here is that typically the separation of the antennas is greater for vertical than it is for horizontal. In addition the multipath spread angle is usually small for a mobile environment, making this technique unattractive for improving the multipath environment for distant mobiles.

5.5.2 Polarization

This form of diversity reception involves utilizing two different branches that are polarized in orthogonal planes. Polarization diversity lends itself to only a two-branch system in practicality. There has been and continues to be work conducted with implementing this type of diversity scheme in a mobile environment. The main advantage that polarization diversity offers a communication system is a reduced number of antennas required for an installation.

5.5.3 Time

The use of time diversity is based on the principle that the signal will undergo different fading characteristics when sent at different times. The required time interval for time diversity is the reciprocal of what is expected to be the fading bandwidth for the channel in question. Time diversity is usually exploited in interleaving and forward error correction. However, one of the drawbacks with utilization of time diversity is the processing time required to collect and correlate all the repeated or interleaved signals.

5.5.4 Frequency

Frequency diversity involves the technique of transmitting the information content on two different frequencies. The frequencies should be such that each will undergo an uncorrelated fade with the other. The primary advantage of using frequency diversity for a communication site is the reduction in the number of antennas needed for the location. The disadvantage is that to achieve diversity utilizing this type of scheme the frequency efficiency of the system is reduced to half of what it was beforehand. In addition frequency diversity requires a second transmitter at the sending location, which is not practical for a mobile environment.

5.5.5   Angle

Angle diversity is another technique utilized for reducing fading in a mobile communication system. Angle diversity has direct applications in the fixed microwave and in mobile wireless communications. The use of angle diversity was exploited in six-sector environments for cellular communication. Angle diversity has the advantage of providing diversity while at the same time reducing the tower loading that would normally accompany space diversity techniques.

This type of diversity relies on the angle of arrival of the signal instead of relying on the physical separation of the antennas to account for different fading characteristics of the same signal. A typical angle diversity system is shown in Fig. 5.26. The signal from the subscriber unit is received equally in both sectors 1 and 2 of the six-sector cell site. Either max ratio or select diversity in the radio itself can be employed to take advantage of this configuration. The chief advantage of this type of configuration is that you can employ only one receive antenna per sector and if it is duplexed a total of six antennas can realize a six-sector cell site whereas using space diversity a minimum total of 12 antennas would be needed.

5.6   Antenna Isolation

Antenna isolation is an important part of a communication system design. Often the isolation from a communication system transmitter to its own

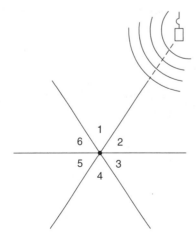

Figure 5.26 Angle diversity.

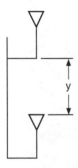

Figure 5.27 Vertical isolation.
y = distance between ends of
antennas.

receiver is integrated into the initial design of the system. However, as is often the case, colocating wireless communication systems at or near the same antenna platform requires adherence to obtaining a minimum level of isolation between a transmitting element and the receive system.

The isolation that is covered here is related to the physical separation of the antennas (transmit to receive antenna). Additional isolation is afforded the communication system by filtering techniques employed at the site. However, physically separating the antennas either vertically, horizontally, or some combination will afford a particular level of isolation. Isolation is normally expressed in terms of dB of attenuation the signal from the transmitter will undergo as it travels toward the receive antenna.

Vertical separation of the antennas achieves the highest single level of attenuation. The amount of attenuation experienced is dependent upon the frequency of operation and the spacing between the antennas (see Fig. 5.27). The calculation for vertical isolation (VI), measured in decibels, is shown in Eq. (5.6).

$$VI = 28 + 40 \log_{10}\left(\frac{y}{\lambda}\right) \text{ dB} \tag{5.6}$$

where $y > \lambda$

$\therefore$ if

$$f = 890 \text{ mHz} \qquad y = 10 \text{ ft} = 120 \text{ in}$$

$$VI = 28 + 40 \log_{10}\left(\frac{120}{13.267}\right)$$

$$= 28 + 38.256$$

$$= 66.25 \text{ dB}$$

Horizontal separation can also be used to achieve isolation between communication systems. However, the amount of physical separation needed to

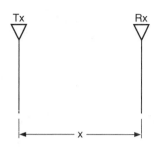

Tx   Rx

Figure 5.28 Horizontal isolation. x = distance between antennas T_x and R_x.

achieve the same level of isolation as that of vertically separating the antennas is much greater. Figure 5.28 illustrates horizontal isolation in an antenna system. The calculation for horizontal isolation (HI), measured in decibels, is shown in Eq. (5.7).

$$(dB) \approx 22 + 20 \log\left(\frac{x}{\lambda}\right) = HI \qquad (5.7)$$

where $\dfrac{x}{\lambda} > 10$

$\therefore$ if

$$T_x = 10 \text{ dB} \qquad f = 890 \text{ mHz}$$

$$R_x = 10 \text{ dB} \qquad x = 10 \text{ ft}$$

$$VI = 22 + 20 \log_{10}\left(\frac{10 \times 12}{13.267}\right)$$

$$= 22 + 19$$

$$= 41 \text{ dB}$$

As in most installations, pure vertical or horizontal separation is not practical and some level of vertical and horizontal separation will be employed. The isolation afforded a slant angle installation is shown in Fig. 5.29. The calculation for slant isolation (SI) is shown in Eq. (5.8).

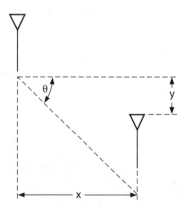

Figure 5.29 Slant isolation. x = horizontal distance. y = vertical distance. θ = slant angle relative to horizontal plane.

$$(VI-HI) \cdot \frac{\theta}{90} + HI \qquad (5.8)$$

$\therefore$ if

$$f = 890 \text{ mHz}$$

$$x = 10 \text{ ft}$$

$$y = 10 \text{ ft}$$

$$\theta = 45°$$

$$SI = (66.25 - 41) \cdot \frac{45}{90} + 41$$

$$= 53.625 \text{ dB}$$

Figures 5.30 and 5.31 show the relative isolation for similar distances both vertical and horizontal as a function of both distance and frequency of operation.

5.7   Beamwidth versus Gain

Since there is no free lunch in anything, the relationship between the antenna's bandwidth and gain needs to be understood. Basically the larger the antenna is physically, in terms of wavelengths, the higher the gain and increased directivity will be for the antenna. However, the wider the bandwidth the antenna is supposed to operate over the less overall gain it will have.

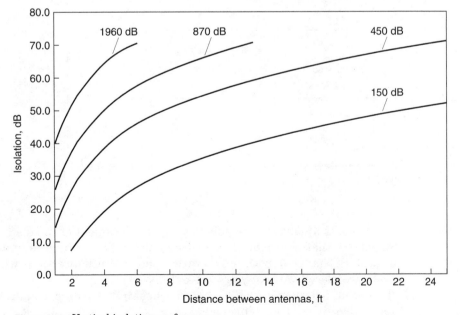

Figure 5.30   Vertical isolation vs. frequency.

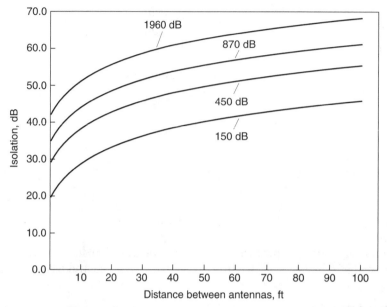

Figure 5.31 Horizontal isolation vs. frequency.

	Dipole arrangement	Elevation	Azimuth
Single			
Two dipoles vertically spaced			
Two dipoles horizontally spaced			

Figure 5.32 Dipole patterns.

Figure 5.32 illustrates the effects that the placement of the elements has on the pattern that is formed. It utilizes a simple dipole antenna and illustrates the pattern change in both the elevation and azimuth directions when two elements are placed vertically or horizontally apart.

Equation (5.9) demonstrates the relationship between gain and beamwidth and further drives the point home.

$$G = 10 \log_{10}\left[\frac{n(41253)}{\theta_E \cdot \theta_h}\right] \tag{5.9}$$

where G = gain of antenna, dBi
  n = efficiency factor (≤ 1)
  θ_E = 3 dB E plane
  θ_h = 3 dB H plane

The relationship shows that if the antenna vertical size is increased by twice the size, the beamwidth of the antenna in the vertical or elevation plane has decreased to half of its original value. In addition, as the beamwidth decreases, the gain also increases by 2 or rather 3 dB.

The antenna beamwidth and gain relationship is shown in Fig. 5.33 and Eqs. (5.10).

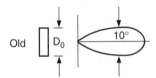

Old D_0 10°

New D_N 8°

Figure 5.33   Antenna beamwidth vs. gain. $D_N = 2 \cdot D_0$.

$$G = \frac{k}{\theta_H \theta_V} \qquad BW_{3\,dB} = \frac{k\lambda}{D} \tag{5.10}$$

$$\therefore \frac{BW_{new}}{BW_{old}} = \frac{k\left(\dfrac{\lambda}{D_N}\right)}{k\left(\dfrac{\lambda}{D_0}\right)} = \frac{D_0}{D_N} = \frac{D_0}{2D_0} = \frac{1}{2}$$

$$BW_{new} = \frac{1}{2} BW_{old} = \frac{1}{2}\theta_{V_{old}} = \theta_{V_{new}}$$

$$\frac{G_{new}}{G_{old}} = \frac{\left(\dfrac{k}{\theta_{V_N}\theta_{H_N}}\right)}{\left(\dfrac{k}{\theta_{V_0}\theta_{H_0}}\right)} = \frac{\theta_{V_0}}{\theta_{V_N}} = \frac{\theta_{V_0}}{\dfrac{\theta_{V_0}}{2}} = 2$$

5.8   Connectors

In wireless communication systems many types of connectors are utilized. Often at a communication site one system will employ multiple types of connectors depending on the application at hand. The connector chosen is electrically an extension of the coaxial cable utilized.

The types of connectors normally utilized at a communication site include the following:

1. N-type

2. 7/16 DIN

3. EIA flange

4. SMA

5. TNC

6. Mini UHF

7. UHF

The two most common wireless communication systems are the N and 7/16 DIN connectors, usually associated with the antenna and transmit path for the system. In the United States, N-type connectors have enjoyed wide use but 7/16 DIN are being more commonly utilized because of their better intermodulation performance. The intermodulation performance is a direct result of the contact area the 7/16 DIN has over the N-type connector; that is, it is greater. The larger the contact surface the better the intermodulation suppression for the connector.

When selecting a connector to utilize at a communication site, the following items are needed:

1. Cable size

2. Frequency range

3. Coupling method

4. SWR

5. Intermodulation suppression

6. Installation

7. Maintenance

5.9   Intelligent Antennas

Intelligent antenna systems are being introduced to commercial wireless communication systems. The concepts and implementation for intelligent antenna systems have been utilized in other industries, primarily the military, for some time.

Intelligent antenna systems can be configured as either receive only or full duplex operations. They can be arranged as either an omni or sector cell site depending on the application at hand.

The fundamental objective is to increase the S/N by reducing the amount of N (noise and interference) and possibly increasing the S (serving signal) in the same process. All the technologies are based on the principle that a narrower radiation beam pattern will provide increased gain and can be directed toward the subscriber and at the same time offer less gain to interfering signals that will arrive at an off-axis angle because of the reduced beamwidth size.

A simple rule of thumb can be used for determining the amount of improved S/N that can be achieved with an intelligent antenna system.

Here N = number of segmentations, and the calculation is shown as Eq. (5.11).

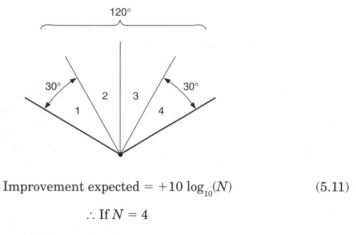

$$\text{Improvement expected} = +10 \log_{10}(N) \qquad (5.11)$$

$$\therefore \text{If } N = 4$$

That is, each segment is 30° out of 120°.

$$\text{Improvement expected} = +10 \log_{10}(4) = 6 \text{ dB}$$

Figure 5.34 illustrates three types of intelligent antenna system; each has positive and negative attributes. All the illustrations shown can be either receive only or full duplex. The difference between the receive only and the full duplex systems involves the number of antennas and potential number of transmitting elements in the cell site itself.

The beam switching antenna arrangement shown is the simplest to implement. It normally involves four standard antennas of narrow azimuth beamwidth, 30° for a 120° sector, and based on the receive signal received, the appropriate antenna will be selected by the base station controller for use in the receive path.

The multiple beam array shown involves utilizing an antenna matrix to accomplish the beam switching. The beam steering array, however, utilizes phase shifting to direct the beam toward the subscriber unit. The direction chosen by the system for directing the beam will affect the entire sector.

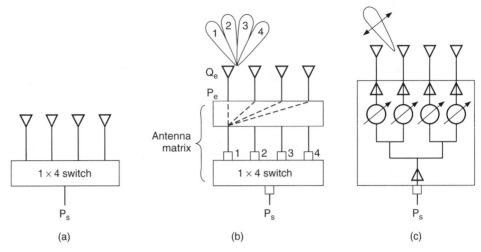

Figure 5.34 Intelligent antenna systems. (*a*) Switched antennas. $P_e = P_s$.
(*b*) Multiple-beam array. $P_e = P_s/4$. (*c*) Steered-beam array. $P_e = P_s/4$.

Normally amplifiers for transmit and receive are located in conjunction with the antenna itself. In addition the phase shifters are located directly behind each antenna element. The objective of placing the electronics in the masthead is to maximize the receive sensitivity and exploit the maximum transmit power for the site.

5.10   Antenna Selection

The antenna selection process can be either straightforward or depending on how it is approached. It is recommended that the antenna picked meet the desired goals for the cell site and the system. However, the antenna selection process should be straightforward if the desired objective is known from the beginning. The antenna selected for the application should be one that meets the following major points as a minimum.

1. Elevation and azimuth patterns meet requirements.
2. Antenna exhibits the proper gain desired.
3. Antenna is available from common stock, company inventory.
4. Antenna can be mounted properly at location, that is, be physically mounted at the desired location.
5. Antennas will not adversely affect the tower, wind, and ice loading for the installation.
6. Visual impact (negative) has been minimized in the design and selection phase.
7. Antenna meets the desired performance specifications.

5.11   Tilting

Antenna inclination, whether uptilting or downtilting, is often misapplied by the engineer designing the network. The primary issue that is often missed in the antenna inclination adjustment phase is the effects of local clutter or terrain on the pattern itself. Often the antenna tilt angle is chosen solely on the basis of the antenna elevation pattern. If the desired result is to, say, remove 3 dB of energy contributed by the site being considered to antenna inclination changes, then the antenna pattern is looked at and wherever the 3-dB point is on the elevation pattern is the desired tilt angle. The negative with simply employing a tilt angle based on the antenna pattern itself is illustrated in Fig. 5.35.

5.12   Sweeping Procedure

The antenna system should have its integrity checked at the time of installation and on a periodic basis. There are several methods of sweeping the antenna system. It is important to have a defined set of objectives for a sweeping procedure, and performance guidelines must therefore be adhered to.

One method of sweeping an antenna system is through use of a time-domain reflectometer (TDR). The TDR is a valuable tool for detecting feedline problems, i.e., crimped cables due to excessive bends or overzealous tightening of grounding clamps. The TDR has the disadvantage of not offering the frequency characteristics of the antenna system.

A second method for sweeping the antenna system is through use of a tracking generator and a service monitor, or spectrum analyzer. This has the unique advantage of determining the frequency characteristics of the antenna system but offers little or no positional data for determining where, if any, a fault lies in the antenna system itself.

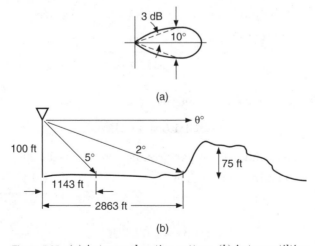

Figure 5.35   (*a*) Antenna elevation pattern. (*b*) Antenna tilting.

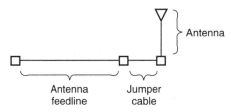

Figure 5.36 Return loss.

A third method is using a network analyzer that has S parameter capabilities. The network analyzer has the ability to do not only frequency-domain but also time-domain analysis of the antenna system. A sweeping procedure using a network analyzer is included in Appendix A. The method used for checking the antenna system can be easily converted to use with another network analyzer.

When sweeping an antenna system, it is important to note that the cable length used at the site can directly affect the conclusions arrived at. For example, if the cable-line loss is 3 dB, a poor antenna can still be considered good if this factor is not included in the analysis.

The performance criteria for sweeping an antenna system (see Fig. 5.36) are derived as follows.

$$\text{SWR } 1{:}1.2 = 20 \text{ dB return loss}$$

$$\text{Feedline: } \tfrac{7}{8} \text{ in at } 1.28 \text{ dB/100 ft}$$

$$\text{Jumper cable: } \tfrac{1}{2} \text{ in LdF at } 3.4 \text{ dB/100 ft}$$

Antenna	20.00	
Feedline (100 ft)	1.28	forward
Jumper (10 ft)	0.34	
Feedline (100 ft)	1.28	return
Jumper (10 ft)	0.34	
	23.24 dB	Return loss

However, if the goal is SWR 1.5 for the antenna system

$$\text{SWR } 1{:}1.5 = 14 \text{ dB}$$

Antenna system performance goal	= 14.00
Feedline and jumper	−3.14
	10.86 dB

$$\text{Antenna} \approx 2.0 \text{ SWR}$$

But

$$\begin{array}{ll}
\text{Antenna SWR (1:1.5)} & = 14.00 \\
\text{Feedline} \cdot \text{jumper} & = 3.14 \\
\hline
& 17.14 \text{ dB} \approx 1{:}1.3 \text{ SWR}
\end{array}$$

If the SWR of the desired antenna system was, say, 1:1.2 or better, this would equate to a 23-dB or better return loss. But if the goal was to have a SWR of 1:1.5, it is possible to have an antenna that has a SWR of 1:2.0 and still meet the system requirements because the cable system masked the real performance of the antenna system. Obviously having an antenna with a SWR of 2.0 or greater would necessitate its removal from the base station antenna system. It is advisable therefore to utilize the SWR value that is referenced for the particular antenna and then calculate in advance what the acceptable range for the return loss should be. I suggest using a tolerance of 10 percent as an upper and lower boundary for the desired results.

An example of a sweeping procedure for an antenna system is included in Appendix B. The procedure can be used directly or simply modified for the particular situation. The results from all sweeping procedures should be cataloged and stored for future comparisons of the particular antenna system.

References

1. Smith, Clint, and Curt Gervelis, *Cellular System Design and Optimization,* McGraw-Hill, New York, 1996.
2. Egenstafer, Frank, "Path Loss and Antenna Gain Elementary Calculations," *RF Design,* February 1995, pp. 52–61.
3. Edward, B. J., and R. S. Webb, "A W-Band Active Phased-Array Antenna," *Microwave J.,* May 1996, pp. 254–262.
4. Schweber, Bill, "Antennas: Critical Links in the Wireless Signal Chain," *EDN* June 9, 1996, pp. 52–62.
5. Cuthbert, Thomas, "Broadband Impedance Matching Methods," *RF Design,* August 1994, pp. 64–71.
6. Browne, Jack, "Advances in Antennas Drive Wireless Systems," *Microwaves & RF,* May 1995, pp. 129–131.
7. Lee William, C. Y., "Vertical vs. Horizontal Separations for Diversity Antennas," *Cellular Business,* December 1991, pp. 56–60.
8. Floh'H, Jean Marie, "Method Determines the Performance of Dipole Arrays," *Microwaves & RF,* September 1995, pp. 104–110.
9. Rosol, George, "Planning an Antenna Installation for a Wireless System," *Microwaves & RF,* October 1995, pp. 116–123.
10. Steyskal, Hans, "Digital Beamforming at Rome Laboratory," *Microwave J.,* February 1996, pp. 100–126.
11. Pozar, David, and Daniel Schaubert, *Microstrip Antennas,* IEEE.
12. ANSI/IEEE 149-1979, *IEEE Standard Test Procedures for Antennas,* IEEE.
13. Waldron, Jim, "If It's Not a Whip, It Won't Work," *Mobile Radio Technology,* July 1990, pp. 22–36.
14. Doty, A. C., John A. Frey, and Harry J Mills, "Capacitive Bottom-loading Tunes Mobile Antennas," *Mobile Radio Technology,* July 1990, pp. 11–18.
15. White, C. M., "Corroded Base Antennas May Cause Radio Noise," *Mobile Radio Technology,* July 1991, p. 16.

16. Nishikawa, N., and Y. Asano, *Vertical Patterns of Mobile Antenna at UHF Frequencies,* IEEE, 1985.
17. Blaese, Herb, "Cellular Antenna Theory with the Installer in Mind," *Mobile Radio Technology,* January 1991, pp. 22–34.
18. Breed, Gary, "Antenna Basics for Wireless Communications," *RF Design,* October 1995, pp. 60–65.
19. Oster, John, and Ed Bachner, *Angle Diversity, a Practical Technique for Reducing Fades,* Andrew Corporation, June 9, 1989.
20. Singer, Andrew, "Use Downtilt Selectively to Improve System Coverage," *Mobile Radio Technology,* November 1990, pp. 62–66.
21. Rosol, George, "Evaluating Base-Station Antennas Performance," *Microwaves & RF,* September 1995, pp. 127–131.
22. Monser, George, "A Reversible Rotman Lens, Useful in Short, Linear, Array Applications," *Microwave J.,* January 1995, pp. 160–164.
23. Singer, Andrew, "The Ultimate Guide to Selecting Base Station Antennas," *Radio Resource,* November 1993, pp. 45–55.
24. IEEE Std 145-1883, *IEEE Standard Definitions of Terms for Antennas,* IEEE.
25. Siva Chebolu, Raj Mittra, and Wiren Becker, "The Analysis of Microwave Antennas Using the FDTD Method," *Microwave J.,* January 1996, pp. 134–150.
26. Schneiderman, Ron, "Antenna Makers Set 'Smart' Goals," *Microwaves & RF,* May 1995, pp. 33–41.
27. American Radio Relay League, *The ARRL 1986 Handbook,* 63d ed., The American Radio Relay League, Newington, CT, 1986.
28. American Radio Relay League, *The ARRL Antenna Handbook,* 14th ed., The American Radio Relay League, Newington, CT, 1984.
29. AT&T, *Engineering and Operations in the Bell System,* 2d ed., AT&T Bell Laboratories, Murray Hill, NJ, 1983.
30. Carr, J. J., *Practical Antenna Handbook,* McGraw-Hill, New York, 1989.
31. Code of Federal Regulations, CFR 47 Parts 1,17,22,24, and 90.
32. IEEE Standard Test Procedures for Antennas, IEEE, New York, 1979.
33. Kaufman, M., and A. H Seidman, *Handbook of Electronics Calculations,* 2d ed., McGraw-Hill, New York, 1988.
34. Lee, W. C. Y., *Mobile Cellular Telecommunications Systems,* 2d ed., McGraw-Hill, New York, 1996.
35. White, Duff, *Electromagnetic Interference and Compatibility,* Interference Control Technologies, Gainesville, GA, 1972.
36. Yarborough, *Electrical Engineering Reference Manual,* 5th ed., Professional Publications, Belmont, CA, 1990.
37. Fink, Donald, and Donald Christiansen, *Electronics Engineers Handbook,* 3d ed., McGraw-Hill, New York, 1989.
38. Fink, Beaty, *Standard Handbook for Electrical Engineers,* 13th ed., McGraw-Hill, New York, 1995.
39. Carlson, A. B., *Communications Systems,* 2d ed., McGraw-Hill, New York, 1975.
40. Johnson, R. C., and H. Jasik, *Antenna Engineering Handbook,* 2d ed., McGraw-Hill, New York, 1984.
41. Pozar, Schaubert, *Microstrip Antennas,* IEEE, New York, 1995.
42. *Reference Data for Radio Engineers,* 6th ed., Sams, 1983.
43. Brewster, *Telecommunications Technology,* Wiley, New York, 1986.
44. DeRose, *The Wireless Data Handbook,* Quantum Publishing, Mendocino, CA, 1994.
45. Dixon, *Spread Spectrum Systems,* 2d ed., Wiley, New York, 1984.
46. Hata, M., "Empirical Formula for Propagation Loss in Land Mobile Radio Services," *IEEE Trans. on Vehicle Technology,* vol. VT-29, no. 3, 317–325, 1980.
47. Low, K., "Comparison of Urban Propagation Models with CW Measurements," 42d IEEE Vehicle Technology Conference, Denver, May 1992, pp. 317–325.
48. Qualcom, *An Overview of the Application of Code Division Multiple Access (CDMA) to Digital Cellular Systems and Personal Cellular Networks,* Qualcom, San Diego, CA, May 21, 1992.
49. Sklonik, M. I., *Introduction to Radar Systems,* 2d ed., McGraw-Hill, New York, 1980.
50. Simo, "IS-95 Based SS-CDMA: Operational ISSUES," Seminar, January 1995.
51. Stimson, *Introduction to Airborne Radar,* Hughes Aircraft Company, El Segundo, CA, 1983.
52. Qualcom, *CDMA Handbook,* Qualcom, San Diego, CA, 1996.

53. Schwartz, Bennett, and Stein, *Communication Systems and Technologies,* IEEE, New York, 1996.
54. Rappaport, *Wireless Communications Principles and Practices,* IEEE, 1996.
55. Webb, Hanzo, *Modern Amplitude Modulations,* IEEE, 1994.
56. Gibson, *The Mobile Communications Handbook,* IEEE, 1996.
57. Steele, *Mobile Radio Communications,* IEEE, 1992.
58. Carlson, A. B., *Communications Systems,* 2d ed., McGraw-Hill, New York, 1975.
59. Lathi, *Modern Digital and Analog Communication Systems,* CBS College Printing, New York, 1983.
60. MacDonald, "The Cellular Concept," *Bell Systems Technical J.,* vol. 58, no. 1, 1979.
61. Rappaport, *Cellular Radio & Personal Communications,* IEEE Press, New York, 1995.
62. Mouly, Pautet, *The GSM System for Mobile Communications,* Mouly Pautet, 1992.
63. Williams, Taylor, *Electronic Filter Design Handbook,* 3d ed., McGraw-Hill, New York, 1995.

6

Filters

6.1 Introduction

Filters play an integral part in the design and operation of a radio and communication system. Many types of filters can be and are deployed. Selection of the types of filters to use is based on their mission statement and cost, in terms of either spectrum or actual monetary issues. Simply put, the purpose of a filter is to allow the desired energy or information to pass undistorted in either phase, amplitude, or time and at the same time completely suppress all other energy.

More often than not the filter characteristics of an existing or new communication system are overlooked by the system design engineer. With the proliferation of wireless communication, especially in the 800-MHz and 1.9-GHz bands, the need to pay particular attention to the filter characteristics of the base station becomes paramount. With this proliferation comes the demand for smaller physical size with increased attenuation characteristics of unwanted signals, at the same time not distorting the desired signal in any fashion.

This chapter introduces several types of filters, their importance to a communication system or radio, and some specific applications for them. Several excellent books written specifically dealing with filters are included in the reference section of this chapter.

6.2 Filter Types

Many types of filters are available for use, and each has its own unique characteristics. The specific characteristics of a filter are driven by its physical construction, which is an important aspect of the selection process, and the types of filters listed are passive filters only. Active filters have many good applications but are not covered in this section.

Four general classifications of filters are used throughout all of radio communications. The filter classifications are listed below. It should be noted that there are many perturbations regarding combinations of the general filter types listed. The specific configuration chosen is entirely dependent upon the application that it is meant to solve and the acceptable trade-offs that come along with the filter choice made.

The general classification for filters falls into one of the following four basic filters:

1. Low-pass filter

2. High-pass filter

3. Bandpass filter

4. Band-reject filter (notch filter)

Ideally a filter would pass without attenuation all the frequencies within a specified pass band and infinitely attenuate all those frequencies outside the pass band. Additionally the time response of the filter would be such that the output is identical to the input with some delay time. In other words the transfer function for the filter should be equal to 1 for the frequencies of interest only (Fig. 6.1). Here Eqs. (6.1) apply.

$$T(s) = \frac{e_{out}}{e_{in}} = \frac{N(s)}{D(s)}$$

where $N(s)$ = zeros
   $D(s)$ = poles   (6.1)

$$T(s) = \frac{N_m S^m + N_{m-1} S^{m-1} + \cdots + N_1 S + N_0}{D_n S^n + D_{n-1} S^{n-1} + \cdots + D_1 S + D_0}$$

where $S = jw$
   highest power of m in numerator is number of zeros
   highest power of n in denominator is number of poles

An example of an ideal low-pass, high-pass, bandpass, and band-reject filter is included in Fig. 6.2.

The specific filter type selected should be determined based on which trade-offs or imperfections can be best tolerated. However, the choice of trade-offs is often a difficult task, since the ideal filter is truly not realizable at this particular time.

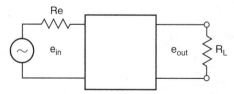

Figure 6.1 Transfer function.

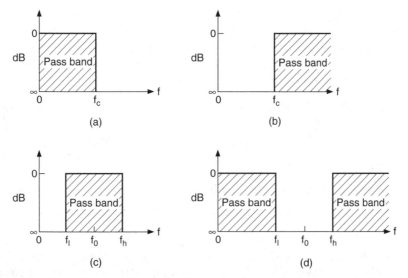

Figure 6.2 Ideal filters. (*a*) Low pass. (*b*) High pass. (*c*) Bandpass. (*d*) Band reject (notch).

6.2.1 Low-pass filters

A low-pass filter simply passes signals from zero frequency (dc) to a certain cutoff frequency and rejects all signals whose frequencies are beyond the cutoff frequency. The simplest low-pass filter is the *RC* circuit shown in Fig. 6.3*a*. The particular transfer function, Expression (6.2), indicates that there is only one pole with this filter.

$$\frac{e_{out}}{e_{in}} = \frac{1}{S + \dfrac{1}{2\pi RC}} \tag{6.2}$$

Figure 6.3*b* is a Bode plot of the low-pass filter. The Bode plot illustrates that the output is equal to the input at low frequencies. However, as the frequency increases the amplitude of the signal decreases to 70.7 percent (3 dB) at the corner frequency, which is the point at which the impedances of the resistor and the capacitor are equal. The corner frequency is

$$f_c = \frac{1}{2\pi RC}$$

Above the corner frequency, the filter's response will continue to roll off at -6 dB/octave. Specifically for each time the frequency doubles (referred to as an octave) the filter's response adds 6-dB attenuation to the signal as it passes through the filter since it is a single-pole filter.

Passive filters like the *RC* filter here have poles located in the left quadrant of the *S* plane plot shown in Fig. 6.3*c*. Here

$$S = \frac{-1}{2\pi RC}$$

The poles for a filter are factors of the denominator in the transfer function. The RC low-pass filter in Fig. 6.3 has one pole centered at f_c. For example, this equates to 6 dB/octave since it is just one pole. However, if the number of poles were, say, 8 this would be 48 dB/octave of attenuation. The method of calculating the attenuation level per octave for a low-pass filter is shown in Example 6.1.

Example 6.1   To determine the attenuation level per octave for a low-pass filter refer to Fig. 6.4.

$$\text{Attenuation/octave} = n \cdot 6 \text{ dB/octave}$$

where n = number poles

Therefore, if the low-pass filter has

$$1 \text{ pole attenuation} = 1 \cdot 6 \text{ dB/octave} = 6 \text{ dB/octave}$$

$$8 \text{ poles} = 8 \cdot 6 \text{ dB/octave} = 48 \text{ dB/octave}$$

The characteristics of a low-pass filter include its insertion loss, the pass band ripples, the attenuation slope, its stopband attenuation, and of course the pass band. Figure 6.5 illustrates the general characteristics for a low-pass filter.

Example 6.2   The calculation of the steepness factor for a low-pass filter is shown in Fig. 6.5.

$$\text{Pass band} = \text{dc to } f_c$$

$$\text{Steepness factor} = A_s = \frac{f_s}{f_c}$$

Steepness factor:

$$f_c = 900 \text{ MHz}$$

$$f_s = 925 \text{ MHz}$$

$$A_s = \frac{925 \text{ MHz}}{900 \text{ MHz}} = 1.02778$$

6.2.2   High-pass filter

The simplest form of high-pass filter is shown in Fig. 6.6a. The high-pass filter will attenuate (reject) all the signals that are below the cutoff frequency and pass all the frequencies above the cutoff frequency.

The high-pass filter is the RC circuit shown in Fig. 6.6a. The particular transfer function is listed next to the figure for the RC high-pass filter. The transfer function indicates that there is one pole and one zero with this filter. The combination of the pole and zero makes the high-pass filter function. Figure 6.6b is a Bode plot of the high-pass filter. The Bode plot illustrates that the output is equal to the input at high frequencies. However, as the frequency decreases the amplitude of the signal decreases to 70.7 percent (3 dB) at the corner frequency, which is the point at which the impedances of the resistor and the capacitor are equal. The corner frequency is represented by the symbol f_c.

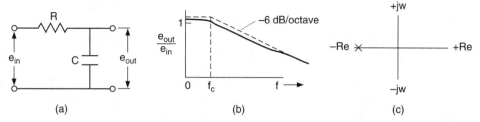

Figure 6.3 (*a*) *RC* low-pass filter. (*b*) Bode plot. (*c*) Low-pass filter polar plot.

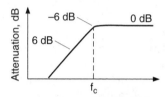

Figure 6.4 Determination of attenuation level per octave for a low-pass filter.

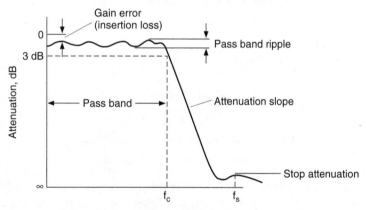

Figure 6.5 Determination of steepness factor for a low-pass filter. f_c = cutoff frequency. f_s = stop frequency.

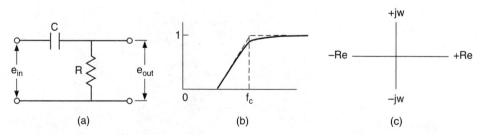

Figure 6.6 High-pass filter. (*a*) *RC* filter; $e_{out}/e_{in} = S/[(1 + S/(2\pi RC)]$. (*b*) Bode plot; corner frequency, $f_c = 1/(2\pi RC)$ (*c*) Polar plot.

Below the corner frequency, the filter's response will continue to roll off at -6 dB/octave. Specifically for each time the frequency doubles (referred to as an octave) the filter's response will add 6 dB attenuation to the signal as it passes through the filter, since it is a single-pole filter. The zeros will also change at the same rate of having 6 dB/octave with each zero. Passive filters like the RC filter here have poles located in the left quadrant of the S plane plot shown in Fig. 6.6c.

The poles for a filter are factors of the denominator in the transfer function. The RC high-pass filter in Fig. 6.6c has one pole centered at f_c. For example, this equates to 6 dB/octave, since it is just one pole. However, if the number of poles were, say, 8 then this would be 48 dB/octave of attenuation.

The filter also has zeros centered at the same location as the poles, and this combination enables the high-pass filter to provide the attenuation to the signal at low frequencies and pass the higher frequencies virtually unattenuated at the higher frequencies. Therefore, if the filter were to, say, have three zeros the attenuation would be -24 dB/octave.

The method of calculating the Bode plot or rather attenuation level per octave for a high-pass filter is as shown in Example 6.3.

Example 6.3

$$\text{Pole attenuation/octave} = n \cdot 6 \text{ dB/octave}$$

$$\text{Zero attenuation/octave} = -m \cdot 6 \text{ dB/octave}$$

where n = number of poles
$\qquad m$ = number of zeros

$$\therefore \text{ if the high-pass filter has 1 pole and 1 zero,}$$

$$\text{Below } f_c \quad \text{attenuation} = 1 \cdot 6 \text{ dB/octave} = 6 \text{ dB}$$

$$\text{Above } f_c \quad \text{attenuation} = -1 \cdot 6 \text{ dB/octave} = -6 \text{ dB}$$

$$\therefore \text{ At } f_c \text{ attenuation} = 6 \text{ dB} - 6 \text{ dB} = 0$$

The characteristics for a high-pass filter are very similar to those for a low-pass filter, with the exception that the roles are reversed for what their objective is at the same cutoff frequency. The steepness factor for a high-pass filter is calculated as shown in Example 6.4.

Example 6.4

$$\text{High-pass steepness factor} = A_{SHP} = \frac{f_c}{f_s}$$

$$f_c = 3000 \text{ Hz}$$

$$f_s = 300 \text{ Hz}$$

$$A_{SHP} = \frac{3000}{300} = 10$$

6.2.3   Bandpass filters

This type of filter passes signal frequencies in the pass band region and attenuates frequencies above and below this region (Fig. 6.7). The bandpass filter is designed as separating the low- and high-pass filter characteristics and then

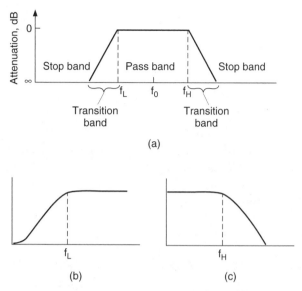

Figure 6.7   Bandpass filter. (a) f_0 = center frequency; f_L = lower cutoff; f_H = higher cutoff. (b) High-pass filter. (c) Low-pass filter.

cascading them as shown in Fig. 6.7. For the bandpass filter the pass band is defined and bounded by the filter's upper and lower -3-dB break frequencies f_H and f_L. The break frequencies occur at the points where the filter's amplitude response passes through the designed -3-dB gain reduction with respect to the pass band gain.

Several key parameters that define a bandpass filter characteristics are shown as Eqs. (6.3).

$$f_0 = \text{center frequency}$$

$$f_0 = \sqrt{f_H \cdot f_L}$$

$$Q = \text{selectivity}$$

$$Q = \frac{f_0}{f_H \cdot f_L} = \frac{f_0}{\text{bandwidth}} \qquad \text{if } Q > 10, f_0 = \frac{f_L + f_H}{2} \qquad (6.3)$$

$$\text{where }\ f_L = \sqrt{\left(\frac{Bw}{2}\right)^2 + f_0^2} - \frac{Bw}{2}$$

$$f_H = \sqrt{\left(\frac{Bw}{2}\right)^2 + f_0^2} + \frac{Bw}{2}$$

$$A_s \text{ (steepness factor)} = \frac{\text{stop band bandwidth}}{\text{pass band bandwidth}}$$

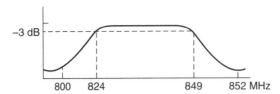

Figure 6.8 Selectivity, Q, of the filter.

The center frequency for the bandpass filter is expressed as the geometric mean between the -3-dB frequencies for the upper and lower boundaries.

The selectivity of the filter (also known as the Q for the filter) establishes the filter's frequency discrimination, or rather sharpness. The selectivity Q of the filter increases as the distance between the lower and higher cutoff frequencies is reduced. See Fig. 6.8 and Eqs. (6.4).

$$f_0 = \sqrt{f_H - f_L} = \sqrt{(849 \text{ MHz}) \cdot (824 \text{ MHz})} = 836.40 \text{ MHz}$$

$$Q = \frac{f_0}{f_H - f_L} = \frac{836.40 \text{ MHz}}{849 - 824} = 34.85 \tag{6.4}$$

$$A_s = \frac{\text{stopband bandwidth}}{\text{pass band bandwidth}} = \frac{52}{25} = 2.08$$

$$A_s L = \frac{824}{800} = 1.03$$

$$A_s H = \frac{852}{849} = 1.004$$

Bandpass filters are either wide-band or narrow-band filters. The narrow-band bandpass filter has a ratio of upper cutoff frequency to lower cutoff frequency of approximately 2 or less. The narrow-band bandpass filters cannot be designed as separate low- and high-pass filters and then cascaded together because the loss at or near the center frequency will become inhibitive as the pass band size decreases in bandwidth. However, the wide-band band pass filter can be separated into a low- and high-pass filter design and treated separately.

6.2.4   Band reject (notch) filter

The band reject filter, also known as the notch filter, is used to selectively remove unwanted signals from the communication system. Specifically the band reject filter attenuates the signal frequencies within the center or notch region and passes frequencies above and below the notch. Figure 6.9 represents a band reject filter.

The filter's upper and lower -3-dB frequencies f_U and f_L are both bound and defined by the width of the notch itself. The width of the notch is referred to

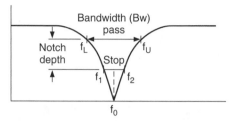

Figure 6.9 Band reject filter (notch). $f_0 = \sqrt{f_L f_U}$
$Q = f_0/\text{BW}$. AS = Pass band/stopband.

as the pass band for the notch. The notch also has a stop bandwidth, defined in Fig. 6.9 as the difference between f_1 and f_2:

$$\text{Center frequency } f_0 = \sqrt{f_L \cdot f_U} \qquad (6.5)$$

$$\text{Selectivity } Q = \frac{f_0}{f_U - f_L} = \frac{f_0}{\text{bandwidth pass band}} = \frac{f_0}{BW_{PB}}$$

$$\text{If } Q > 10 \; f_0 = \frac{f_L + f_U}{2}$$

$$\text{Notch depth} = \text{pass band (db)} - \text{stop band (dB)}$$

$$\text{Steepness factor} = \frac{\text{pass band bandwidth}}{\text{stop band bandwidth}}$$

For a notch filter the filter's selectivity is defined as Q. The selectivity of the filter will increase as the difference between the upper and lower frequencies in the pass band is reduced. Another key parameter for the notch filter is the notch attenuation, or rather notch depth. It is defined as the gain reduction in dB relative to the pass band of the notch filter.

A notch filter is considered wide band if the ratio between its upper and lower cutoff frequencies is an octave or more and therefore can be separated into low- and high-pass filters. If the notch filter is considered wide band, this will enable a high- and low-pass filter design to be superimposed upon each. However, the definition of an octave separation defining the break point for separating the design is not locked in concrete. The rest of the notch filters are considered narrow-band filters. See Example 6.5 and Fig. 6.10.

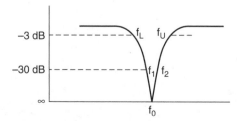

Figure 6.10

Example 6.5   See Fig. 6.10.

$$f_L = 845 \text{ MHz} \qquad f_U = 846.5 \text{ MHz}$$
$$f_1 = 845.5 \qquad f_2 = 846.0 \text{ MHz}$$
$$f_0 = 845.75 \text{ MHz}$$
$$Q = \frac{f_0}{f_U - f_L} = \frac{845.75}{846.5 - 845} = 563.83$$
$$\text{Notch depth} = -3 \text{ dB} - (-30 \text{ dB}) = 27 \text{ dB}$$
$$\text{Steepness factor} = \frac{846.5 - 845}{846 - 845.5} = 3$$

6.2.5   Crystal filter

A crystal filter is a narrow bandpass filter, commonly referred to as a quartz filter. The crystal filter is used primarily in the IF circuits for a radio and is equivalent to an LC tank circuit. Specifically the crystal filter will pass frequencies at either side of the crystal frequency but reject all the others. Figure 6.11a is a representation of a crystal filter. The filter shown in 6.11b exhibits a high Q while the one in 6.11c exhibits a low Q. The difference between these is an inductor and capacitor are added to Fig. 6.11c, which widens the pass band for the crystal filter itself but at the same time makes it tunable over a range of frequencies.

6.2.6   Harmonic filter

The harmonic filter is either a low-pass or bandpass filter depending on its application. The objective of the harmonic filter is to remove all the harmon-

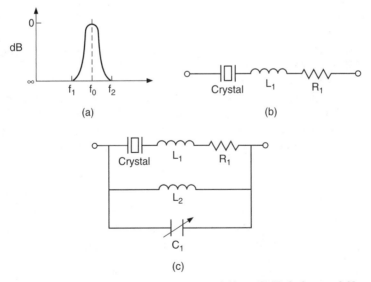

Figure 6.11   (*a*) Frequency response to crystal filter. (*b*) High Q crystal filter. (*c*) Low Q crystal filter.

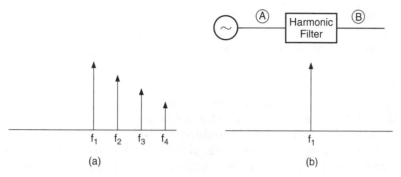

Figure 6.12 Harmonic filter. (a) f_1 = 880 MHz; $2f_1$ = 1760 MHz = f_2; $3f_1$ = 2640 MHz = f_3; $4f_1$ = 3520 MHz = f_4. (b) Filter output.

ics except the carrier and its sidebands from transmission. Figure 6.12 is a simple representation of a harmonic filter utilizing a low-pass filter.

6.2.7 Surface acoustic wave (SAW) filter

The utilization of SAW filters in a communication system is a chief reason for the reduction in physical size of the devices. SAW filters can come as low-pass, high-pass, bandpass, and notch filters. The advantage with using SAW filters is primarily their extremely small size, low manufacturing cost, and relatively sharp bandpass response coupled with the low group delay across the band. Their main drawback is their large insertion loss. However, based on where they are placed within the communication system, usually the IF region, the insertion loss can be overcome with amplifier gain.

6.2.8 General characteristics

The following is a general selection process for determining the type of general filter needed for the application. However, for certain applications several filters can be cascaded together to solve a particular design issue.

The general rules are as follows:

Low-pass filters: high-frequency interference rejection; band limiting; harmonic suppression

High-pass filters: band limiting; noise reduction; interference elimination; broadcast signal conditioning

Bandpass filters: band limiting; comb filter; interference elimination

Notch filter: Selective frequency rejection; noise reduction; interference elimination

6.3 Filter Classifications

Once the decision has been made to select the type of filter a choice must be made as to the type of filter design necessary. There are several general choices for filter design; each has its positive and negative trade-offs. Filter classifications that will be covered here involve Butterworth, Tchebysheff (Chebyshev), Bessel, and elliptic filter types.

6.3.1   Butterworth

The Butterworth filter is one of the common filter types utilized. It has a transfer function that provides a flat amplitude response in the pass band and rolls off at -6 dB/octave beyond the cutoff frequency. This type of filter has the poles uniformly spaced on a unit circle about the origin in the left half of the S plane, which will yield a maximally flat frequency response (Fig. 6.13). Table 6.1 indicates the attenuation that is expected depending on the filter type selected for the design. The Butterworth filter is usually selected when pass-band gain accuracy is of extreme importance. The Butterworth filter also exhibits a moderate amount of overshoot in the phase response, as shown in Eqs. (6.6).

TABLE 6.1   Butterworth Filter

Filter type	Ω
Low pass	W_x/W_c
High pass	W_c/W_x
Bandpass	$BW_x/BW_{3\text{ dB}}$
Notch	$BW_{3\text{ dB}}/BW_x$

Attenuation $A_{\text{dB}} = 10 \log_{10} (1 + \Omega^{2n})$.
Ω = dimensionless ratio of frequencies
W_x = stop frequency
W_c = cutoff frequency
BW_x = stop band
$BW_{3\text{ dB}}$ = pass band

For example, [2]

$$n \cdot (-45°) \qquad \text{at 3-dB frequency}$$

$$n \cdot (-90) \text{ stop} \qquad \text{where } n = \text{number of poles (complexity)}$$

If $n = 3$

$$\text{Phase shift} = 3 \cdot -45 = -135° \text{ (3 dB)}$$

$$3 \cdot (-90) = -270 \text{ (stop band)} \qquad (6.6)$$

$$\phi_{\text{pass band}} = \frac{-45 \, nf_x}{f_c{}^2}$$

where f_x = any frequency
f_c = 3-dB cutoff frequency

$$\text{Phase delay} = T_{pd} = \frac{\phi}{360} \cdot \frac{1}{f_x} \qquad \text{or} \qquad T_{pd} = -\frac{\beta}{W}$$

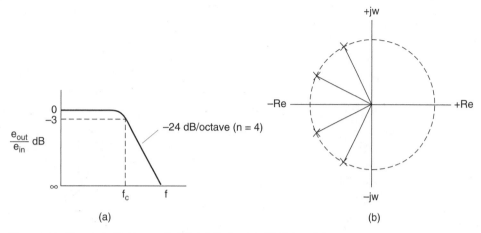

Figure 6.13   Four-pole Butterworth LP. (*a*) Bode plot. (*b*) Polar plot.

where β = phase shift radius
$$w = 2\pi f_x$$

The phase delay for the Butterworth filter is shown above. The amount of phase delay is dependent upon the complexity of the filter, either poles or zeros. In addition the phase delay is directly related to the frequency response of the filter itself. As the filter's response gets narrower and narrower, the phase delay will increase, driven by a combination of the complexity and the steepness of the filter.

6.3.2   Tchebysheff (Chebyshev)

The Tchebysheff filter types are selected when the roll-off rate between the pass band and the stop band, that is, the transition band, is of primary importance and the gain variations (amplitude differences) within the pass band are of secondary concern.

The transfer function for a Tchebysheff filter achieves a sharp transition band at the expense of allowing ripple in the pass band.

The Tchebysheff low-pass filter shown in Fig. 6.14 has its poles distributed on an ellipse about the origin. This type of filter has an equiripple pass band where the ripple's magnitude is dependent upon the deviation of the ellipse from a circle and steeper monotonic roll-off above the corner frequency than, say, a Butterworth filter with the same number of poles.

$$A_{dB} = \text{attenuation} = 10 \log_{10} [1 + \varepsilon^2 C_n^2(\Omega)]$$

$$\varepsilon = \sqrt{10^{\alpha_0} - 1}$$

where α_0 = pass band ripple, dB
C_n = Chebyshev polynomial

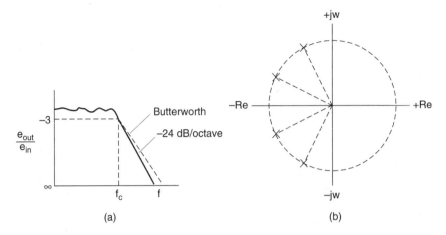

Figure 6.14   Four-pole Tchebysheff low-pass filter. (*a*) Bode plot. (*b*) Polar plot.

6.3.3   Cauer filter

These filters, also called "antialiasing filters," offer the steepest roll-off above the corner frequency. They have good pass band flatness and low wide band noise characteristics, but their nonuniform group delay can cause some overshoot or ringing. The Cauer low-pass filter (Fig. 6.15) has poles distributed on an ellipse about the origin and zero pairs on the imaginary axis above the corner frequency. The filter's response is equiripple in the pass band and equiripple stop band floor. This filter produces a sharper roll-off rate than either the Butterworth or Tchebysheff but at the expense of the stop band floor.

6.3.4   Bessel filter

Bessel filters have specific digital applications. A Bessel filter will delay all frequency components in the pass band of the filter by the same amount of time. This type of filter is used in the reconstruction of waveforms and digital radio applications. It has a uniform group delay with no real ringing or overshoot, but it has a slow roll-off.

According to Fourier analysis, a step input consists of many frequencies related in phase and amplitude. A Bessel filter has constant group delay and will preserve the step response delayed in time, but with a rise time inversely related to the corner frequency. With a Bessel filter no waveform ringing will occur.

The Bessel transfer function approximates a constant time delay in the pass band. The phase delay increases linearly from 0 to $N\,\pi/4$ radians, and amplitude response is -3 dB down at the cutoff frequency f_c. The fast settling time and nominal overshoot are other key characteristics of this type of filter.

6.4   Filter Performance Criteria

Many aspects and criteria define a filter's performance. The first and most important is the mission statement the filter itself is trying to resolve. Specifically if the design calls for a filter to pass frequencies between 1930 and 1945 MHz, a bandpass filter might be the best general type to utilize. However, the other attributes for the filter will determine just how well it performs in the application. For example, the ability to dissipate power might not be a criterion for a receive filter but it will be very important for a transmit filter.

When selecting a filter, the following are some of the criteria that you should define during the design phase for the communication system.

Frequency response.   The frequency response of a filter defines which frequencies will be passed and which will be attenuated. The components that make up the frequency response characteristics of a filter involve the pass band, the cutoff band, the transition band, and the stop band as shown in Fig. 6.16.

Pass band.   This is one of the most important filter criteria. The pass band defines which frequencies will pass through the filter and which frequencies

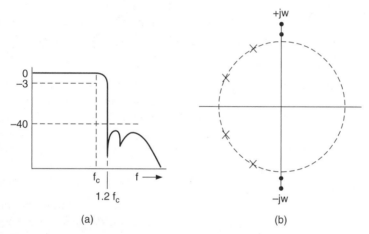

(a)   (b)

Figure 6.15   Four-pole/zero Cauer low-pass filter. (*a*) Bode plot. (*b*) Polar plot.

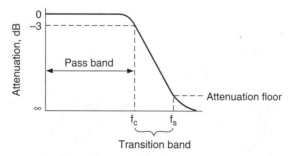

Figure 6.16   Depiction of filter's ability to reject out-of-band energy. Shape factor SF $= f_s/f_c$, *where* $f_c =$ cutoff frequency. $f_s =$ stop frequency.

will be discriminated against (attenuated). The pass band is normally defined as the area of the filter which experiences the lowest level of attenuation, ideally 0 dB, and has a characteristic low-level ripple.

Cutoff frequency.   This is the frequency where the pass band is at 3 dB of attenuation, or rather the end of the desired pass band.

Transition band.   This is the portion of the filter response that is between the cutoff frequency and the stop frequency. It is the part of the filter response where the greatest attenuation change occurs.

Stop band edge.   This is the highest-frequency f_r at which the pass band ripple occurs. The stop band edge is also the transition point where a small increase in frequency gives a large increase in attenuation. Receiver front ends usually specify the stop band attenuation at the upper and low frequencies it operates over. For a transmitter the stop-band frequency is normally specified with respect to the receive band for the receiver.

Insertion loss.   The insertion loss or attenuation in the pass band is important, since this defines how much loss the filter will impose upon the signal as it traverses through the filter itself. Ideally the insertion loss through a filter should be 0 dB. However, a passive filter will have some insertion loss, which also equates to a noise figure for the filter itself.

In addition to insertion loss there is also a factor called gain loss which makes up part of the insertion loss the filter imposes upon the signal. The gain error is the difference between the specified and actual pass band gain (insertion loss). For a typical filter the gain error can be as high as a few percent. This value can be referenced to many things including the frequencies below the cut-off for a low-pass filter or the entire pass band. It can even refer to the band reject portion of the filter. However, it normally refers to the entire pass band.

Pass band ripple.   The pass band ripple is the variation in gain (insertion loss) over the pass band, also referred to as the in-band variation of the signal. The plot of gain vs. frequency response for the filter in Fig. 6.17 shows a ripple across the pass band instead of a flat response [Eq. (6.7)].

$$\text{Gain error} = 5\% = (1\text{-dB insertion loss}) \cdot 1.05 = -1.05 \text{ dB} \qquad (6.7)$$

$$\text{Insertion loss (dB)} = \text{amplitude before filter} - \text{amplitude after filter}$$

Typically a filter will have 1 to 2 percent ripple over most of the filter's pass band. Figure 6.18 is another example of pass band ripple as it applies to a band pass filter.

Attenuation floor.   This is the highest attenuation level for the filter at the stop frequency. The attenuation floor reference F_s is where the roll-off response of the filter's transition band crosses the attenuation floor (Fig. 6.16).

Shape factor.   The shape factor for a filter is a measure of the filter's attenuation steepness for the transition band. The steepness factor will

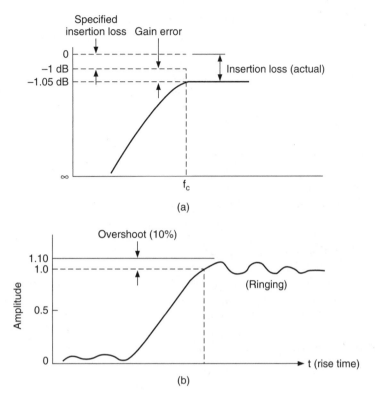

Figure 6.17   (a) Pass band ripple. (b) Low-pass band filter.

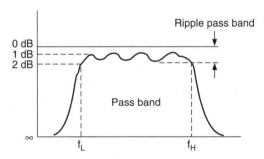

Figure 6.18   Pass band ripple.

increase in value as the number of poles and/or zeros increases, or rather as the order of the filter increases as shown in Eq. (6.8).

$$SF = f_s / f_c \qquad (6.8)$$

Ideally the steepness factor should be unity, that is, $SF = 1$.

Phase error.   Phase error is also referred to as phase linearity. It is the linearity of the phase shift vs. frequency. If there is no phase error, the phase linearity line should be a straight line. If there is no phase error, there is no group delay, since the derivative of a constant is 0 and the group delay is the derivative of the phase error.

Group delay.   Group delay is defined as the time delay through a filter for a finite length of time for a burst (pulse). Ideally the group delay for a filter should be constant across the entire pass band of the filter. A group delay which is not constant across the pass band can cause overshoot or ringing in the pass band itself.

Since the group delay for a filter is a derivative of the filter's phase shift, it is mathematically represented as phase vs. frequency response. It is important to note that the actual magnitude of the group delay is not important, but rather its flatness across the pass band is the key to the filter's performance and is a measurement of distortion.

If the group delay is not constant over the bandwidth of the pass band for the filter where the desired modulated signal resides in frequency, some form of waveform distortion will take place. The narrower the desired signal in bandwidth the less likely it will undergo any noticeable group delay. However, the wider the bandwidth of the signal the more susceptible it is to distortion due to group delay. Specifically the 30-kHz channel used for DAMPS in the United States is not really susceptible to group delay problems. However, the CDMA signal in IS-95, which is 1.288 MHz wide, is more susceptible to group delay, defined as Eq. (6.9).

$$T_{pd} = \frac{-\mathrm{dB}}{\mathrm{dW}} \tag{6.9}$$

The tolerance in group delay over the entire filter pass band is sometimes called group delay ripple and is expressed in units of time such as milli- or microseconds. The absolute value is normally not mentioned, only the deviation from some fixed value.

Lastly any impedance mismatch in the communication system will also invoke some group delay due to SWR problems. To illustrate what group delay can look like for a digital signal, Fig. 6.19b is an eye diagram of a QPSK signal that is experiencing group delay.

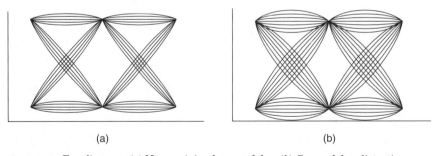

(a)   (b)

Figure 6.19 Eye diagram. (a) No or minimal group delay. (b) Group delay distortion.

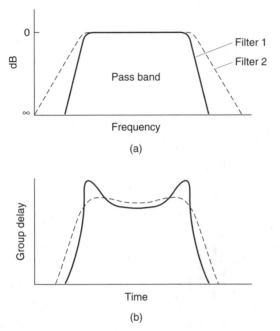

Figure 6.20 Group delay. (*a*) Frequency response. (*b*) Time response.

Group delay is often ignored in favor of amplitude response, or rather insertion loss for a filter's performance. Filter skirts often exhibit some form of group delay. Unless some form of compensation is built into the filter, the steeper the filter skirts the larger the group delay. Figure 6.20 is an example of a group delay response vs. frequency for a band pass filter.

To overcome the group delay, some form of delay compensation is employed in the filters to overcome the effects of selective group delays in the pass band. One of the compensation methods is to move the sharp rise in group delay to a portion of the spectrum that is not intended for use by the communication system. The movement of the group delay by changing its location is usually accomplished by relaxing the selectivity of the filter itself.

Another component of delay is the carrier delay, directly related to the group delay of a filter. If the filter's phase characteristics are linear with respect to the pass band frequency, both the carrier and the sidebands that accompany the carrier will have the same delay when passing through the filter itself. The same delay, if possible, will ensure that the output of the filter is the same as the input leading to a transfer function of 1 for the filter itself. An example of carrier delay is shown in Fig. 6.21 for reference with an AM signal.

Selectivity Q. The selectivity of the filter is another key attribute for the filter. The higher the selectivity of the filter the better it rejects unwanted signal from being passed through the filter unattenuated. Ideally the filter should be extremely selective and allow only the desired signals through without

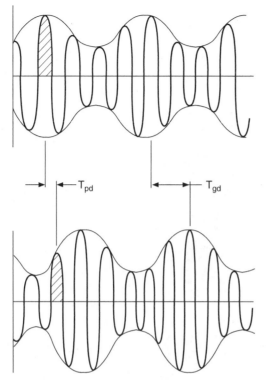

Figure 6.21 Nonlinear phase of AM signal. Carrier
delayed by T_{dp}; envelope delayed by T_{gd}.

distorting the desired signal. The Q of a filter is defined as the ratio between
the center frequency and the bandwidth of the filter.

Temperature stability.   The temperature stability of a filter is a very important
criterion. Specifically the filter should be defined in terms of its tolerance in
parts per million per degree of temperature change, ppm/°C. Ideally a filter
should retain its filter characteristics over the temperature range that it will
be subjected to, but since the filter is constructed of various components that
will change their physical dimensions with temperature, the chance for a
change in frequency response for the filter is great. A common technique
employed in the filter construction process is to utilize materials that will
offset each other's change in physical characteristics as temperature changes.

6.5   Control Theory

Control theory has a direct application for describing the characteristics of
a filter system. The placement and selection of the poles and zeros for a filter directly affect how it will perform. Naturally the more poles and zeros
implemented into the filter design the more complex the filter becomes. See
Fig. 6.22 and Eqs. (6.10).

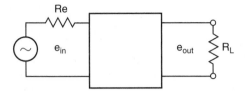

Figure 6.22 Transfer function.

$$T(s) = \frac{e_{out}}{e_{in}} = \frac{N(s)}{D(s)}$$

where $N(s)$ = zeros
$D(s)$ = poles   (6.10)

$$T(s) = \frac{N_m S^m + N_{m-1} S^{m-1} + \dots + N_1 S + N_0}{D_n S^n + D_{n-1} S^{n-1} + \dots + D_1 S + D_0}$$

where $S = jw$
highest power of m in numerator is number of zeros
highest power of n in denominator is number of poles

The transfer functions for both a low-pass and a high-pass filter are shown in Fig. 6.23. It is interesting to note that the low-pass filter is an integrator circuit and the high-pass filter is a differentiation circuit in control theory.

Figure 6.24 shows a simple control circuit that has a feedback loop. The transfer equation (6.11) represents a second-order system with two poles and one zero. The denominator for the transfer function represents the frequency components of the control circuit. The frequency components are identified as the natural frequency for the system and the dampening factor, both shown.

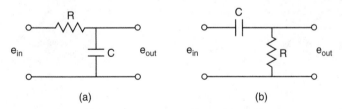

Figure 6.23 First-order functions. (a) Loss pass (integration);
$e_{out}/e_{in} = 1/(RC_s + 1)$. ($b$) High pass (differentiation);
$e_{out}/e_{in} = RC_s/(RC_s + 1)$.

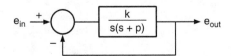

Figure 6.24 Simple control circuit with a feed-back loop.

$$\frac{e_{out}}{e_{in}} = \frac{k/S(S+p)}{1 + k/S(S + p)} = \frac{k}{S^2 + pS + k} \tag{6.11}$$

where w_n = natural frequency
  ε = damping factor
  ε = 1 (critical damping)
  > 1 overdamped
  < 1 underdamped

$$\therefore S^2 + \underbrace{p}_{zE w_n S} + \underbrace{k}_{w_n^2}$$

The location of the poles and zeros determines the location of the knees of the Bode plot. However, the dampening factor identified indicates the performance of the filter in terms of how it will respond to an input. If the dampening factor (DF) is unity, this is an ideal filter and is referred to as critically damped. If the DF is less than unity, the filter is underdamped, and this indicates possibly the filter does not reject unwanted signals as well as it should. If DF is greater than unity, overdamping occurs and the filter could cause ringing and/or attenuate the designed pass band (Fig. 6.24).

Figure 6.25 is a simple example of a filter which has one zero and two poles in it. Here $e_{out}/e_{in} = (S - 1)/(S - 3)(S - 7)$. The Bode plot, which is a representation of the amplitude or phase vs. frequency, is shown in Fig. 6.25. The transfer function is a representation of a band pass filter.

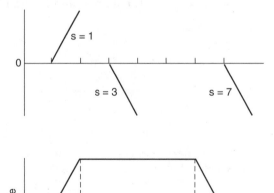

Figure 6.25   Bode plot. Slope = 20 × (n) dB/decade.

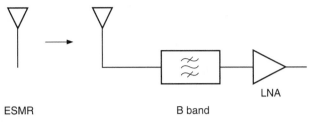

ESMR   B band

Figure 6.26 Receive system.

6.6 Filter Selection

The success or failure of a communication system that operates in the presence of noise, either artificially made or thermal, is improved with the proper selection of filters in its design. Filters in a communication system exist in both the transmit and receive portions. The objective of the cell sites transmit filter is to suppress the emissions that come from the modulator that are later amplified. The unwanted signals needing suppressing could be transmit emissions which fall into the receive band of either your system or another operator's system. The receive filters at the cell site are meant to prevent receiver overload, suppress the third-order intermodulation products (IMP) below the receiver noise floor, and improve the cell sites selectivity.

6.6.1 Communication system requirements

The filter requirements for a communication site should be sufficient to ensure that the maximum sensitivity and selectivity for the system is achieved. The selectivity should be sufficient to pass the desired signal and reject the unwanted signal. In addition sensitivity should always be factored into a communication site's filter requirements since there are many ways to remove unwanted signals from a system but many of these methods involve increasing the attenuation of the desired signals in the pass band.

The transmit filter for a communication site should ensure that the transmitter emissions remain within the FCC-defined emission mask. In addition it is not uncommon for a communication system to install a more selective transmit filter to further reduce the out-of-band energy emissions into other systems even though they are meeting the FCC emission mask. The receive filter for a communication site should ensure that the receiver is protected from adjacent and out-of-band interference, overload, and noise suppression.

A simple example of filter requirements for a communication system involving receiver overload is shown in Fig. 6.26 and Example 6.6.

Example 6.6 Refer to Fig. 6.26.

$$\text{ESMR Tx} = 100 \text{ W ERP} \qquad \text{LNA} = -27 \text{ dBm (input)}$$
$$1 \text{ dB}$$

$$
\begin{array}{lll}
\text{ESMR Tx} & = & 50 \text{ dBm} \\
\text{Antenna isolation} & = & 30 \text{ dB} \\
\text{Receive line loss} & = & \underline{2 \text{ dB}} \\
& & +18 \text{ dBm} \qquad \text{into filter}
\end{array}
$$

$$
\begin{array}{l}
+18 \text{ dB} \\
\underline{-27 \text{ dBm}} \\
\Delta 45 \text{ dB}
\end{array}
$$

45 dB attenuation needed from receive filter

It involves a colocation of a B-band cellular operator with an ESMR operator. The example does not factor into the issue of composite energy and phase noise in a receiver.

6.6.2   Size

The physical characteristics of the filter need to be factored into the selection process. Specifically if a band pass filter with, say, 11 poles is selected, the situation where it will be physically mounted must be taken into account. For instance, if the filter is part of a tower top amplifier that is located immediately following the antenna, the size, weight, and mounting characteristics need to be factored into the selection. In this example if the filter was initially designed to be in a temperature-controlled environment, placing it in a non-temperature-controlled environment might cause a change in the frequency characteristics of the filter itself. Also the type of mounting needed might add undue weight and wind loading to the tower, based on the location and number of units.

6.6.3   Cost

The cost of the filter needs to be factored into any design. If the filter selected for the situation is, say, the best filter and approaches the ideal filter, this might be an ideal situation from a pure design aspect but the cost associated with it would now become a major component in the cell site infrastructure cost.

6.7   New Technology

The proliferation of wireless communications in the marketplace has resulted in the need to isolate one communication system from another. However, with the increased demand for wireless communications the demand to utilize existing wireless facilities for multiple users has also increased. The increase in wireless communications has directly resulted in increased levels of interference. New technologies have emerged in the wireless community that promise the ability to reject all the unwanted signals while at the same time improving the existing signals. Two types of filter technologies that are being pursued involve superconducting filters, both thick and thin film, and adaptive filters.

6.7.1 Superconductors

With the discovery of high-temperature superconductors the promise of achieving the ideal filter for a communication system may become a reality, the ideal filter being one that has no loss in the pass band and infinite loss everywhere else. High-temperature superconductors offer the ability to have filters obtain a near infinite number of poles with virtually no loss. The filters can be deployed utilizing existing cooling systems since the ideal temperature for high-temperature superconductors is around 77 kelvins. Currently the superconducting efforts are centered in two camps, thick film and thin film. Each has its advantages and disadvantages. The thick film filter superconducting filters are able to handle more power and have better intermodulation specifications. However, they are larger and tend to limit themselves to a macrocellular environment. Efforts are currently underway to minimize the thick film filter size so it can be deployed in a micro base station.

The thin film superconductors, while not being able to handle as much power, do enjoy the advantage of physical size. The thin film technology lends itself immediately to mobile applications and micro base stations.

6.7.2 Adaptive filters

An adaptive filter is the practical implementation of an ideal filter with the exception that it adjusts to a changing signal condition. The ability of the adaptive filter to adjust its response makes it an ideal candidate for interference rejection, equalization, echo and noise cancellation, and many more such applications.

For a spread-spectrum system the capacity of the system is normally limited to the mutual interference between the intended user and the unintended user. The implementation of an adaptive filter for a heavy reuse environment has the potential for increasing capacity in the network by suppressing the unintended signal, or rather jamming signal.

The adaptive filter can despread the jamming signal through its processing gain. Implementation of adaptive filters can also be used to separate unwanted signals by nulling them out when they arrive at different times than the desired signal.

When using an adaptive filter for interference rejection, the following must be taken into account:

1. The desired response of the signal being processed by the adaptive filter must be known.

2. The filter's implementation must be physically realizable and must work in real time.

In most cases the actual desired response of the filter is not known, since the desired signal will most likely be affected when the interference is removed.

Three types of adaptive filter implementations are shown in Fig. 6.27.

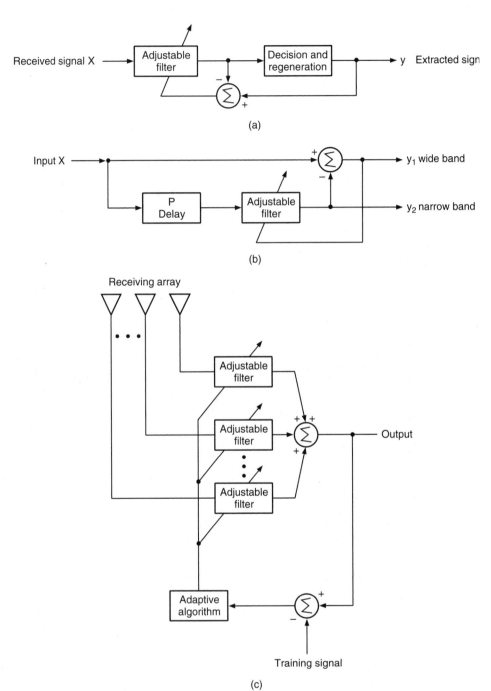

Figure 6.27 (*a*) Decision-directed signal extraction. (*b*) Adaptive whitening filter. (*c*) Adaptive null steering antenna.

References

1. Smith, Clint, and Curt Gervelis, *Cellular System Design and Optimization,* McGraw-Hill, New York, 1996.
2. Williams, Arthur, and Fred Taylor, *Electronic Filter Design Handbook,* 3d ed., McGraw-Hill, New York, 1995.
3. Weston, D. A., "Filter Design Reduces RF Output Noise," *Microwaves & RF,* March 1995, pp. 91–94.
4. Aparin, V., and P. Katzin, "Electronically Tuned L-S Band Filters," *Applied Microwave & Wireless,* Fall 1994, p. 48.
5. Zargar, Rosita, "Filters Eliminate the Aliasing Problem," *Evaluation Engineering,* September 1995, pp. 32–37.
6. Chodora, Jason, "Group Delay Characterization of Frequency Converters," *Microwave J.,* September 1995, pp. 102–116.
7. Chen-Chyi You, Cheng-Liang Huang, and Chung-Chuang Wei, "Single-Block Ceramic Microwave Bandpass Filters," *Microwave J.,* November 1994, pp. 24–35.
8. Mayle, Eugene, "A UHF Delay Equalizer," *RF Design,* April 1995, pp. 24–29.
9. Browne, Jack, "SAW Filters Redefine Loss Characteristics," *Microwaves & RF,* September 1995, pp. 135–138.
10. Ayers, Stan, "SAW Filters Save Space and Reduce Costs in Wireless Products," *EDN,* Sept. 1, 1995, pp. 153–158.
11. Ghose, Rabindraw N., *Interference Mitigation,* IEEE, New York.
12. Reed, Jeffrey, "Adaptive Filter Technology," *Defense Electronics,* May 1989, pp. 83–90.
13. American Radio Relay League, *The ARRL 1986 Handbook,* 63d. ed., The American Radio Relay League, Newington, CT, 1986.
14. Dixon, *Spread Spectrum Systems,* 2d ed., Wiley, New York, 1984.
15. Kaufman, M., and A. H. Seidman, *Handbook of Electronics Calculations,* 2d ed., McGraw-Hill, New York, 1988.
16. Lee, W. C. Y., *Mobile Cellular Telecommunications Systems,* 2d ed., McGraw-Hill, New York, 1996.
17. Stimson, *Introduction to Airborne Radar,* Hughes Aircraft Company, El Segundo, CA, 1983.
18. White, Duff, *Electromagnetic Interference and Compatibility,* Interference Control Technologies, Gainesville, GA, 1972.
19. Yarborough, *Electrical Engineering Reference Manual,* 5th ed., Professional Publications, Inc., Belmont, CA, 1990.
20. Fink, Donald, and Donald Christiansen, *Electronics Engineers Handbook,* 3d ed., McGraw-Hill, New York, 1989.
21. Schwartz, Bennett, and Stein, *Communication Systems and Technologies,* IEEE, New York, 1996.
22. Rappaport, *Wireless Communications Principles and Practices,* IEEE, 1996.
23. Fink, Beaty, *Standard Handbook for Electrical Engineers,* 13th ed., McGraw-Hill, New York, 1995.
24. Steele, *Mobile Radio Communications,* IEEE, 1992.
25. Jakes, W. C., *Microwave Mobile Communications,* IEEE Press, New York, 1974.
26. Carlson, A. B., *Communications Systems,* 2d ed., McGraw-Hill, New York, 1975.
27. Lathi, *Modern Digital and Analog Communication Systems,* CBS College Printing, New York, 1983.
28. MacDonald, "The Cellular Concept," *Bell Systems Technical Journal,* vol. 58, no. 1, 1979.
29. *Reference Data for Radio Engineers,* 6th ed., Sams, 1983.

Receivers

7.1 Introduction

The receiver and the receive system utilized by a wireless system is a crucial element of the network. Specifically the receiver's job is to extract the desired signal from the plethora of other signals and noise that exist in its environment. The basic receiver block diagram is listed in Fig. 7.1.

The receiver as discussed in this chapter involves the portion of the radio system that includes everything in the receive path starting with the antenna system itself. When talking of a receiver in the more classical sense, it involves only the portion of the network that is directly involved with downconverting and demodulating the signal to extract the initial information content. The rationale, however, with including the rest of the components that make up the receive system is to treat it as a whole system since all the components in the receive path directly influence the ultimate performance of the receiver.

This chapter attempts to cover more of the practical aspects of the receive system for a wireless network. Its objective is to discuss the more common types of receivers that are utilized (their fundamental architecture) and to define performance criteria. The chapter intertwines the use of examples with the material to help reinforce the material presented. For a more detailed and

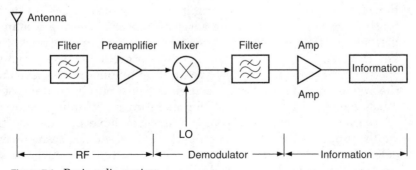

Figure 7.1 Basic radio receiver.

theoretical approach to receiver architecture, many of the references listed at the end of the chapter can be utilized.

7.2   Types of Receivers

Many types of receivers are utilized for wireless communication. The type of receiver utilized for the communication system should be selected so the information content desired to be received is done in the most efficient method, which incorporates not only financial but also spectral efficiencies where the desired information content is delivered to the receiver.

Normally a communication system receiver must deal with a signal spectrum that contains more than just the desired signal. The multitude of signals that the receiver must simultaneously deal with puts a price on device linearity for each of the stages in the receiver's path. The receiver in its operation must select the desired carrier (signal) from a multitude of other signals, amplify the weak desired signal, and then demodulate it.

The top electrical performance and cost drivers for a receiver typically are

1. Frequency range

2. Dynamic range

3. Phase noise

4. Tuning resolution

5. Tuning speed

6. Sensitivity

7. Distortion (gain and phase)

8. Noise

9. Others

The receiver design must incorporate the desired performance criteria and at the same time minimize the number of stages between the RF and IF portions of the receiver system, which is dependent upon the modulation scheme selected.

The information content to be sent and received has a direct role in the selection of the type of modulation format utilized, which ultimately determines the type of receiver utilized in the wireless network. There are many aspects to determining the type of modulation format desired in a network. However, one key criterion is the information bandwidth needed as compared to available spectrum. Specifically, if you have an infinite amount of spectrum to utilize, the type of modulation format chosen is based purely on the cost constraints imposed on the system operator for operating a profitable business.

However, as a matter of course you never have infinite spectrum (bandwidth) to utilize for the wireless system. Therefore, a trade-off is needed to determine what modulation format will be used to maximize the spectral and cost constraints of the system design [Eq. (7.1)].

$$E(t) = A \sin (2 \pi f_c t + \phi) \qquad\qquad (7.1)$$

where A = amplitude
$\quad\quad\;\; f_c$ = carrier frequency
$\quad\quad\;\; \phi$ = phase
$\quad\quad\;\; t$ = time
$\quad\quad\;\; E$ = instantaneous electric field strength

A more detailed look of modulation formats that are available and some of their variants is included in Chap. 4. However, the three general types of modulation types are shown in Fig. 7.2 for both analog and digital. Through manipulation of Eq. (7.1) all the modulation schemes are achievable.

There are three general types of receivers, amplitude modulation (AM), frequency modulation (FM), and phase modulation (PM). Many receivers utilize a combination of the three types of basic receiver elements based on the technology platform utilized for the wireless system. The three basic types of receivers are discussed below.

7.2.1 Amplitude modulation (AM)

The most basic type of radio receiver involves amplitude modulation, or rather amplitude demodulation. There are many variants for demodulating an AM signal; however, the most basic method is through use of an envelope detector. Specifically, the envelope detector in Fig. 7.3 is extremely easy to construct.

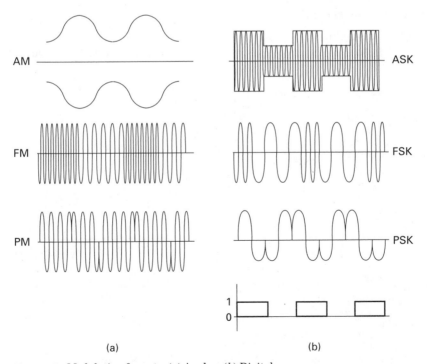

(a)   (b)

Figure 7.2 Modulation formats. (*a*) Analog. (*b*) Digital.

However, AM is not as spectral-efficient as either FM or PM in occupied spectrum for the information content or in terms of power per information content needed to deliver the message. But the receiver is extremely simple to produce, lending itself to a low cost for users. In AM modulation the modulated wave consists of a carrier wave with frequencies equal to the sum and difference between the carrier and the modulating frequency, i.e., sidebands. But one key drawback with utilizing AM modulation is its susceptibility to noise. A direct example of susceptibility to noise is experienced when the electronic ignition noise, if there is a shielding problem, can be heard on the AM radio as a whining sound that either increases or decreases in pitch depending on the speed of the engine. Referring to Fig. 7.3, the ability to determine the specific frequency that will be received and ultimately demodulated is the simple filter that comprises the front end of the AM receiver itself.

Figure 7.4 illustrates the AM demodulation in a pictorial method.

Another type of AM receiver is the single sideband receiver (SSB), which involves AM modulation and demodulation, but SSB is more spectral- and power-efficient than its counterpart, the envelope detector. An example of a SSB receiver is included in Fig. 7.5. The configuration for USD and LSB receiver is functionally the same as the SSB receiver with the exception that a low- or high-pass filter is utilized in the intermediate-frequency (IF) portion of the receiver. AM can have either data or voice as the information content utilized for this form of transport technology.

7.2.2 Frequency modulation (FM)

FM demodulation is one of the more common forms of receiver design. FM is utilized extensively for cellular communication and is often referred to as ana-

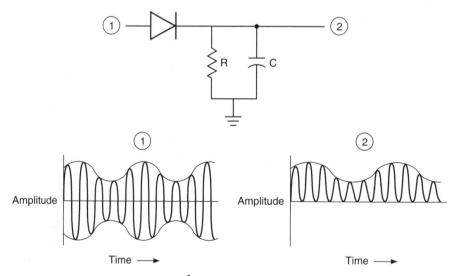

Figure 7.3 Envelope detector. $f_c = \dfrac{1}{2\pi RC}$.

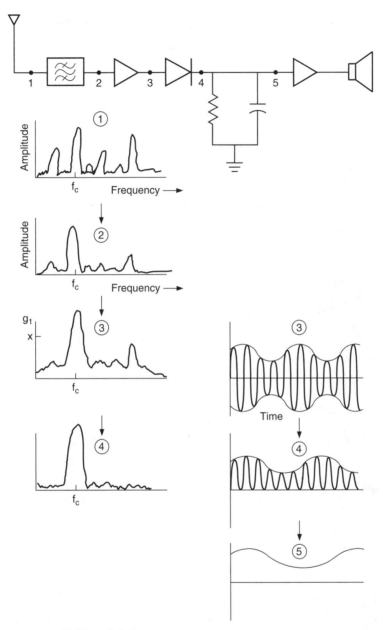

Figure 7.4 AM demodulation.

log, although the term is only partially correct in this context. Utilization of FM for a radio receiver has the advantages of being more noise-immune, but this comes at the expense of receiver complexity.

The most common FM demodulation technique is the superheterodyne FM reception. Specifically using two downconverters to arrive at the desired information content provides excellent sensitivity and selectivity for a radio system.

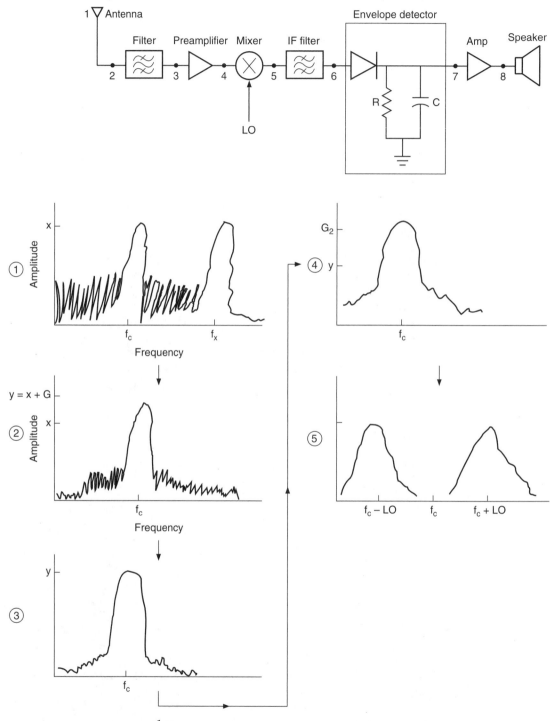

Figure 7.5   AM receiver. $f_c = \dfrac{1}{2\pi RC}$ × G = antenna gain. G_1 = preamplifier gain. G_2 = amplifier gain.

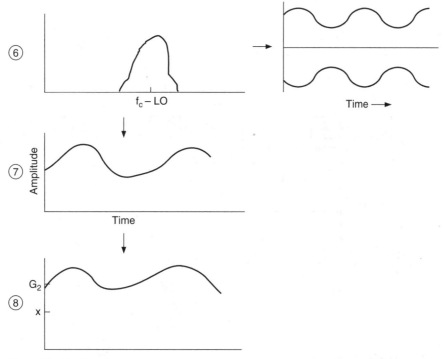

Figure 7.5 *Continued.*

Many types of radio systems employ FM for transporting information content. Some communication systems utilizing FM are AMPS, TACS, N-AMPS, SMR, and various forms of two-way communication systems besides commercial FM radio services. The usual bandwidths for an FM system can range from 200 kHz for a commercial FM broadcast system to 30 kHz for AMP all the way to 10 kHz for N-AMPS. Again the bandwidth that is utilized for the system is dependent upon numerous issues involving available spectrum and information content to be sent and received. The available spectrum and information content have a direct impact on the cost of the system.

The FM receiver configuration is shown in Fig. 7.6. It is important to note that the two downconversions that take place improve the overall performance of the radio.

FM radios can demodulate either voice or data or both depending on the type of information to be sent. For example, FSK and GMSK send digital information utilizing an FM modulation and demodulation method. However, FM is more commonly associated with analog voice communication.

7.2.3 Phase modulation (PM)

Phase modulation receivers offer many advantages over pure AM or FM receivers. Specifically the PM receiver is able to be more spectral-efficient

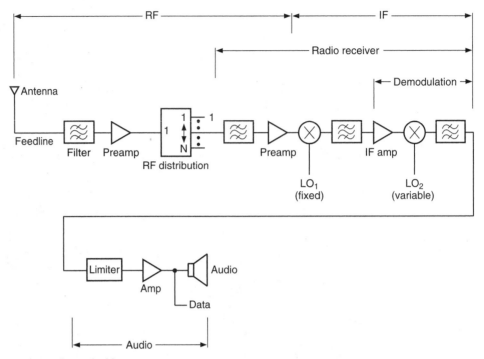

Figure 7.6   Basic double conversion receiver.

owing to its modulation method. The specifics of the modulation methods are included in Chap. 4.

A generalized block diagram of a PM receiver is shown in Fig. 7.7, a rather straightforward approach to PM demodulation. It is important to note that the in-phase and quadrature signals are extracted from the same received signal that enters the antenna system of the receiver. The in-phase and quadrature portions are separated by mixing the signal with the same local oscillator frequency but shifted 90°. Both the in-phase and quadrature portions of the signal are then sent for further postprocessing. However, the ability to demodulate the PM signal is based on the ability of the receiver to properly separate the in-phase and quadrature portions of the signal.

When the receiver has difficulty differentiating the in-phase and quadrature portions of the signal, bit errors occur (BER) and some or all of the information content on that channel is lost or corrupted.

PM lends itself to digital communication and is one of the primary communication formats for IS-54 and IS-136, to mention but a few.

7.3   Receiver System Blocks

The basic building blocks of a radio receiver system are shown in Fig. 7.8. They are virtually the same for all the types of modulation formats chosen. The chief differentiator for all the types of receive systems lies in the demodulation portion of the receiver itself. This section covers the various aspects of the building blocks for a receive system.

7.3.1 Antenna system

The antenna system is the first stage in the receive path for the receiver. The antenna's purpose is to decouple the electromagnetic energy from the atmosphere and transfer this to the feedline for the communication site. The many attributes of a receive antenna are listed in Chap. 5. Here it is assumed that the antenna system is properly matched and operates in the frequency band of interest with uniform gain of the whole bandwidth.

7.3.2 Feedline

The feedline physically connects the antenna to the rest of the receive system. The feedline is comprised of cable and associated jumpers that normally connect the antenna to the receive filters. The feedline is an important element in the receive system and has a direct role in determining how well it will operate.

   The feedline in Fig. 7.9 consists of a jumper cable, which connects the antenna to the coaxial cable, commonly known as the feeder or feedline. The jumper cable is used to connect the feedline to the antenna purely because of physical constraints where bending a $\frac{7}{8}$-in cable is much more

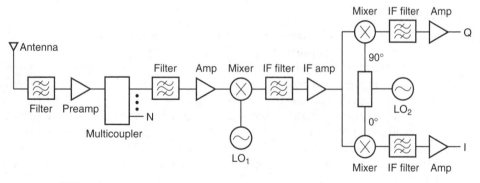

Figure 7.7 PM receiver.

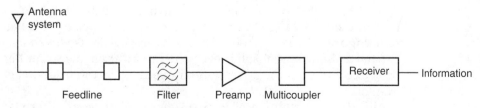

Figure 7.8 Radio system block diagram.

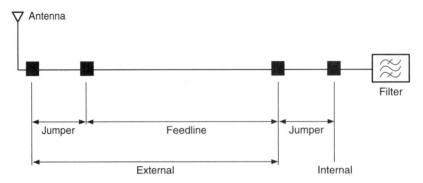

Figure 7.9   Feedline.

difficult than bending a $\frac{1}{2}$-in cable. The jumper, simply put, is meant to aid in the installation and maintenance portion for the antenna system.

In Fig. 7.9 the feedline usually enters the communication facility at the bulkhead for a shelter or external cabinet type of installation. The feedline is then terminated via a connector which again allows for another jumper cable to be installed. The jumper cable that is used between the feedline and the receive filters is normally $\frac{1}{2}$-in superflex or LDF. The objective again is for ease of installation and maintenance of the facility. The chief disadvantages with utilizing additional jumpers are the additional losses, intermodulation, and potential connector problems that can arise.

7.3.3   Filter

The filter or filters utilized for a communication system are a key component of a communication system. There are many types of filters, and the majority of them are included in Chap. 6. Normally several filters are employed in a receive path for a communication system. Specifically the filter should pass only the frequencies of interest, reject all the other frequencies, and do so with no loss in amplitude for the desired signals. Obviously this is not practical to implement when space constraints and cost enter into the decision matrix. However, the filter that is used in the communication receive path has a large role in determining just how well the receiver will ultimately perform.

The receive bands for cellular and PCS in the United States are shown in Fig. 7.10. The general front-end filter configurations are shown for cellular A- and B-band systems. Figure 7.11 is a plot of the cellular A-band receive filter system, and Fig. 7.12 is an illustration of the cellular B-band filter system. Figure 7.13 is a receive filter system that is used also for either A- or B-band cellular systems; however, the filter system shown in Fig. 7.13 offers no selectivity from the other cellular operator for the front end of the receive system.

With the new addition of PCS systems a plethora of filtering systems are needed depending on which portions of the spectrum the operating company is

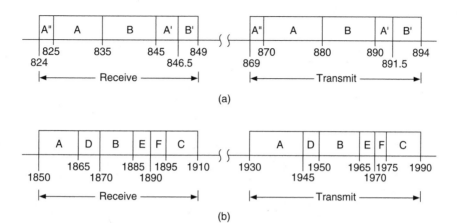

Figure 7.10   (a) U.S. cellular and (b) PCS spectrum chart.

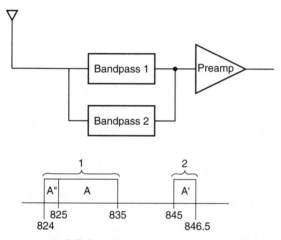

Figure 7.11   Cellular A band.

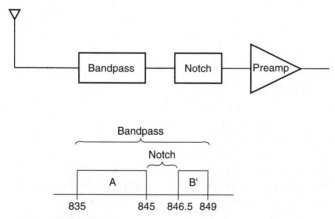

Figure 7.12   Cellular B band.

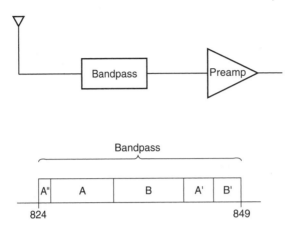

Figure 7.13 Cellular A- and B-band filter.

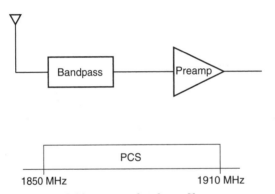

Figure 7.14 PCS spectrum bandpass filter.

operating at. With the auction process it is possible to obtain several different bands in the same market, which necessitates the use of special filters. Obviously one choice is to have a wide front-end filter allowing all the radio spectrum to enter into the receiver's preamplifier (Fig. 7.14). However, this situation would present front-end problems, especially in a mixed technology environment that the PCS auction process has guaranteed. Therefore, the filter schemes shown in Figs. 7.15 to 7.22 represent the configurations that are most likely to take place.

Obviously there are many perturbations to the above list of potential frequency blocks, but the fundamental concepts still remain the same. In addition, if an operator wishes to segregate a spectrum for dual or even trial technologies, it might be necessary to employ more selective filtering in the network to ensure the different base stations operate properly within a network.

For example, if a TDD and GSM system were deployed by the same operator in, say, the A block of a particular MTA, it might be necessary to utilize a different filtering system for the GSM portion as compared to the TDD section.

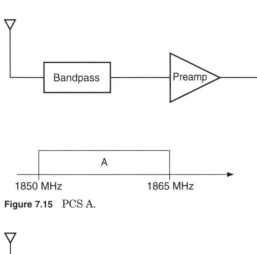

Figure 7.15   PCS A.

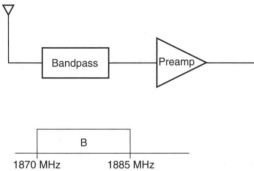

Figure 7.16   PCS B.

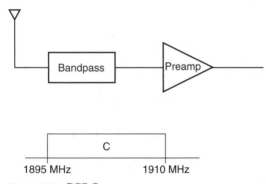

Figure 7.17   PCS C.

7.3.4 Preamplifier

The preamplifier is usually the first active component in a communication system's receive path. The basic function of any RF preamplifier is to increase the signal-to-noise ratio of the received signal. The preamplifier receives the desired signal at the lowest level in any of the receive stages for the communication site. Since the RF preamp receives the desired signal at the lowest

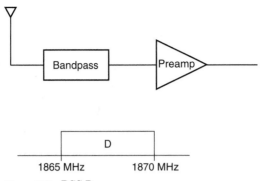

Figure 7.18 PCS D.

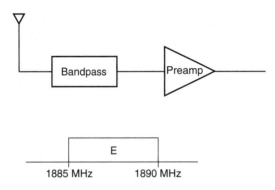

Figure 7.19 PCS E.

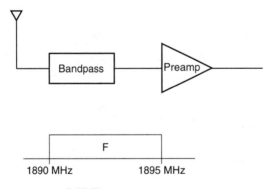

Figure 7.20 PCS F.

level of any receive stage in the cell site's receive path, any noise or other disturbances introduced in this stage have a proportionally greater effect.

The performance of the cell site's receiver with respect to weak signals depends on the performance of the preamplifier, or rather the signal-to-noise ratio of its output. The key issue is that amplifiers do not discriminate between what is the signal and what is noise or interference within the amplifier's pass band. In fact the preamplifier will amplify the desired signal plus any noise equally.

The preamplifier shown in Fig. 7.23 sometimes has a degree of filtering incorporated into the preamplifier itself or is just a straight amplifier. The preamplifier must have a sufficient gain to assist the receiver in its sensitivity. However, too much gain has the adverse effect of creating more intermodulation products

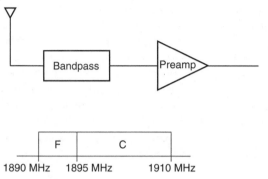

Figure 7.21 PCS C and F.

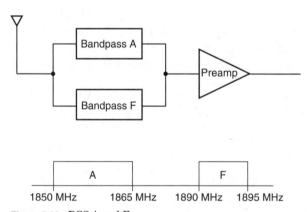

Figure 7.22 PCS A and F.

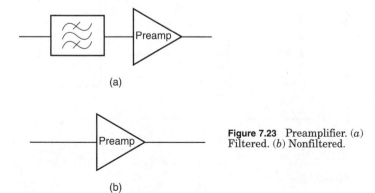

Figure 7.23 Preamplifier. (*a*) Filtered. (*b*) Nonfiltered.

when in the presence of strong signals. Additionally any amplifier has a power budget, regardless of its gain and robustness. The preamplifier for the cell site will amplify the out-of-band emissions that made it past the cell site filters. Depending on the amount of the out-of-band emissions allowed to pass into the receive path for the cell site, a decrease in overall receiver gain can be experienced.

The preamplifier shown in Fig. 7.24 is an amplifier with redundancy built into the path. The objective with this design is to ensure that if one leg of the receive path is damaged for a multitude of reasons the other path will ensure the receiver signal is still allowed to pass to the rest of the receive system. The design shown in Fig. 7.24 enables the designer to improve the overall gain while keeping the noise figure of the preamplifier at a very low value. The design also improves the mean time between failures, and in the event one path fails only a 3-dB reduction in power is experienced over what it would be if both branches were functional. The design in Fig. 7.24 assumes that there is a separate power supply and other components that ensure complete redundancy for this part of the receive system.

7.3.5 Multicoupler

The multicoupler is a device which ensures that received signals are routed to the appropriate receivers. Usually the multicoupler has several stages of splitting as shown in Fig. 7.25. The multicoupler itself normally has the preamplifier included as part of the configuration and is therefore included here. If the preamplifier is not included, the multicoupler contains only a combination of RF splitters. The configuration shown enables one receive antenna to be connected to many radios. The key advantage with utilizing a multicoupler is the reduction in the number of antennas and feedlines required for a cell site. By

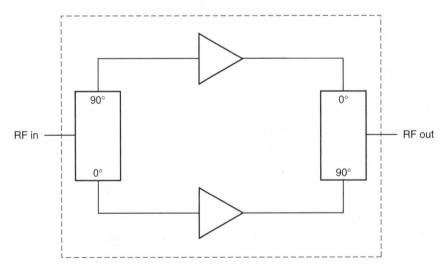

Figure 7.24 Preamplifier.

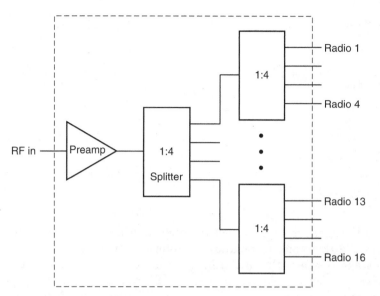

Figure 7.25 Multicoupler.

utilizing a multicoupler many radios within a communication site can share the same receive antenna.

The multicoupler shown in Fig. 7.25 implies that the receive path from the multicoupler itself to each individual radio is the same. However, if the path is different for, say, radio 1 and radio 16 because of internal RF plumbing issues, it might be necessary to include additional gain or padding for the receive branch or group of radios that has the imbalance.

7.3.6 Radio receiver

The radio receiver itself is referred to here as the physical device that converts the RF energy into a usable form. The radio receiver can have from one to multiple receive paths connected to it. Usually two paths are connected to the radio receiver in a cell site and only one path for a mobile or portable unit.

The basic radio receiver is shown in Fig. 7.26. Only one receive path is shown for ease of illustration since the receive path would be the same for each one. However, if more than one receive path were utilized for the radio, a form of diversity reception would be utilized to maximize the receive path's fade margin protection discussed later.

The radio receiver in Fig. 7.26 involves receiving the RF energy from the multicoupler, which enters the radio though a low-loss cable connecting the receiver to the multicoupler. The RF energy is then passed through a filter for additional selectivity and then amplified for additional gain. The RF energy is then put through a mixer which enables the signal to be downconverted to an intermediate frequency. The intermediate frequency is then filtered and then amplified again.

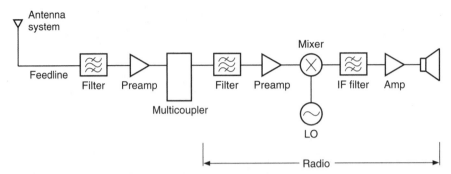

Figure 7.26 Radio receiver system.

The IF signal is now passed through another mixer and filter which now places the IF signal in the audio or information stage. The audio or information stage is where the initial information content is conveyed to the desired receiver, either a person or an electronic terminal device.

Obviously the above example is a simplified version of the events that take place in the receive path. The specific sequence of events that take place to the signal that goes through the receive path chain is dependent upon the type of modulation and information content that is to be interrupted by the end user or device.

Filter. The filter in a receiver itself is meant to accomplish several functions. The first objective is to improve the radio's selectivity by having the filter eliminate all the unwanted energy that has been allowed to pass and generate in the receive path of the communication system. The second objective, and the most important, is to protect the amplifier in the radio from desensitization and overload due to out-of-band emissions that could occur.

The filter used in the radio is a bandpass filter that usually can operate over the entire spectrum band of interest. Normally a radio receive filter will be able to pass RF energy for the A band equally as well as it will pass the B-band frequencies in, say, a cellular system. This enables manufacturers to construct radios that will operate in both portions of the desired band. The point that the radio receive filter is effectively wide places even more emphasis on the need to have a more selective filter at the beginning of the receive path system.

Preamplifier. The preamplifier for the radio itself is the first amplification stage in the actual radio. The purpose of the amplifier in the receive path of the radio receiver is to help set the noise figure and sensitivity of the radio itself. In addition the amplifier is meant to overcome any conversion losses experienced in the receive path as a result of the filter or mixer. The location of the preamplifier in the radio is shown in Fig. 7.26.

Mixer (downconversion). The receiver mixer has a critical role in converting the incoming RF spectrum containing the information content into a IF output ideally without adding noise or intermodulation products along the way. In most modern radios two mixers are employed as part of the downconversion process. However, for simplification only one mixer is shown in Fig. 7.26.

Mixers are sometimes described in textbooks as multipliers. A mixer normally has three ports and is a vital component in either up- or downconverting the information into another frequency. For a receiver the mixer can either up- or downconvert the frequency; however, the standard method utilized is to downconvert the frequency so it can be processed at an intermediate-frequency band instead of the RF band.

The downconversion process for the receiver is usually accomplished with a mixer, represented in system block diagrams as a circle with a cross through it. The mixer's function is to translate the incoming or source signal with that of another signal, usually the local oscillator frequency. A brief diagram of a mixer is shown in Fig. 7.27.

In most receivers two downconversion processes take place. The first involves reducing the initial RF signal to a level that can be processed better in the receiver. This is accomplished by converting the RF signal to an intermediate frequency (IF) and then mixing it again to reduce it even further to a second IF level. The rationale behind this method is to improve the overall performance of the receiver.

For a mixer it is very important to ensure that the device is matched to not only the f_1 portion, usually the RF or source signal, and also f_2, the local oscillator port. Proper matching ensures that phase and gain matching over the desired bandwidth takes place without adding distortion and/or phase noise to the signal.

An example of the mixing, downconverting process is shown in Fig. 7.28, which shows a source signal coming into the mixer, which is f_1 and is represented by the

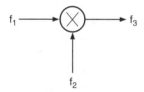

Figure 7.27 Mixer. f_1 = source signal. f_2 = local oscillator frequency. f_3 = resultant frequency.

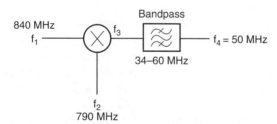

Figure 7.28 Receiver mixer. f_1 = source frequency. f_2 = local oscillator. f_3 = resultant frequency. f_4 = intermediate frequency.

signal entering the mixer on the left. The source signal is combined with the second frequency f_2, which is the local oscillator frequency. The combination of the two frequencies is the output f_3 shown in Fig. 7.28:

$$f_3 = f_1 - f_2 = 50$$

$$f_1 + f_2 = 1630$$

$$2f_1 - f_2 = 890$$

$$2f_1 + f_2 = 2470$$

$$f_1 - 2f_2 = \text{NA}$$

$$f_1 + 2f_2 = 2420$$

$$\vdots$$

It is interesting to note that the mixer has many products as a result of the two signals combining. In order to remove the undesired products, the output of the mixer (f_3) is then passed through a bandpass filter. A low-pass filter could also be used for this postprocess effort also; however, the bandpass filter employed will help reduce any unwanted noise energy that could be present. The output frequency is the resultant frequency, some lower frequency usually referred to as an intermediate frequency (IF), and is represented here as f_4. The IF frequency has only one component output from the downconverting process in this example.

The mixer has several key elements for performance that need to be met in order to ensure that its performance in the receiver communication path does not cause any impairments to the initial signal. The performance criteria involve conversion loss (CL), noise figure (NF), saturation level (SL), image frequency (IMF), and dynamic range, to mention some of the more important aspects.

The dynamic range of the mixer itself is very important since it determines the effective operating range in terms of signal strength the mixer is able to operate over. The dynamic range for the mixer is normally limited at the lower end by its noise figure, carrier-to-noise ratio (C/N), and the receiver bandwidth. It is important to note that the receiver bandwidth is dependent upon the technology platform chosen for the communication system. The saturation point for the mixer is approximately 3 dB above the 1-dB compression level for the mixer itself.

The particular frequency used for the local oscillator is critical for ensuring good performance from the receiver itself. A problem that can enter into the receive system is an image frequency, a particular frequency that is not the desired signal but based on the IF utilized it may be downconverted for postprocessing in the IF stage of the receiver itself.

An example of the image frequency is shown in Fig. 7.29, where the desired signal is in the presence of an undesired signal. The process of downconverting is the same here as that in Fig. 7.28. Based on the local oscillator (f_2), however, the undesired signal f_{1_B} can be passed through the bandpass filter that

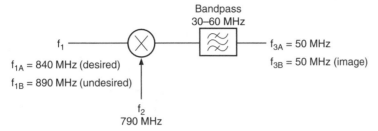

Figure 7.29 Image frequency.

normally is after the mixer itself. Several methods are used to eliminate the image frequency: change the local oscillator frequency and/or employ better front-end filtering before the preamplifier to eliminate the source of the undesired signals.

Noise is a factor that must be accounted for in any system, and it is ultimately the receiver's job to discriminate the desired signal and information content from that of noise in the environment. However, the communication system contributes to the overall noise in the receive path, and the mixer is no exception.

An example of the effect that the mixer has in contributing to the noise of the system is shown in Fig. 7.30. This example utilizes only a signal frequency for clarity of the example shown. The same desired and local oscillator frequencies are used as in the previous examples. The difference in this example is that noise and power are now incorporated in the downconverting process. In Fig. 7.30 the desired signal enters the mixer at a determined frequency, power level, and noise component. The noise component is a representation of the noise power that exists over the receiver's bandwidth. The noise power is assumed to be homogeneous for the entire receive band to simplify the example.

Signal f_1 enters the mixer which applies the local oscillator to it, having its own power level and specific frequency. The noise component N_2 represented here is the noise figure for the mixer itself. The conversion loss (CL) is shown in this example as being 15 dB. Therefore, the initial desired signal will be downconverted by the mixer, but this comes at the price of losing power, or conversion loss. The conversion loss also has the pleasure of reducing the noise power that enters the system by the same amount; however, the mixer adds noise to the signal because of its noise figure. Therefore, resultant noise that exits the mixer is −109.5 dBm and not −110 dBm.

The signal is then passed through a bandpass filter which also attenuates the desired signal to get the resultant frequency f_4. Through the process the S/N ratio has degraded by 0.6 dB from 35 dB at the input of the mixer, f_1 to the output of the bandpass filter, f_4.

Figures 7.31 through 7.33 represent different types of mixer configurations. These are not the only mixer configurations available but give a brief glimpse into the configuration they are composed of.

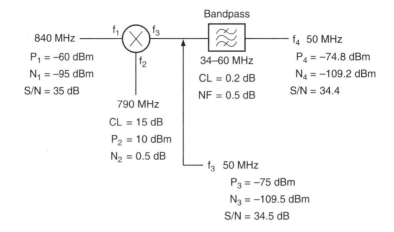

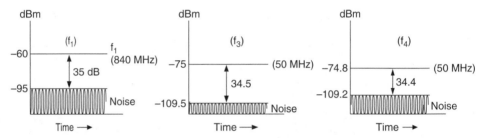

Figure 7.30 Mixer example.

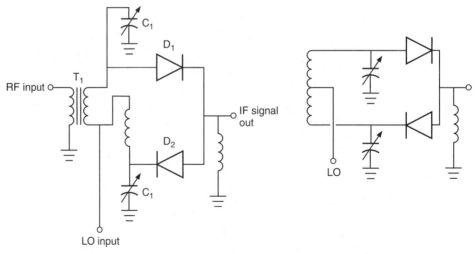

Figure 7.31 Singly balanced mixer. (*Milton Kaufman, Radio Operator's License Q&A Manual, 10th ed., Hayden, Hasbrouck Heights, NJ.*)

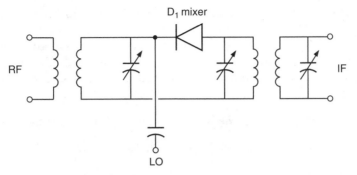

Figure 7.32 Single-ended diode mixer circuit.

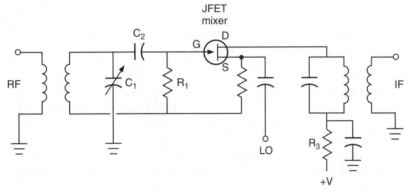

Figure 7.33 Single-ended active mixer.

IF stage. The intermediate (IF) stage of the receiver normally undergoes an additional downconversion so the second intermediate frequency is at a lower frequency for ease of postprocessing. Most of the amplification in the receiver takes place at the IF level. The IF stage is an important part of the receive chain since this is where the postprocessing of the information takes place. The IF stage can take on several variants depending on the technology selected. However, the basic premise is the same in that the signal is now at a lower-frequency range where it can be postprocessed more easily. Several important criteria define the process. They involve IF selectivity and image rejection, to mention two important figures of merit.

IF selectivity is probably the most or one of the most important specifications for a receiver. To prevent interference between channels at the receiver, IF selectivity is used to obtain the necessary interference protection. The selectivity of an IF section is a measure of the total response of all the IF stages, if several are involved with the process. The selectivity of the receiver must be sufficient to allow desired modulated signal to be amplified uniformly across the desired band but yet reject all the unwanted energy. The selectivity of a receiver is usually defined in terms of its Q.

The image rejection is an unwanted signal specification, usually a filter specification, and is the level in dB between the desired signal and the image signal's power. It is measured by applying a signal which is at an image frequency and increasing its signal strength until it is detected [Eq. (7.2)]. Then the desired signal is applied until it is detected. The difference is the image rejection level.

$$\text{Image rejection} = p \text{ received} - p \text{ image detected} \tag{7.2}$$

where p received $= p$ desired $= p$ image before filtering

Ideally the image rejection should be infinite, or rather the same value as the power received, meaning that no image energy is detected. Remembering the image example presented before the source signal which will also produce an image that falls within the pass band of the receiver, IF filters should be attenuated by the cell site filters and/or the receiver filter. The attenuation of the signal and the selection of the local oscillator should be selected to ensure that no adverse problems occur in the receiver itself.

Audio/information stage. The audio, or rather information, stage of the receive process is where the initial information content is extracted and utilized by the terminal or users at the end of the communication link. The audio stage is where the signal is eventually demodulated and transformed into information. The demodulation of the signal is dependent upon the modulation and information content desired for the system. Demodulation is covered in more detail in Chap. 4.

7.4   Diversity

Diversity is technique utilized to minimize the negative effects of fading in a mobile environment. The smaller the bandwidth for the information content to utilize, the more susceptible to fading a communication system will be. To overcome some of the negative effects of fading, several diversity techniques are utilized. Each technique has its advantages and disadvantages. Several excellent papers on diversity (Jakes[43] and Lee[29]) can be used for additional supplemental reading.

7.4.1   Horizontal

Horizontal diversity is the most common form of diversity utilized currently in mobile communications, primarily because of its ease of implementation. It is typically referred to as space diversity; however, many other forms of diversity could also be referred to as space diversity. Horizontal diversity is utilized for reducing the fading that normally takes place in a mobile environment. The horizontal separation needed is dependent upon the height of the antenna above the AGL (the angle of arrival of the signal) and the correlation desired between the signals.

Figure 7.34 is an example of horizontal diversity. It shows the general equation utilized to determine the amount of physical separation between the

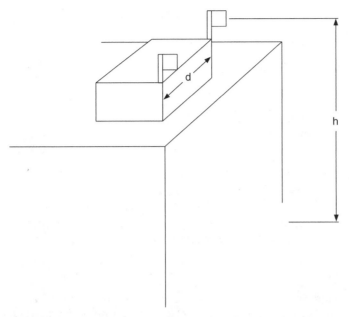

Figure 7.34 Horizontal diversity. $d = h/11$, where d = antenna spacing (ft), h = height AGI (ft), and $p = 0.7$.

antennas. The equation used is for a two-branch receive system. A slight modification needed when employing more branches can be found in Jakes.[43]

> **Example 7.1** A rooftop installation is chosen for the radio communication site. The desired correlation factor for the diversity branch is 0.7. The AGL for the cell site is 100 ft (30 m) and the ASML is 10 ft (3 m). What horizontal separation is needed for this cell site at 930 MHz?
>
> $$d = \frac{h}{11}$$
>
> where d = antenna spacing, ft
> h = height AGI, ft
>   ($p = 0.7$)
>
> $$d = \frac{h}{11} \cdot f_c = \frac{100}{11} \cdot \frac{835}{930} = 8.2 \text{ ft} \qquad f_c = \frac{835}{f_0}$$
>
> If $f_0 = 1860$
>
> $$d = \frac{h}{11} \cdot f_c = \frac{100}{11} \cdot \frac{835}{1860} \approx 4 \text{ ft}$$

7.4.2 Vertical

Vertical diversity is another diversity scheme which, when applied to the mobile communication field, requires the antennas to be separated by a defined vertical separation. Vertical diversity involves stacking the receive antennas on top of each other to overcome fades in the network. Vertical separation can be

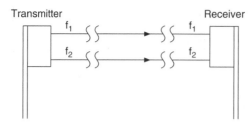

Figure 7.35 Frequency diversity.

used as a method for diversity reception for mobile communications. However, the drawback here is that typically the separation of the antennas is greater for vertical than it is for horizontal. In addition the multipath spread angle is usually small for a mobile environment, making this technique not attractive for improving the multipath environment for distant mobiles.

7.4.3 Polarization diversity

This form of diversity reception involves utilizing two different branches that are polarized in orthogonal planes. Polarization diversity lends itself to only a two-branch system in practicality. Work has been and continues to be conducted with implementing this type of diversity scheme in a mobile environment. The main advantage that polarization diversity offers a communication system is a reduced number of antennas required for an installation.

7.4.4 Frequency

Frequency diversity involves utilizing two distinct frequencies to convey the same information content. The concept is that while one frequency is undergoing a deep fade the other frequency is not experiencing the same fade depth. This will therefore ensure that the communication link remains at the desired level of service. However, frequency diversity has the unique disadvantage of requiring twice the resources both physically and also spectrally to deliver the same information content. The advantage frequency diversity has over space diversity techniques is that this method can utilize one antenna branch and therefore reduce the physical real estate on a tower. An example of frequency diversity is shown in Fig. 7.35.

7.4.5 Angle diversity

Angle diversity utilizes different antennas, but each antenna is oriented in a different direction. This is meant to reduce the number of antennas required for a cell site when multiple sectors are utilized. In addition the signal from a directive antenna has less severe fading characteristics. One application of angle diversity is used in a six-sector cell site where the number of antennas are reduced in count (Fig. 7.36). Each sector has one antenna oriented in a different direction. The angle of arrival for the signal from the subscriber unit is

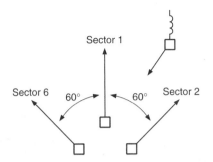

Figure 7.36 Angle diversity.

where the advantages of angle diversity take place. However, the maximum number of branches that can be employed by angle diversity is three.

7.4.6 Diversity combining

Several methods are employed for improving the received signal level at the receiver independent of the antenna diversity scheme. However, the diversity scheme utilized for the antenna has a direct impact on how effective the diversity combining technique will be.

Two diversity combining techniques are max ratio and select diversity. Both have advantages and disadvantages.

Max ratio. Max ratio combining is a technique that is used to improve the signal level at the receiver. The received signal from one branch is combined with another branch to increase its signal level. The number of branches used for max ratio combining is normally two but can at times include three. An example of max ratio combining is included in Fig. 7.37, and this is done at the IF level. The advantage with this technique is that real gain is achieved with this method since the signals are combined. A maximum of 3 dB is realizable with max ratio combining using two branches. The disadvantages, however, involve complexity of the receiver design.

Select. Select diversity involves a technique where the strongest branch in the receive system is chosen as the path for eventual demodulation and use by

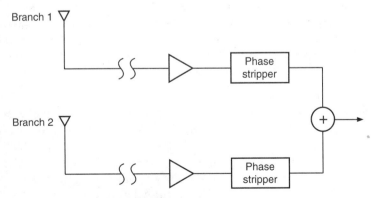

Figure 7.37 Max ratio combining.

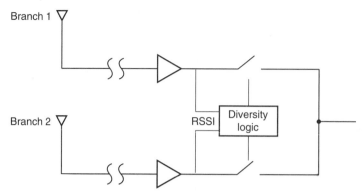

Figure 7.38 Select diversity.

the receiver itself. The objective is to select the branch with the strongest received signal strength (RSSI) and utilize that branch. The selection process is continuous and is meant to overcome the fades that take place. The number of branches used for select diversity is usually two but at times includes up to three. The number of branches used is driven by cost, since the majority of the improvement is experienced with two branches.

An example of select diversity combining is included in Fig. 7.38. The advantage of this technique is that it enables the strongest signal to be selected and therefore eliminates the deep faded signal from being combined as in the case of max ratio. However, select diversity also has the disadvantage of not having a real gain advantage, only a deep fade protection.

7.5   Performance Criteria

The performance criteria for a radio receiver are the key element to determining if the radio selected to extract the information content into a usable form will do so with success or failure depending on the environmental conditions it is placed within.

The performance criteria covered in this section do not pertain to the physical environmental issues, power consumption, heat-exchange requirements, and mean time between failure (MTBF), to mention a few. Instead the objective of this section is to provide a reference to the radio performance criteria that a design engineer should utilize in selecting a radio for the network.

7.5.1   Sensitivity

The ability for a receiver to detect a weak signal is determined by its sensitivity. Receiver sensitivity is a very important figure of merit for a receiver. The receiver sensitivity must be such that it can detect the minimal discernible signal (MDS) from the background noise. The MDS is a measure of sensitivity which incorporates the bandwidth of the system. It will differ from one receiver to another based on the bandwidth of the signals received.

A receiver's sensitivity is best defined by referencing the input signal level to the noise the receiver generates. The MDS is measured by turning off the AGC, applying a signal, with the correct bandwidth, and increasing the signal output from the generator until the signal plus noise is 3 dB higher than when there was no signal and only noise.

Specifically there is a relationship between thermal noise, the receiver's noise figure, and the bandwidth of the signal that the receiver is trying to detect. The relationship for receiver sensitivity is defined below in Eq. (7.3).

$$\text{Sensitivity} = 10 \log_{10} (kTB) + 10 \log_{10} (\text{bandwidth, Hz}) + NF \qquad (7.3)$$

$$f_c = 840 \text{ MHz} \qquad T = 25°C \qquad k = 1.38 \times 10^{-23} \text{ J/K}$$

$$10 \log_{10} (kTB) = -174 \text{ dBm/Hz}$$

$$\text{Sensitivity} = -174 \text{ dBm/Hz} + 10 \log_{10} (\text{bandwidth}) + NF$$

where k = Boltzmann's constant
T = temperature, K
B = bandwidth, Hz

Equations (7.4) show how to calculate a receiver's sensitivity for both an AMP voice channel and a CDMA channel. The AMPS channel is 30 kHz wide and the CDMA channel is 1.23 MHz wide.

Here, NF = 4 dB, assuming no processing gain.

$$\text{Sensitivity}_{\text{analog (30 kHz)}} = -174 \text{ dBm} + 10 \log_{10} (30 \text{ kHz}) + 4$$

$$= -125.22 \text{ dBm}$$

$$(7.4)$$

$$\text{Sensitivity}_{\text{CDMA (1.23 MHz)}} = -174 \text{ dBm} + 10 \log_{10} (1.23 \text{ MHz}) + 4$$

$$= -109.10 \text{ dBm}$$

Another method of defining sensitivity for a receiver is by referencing it to a specific S/N ratio (SNR). The SNR is defined in Eq. (7.5) and using a brief example an S/N of 17 dB is chosen.

$$\text{SNR} = 10 \log_{10} (kTB) + 10 \log_{10} (\text{bandwidth}) + NF + SNR \text{ (dB)} \qquad (7.5)$$

Here, NF = 4 dB, assuming no processing gain.

$$\text{Sensitivity}_{\text{analog(30 kHz)}} = -174 \text{ dBm} + 10 \log_{10} (30 \text{ kHz}) + 4 + 17$$

$$= -108.23 \text{ dBm}$$

Often sensitivity is defined as either negative decibels per milliwatt or in terms of microvolts. The conversion between decibels per milliwatt and micro-volts is shown in Eq. (7.6).

$$0 \text{ dBm} = 1 \text{ mW} = 1 \times 10^3 \text{ } \mu\text{W} \qquad (7.6)$$

$$\mu\text{V} = (50 \text{ } \mu\text{W})^{1/2} \text{ referenced to } 50 \text{ } \Omega$$

$$\therefore -116 \text{ dBm} = 2.511 \text{ pW}$$

$$= 0.224 \text{ } \mu\text{V}$$

7.5.2   Selectivity

Receiver selectivity provides a measure of the protection afforded the radio from off-channel interference. The degree of selectivity is largely driven by the filtering system within the receiver. The IF portion of the receiver affords the most benefit for selectivity. The greater the selectivity the better the receiver is able to reject unwanted signals from entering it. However, if the receiver is too selective, it may not pass all the desired energy.

7.5.3   Dynamic range

The dynamic range is a very important figure of merit and defines the range of signals that the receiver can handle within the specified performance of the receiver. There are several ways to specify the dynamic range of the system. One way is to define it as the range from the MDS to the 1-dB compression point of the receiver. This is often called the blocking dynamic range. Another method is to define the range from the MDS to where the third-order IMD equates the MDS signal and is referred to as spurious free dynamic range (SFDR), or it can be specified as the difference between MDS and a specified IMD level.

Figure 7.39 can be used to determine the dynamic range of a radio system. Note that the signal slope in the chart has a slope of 1:1 while the third-order intercept has a slope of 1:3. The actual intercept point is a calculated value only.

SFDR is a very important specification when the site is near other radio transmitters, since it is a direct indication of how the signal interferes with adjacent channels. SFDR provides a measurement of the performance of the radio as the desired signal approaches the noise floor of the receiver, providing an overall receiver SNR or BER. For example, if a radio can accurately digitize signals from -13 to -104 dBm in the presence of multiple signals, the DR = -91, and it implies a SFDR of 95 to 100 dB.

It is interesting to note that the SFDR can be improved when the signal level is reduced from the full scale, and this can improve the actual dynamic range even with the reduction in signal amplitude.

Dynamic range is usually defined as the range over which an accurate output will be produced. The lower level is called sensitivity and the upper level is called the degradation level. The sensitivity is determined by noise figure, IF bandwidth, and the method of processing. However, the degradation level is determined by whichever of the components in the receiver reaches its own degradation level first. It is therefore important to understand all the components in the receiver path since the first component to degrade will define the dynamic range of the system.

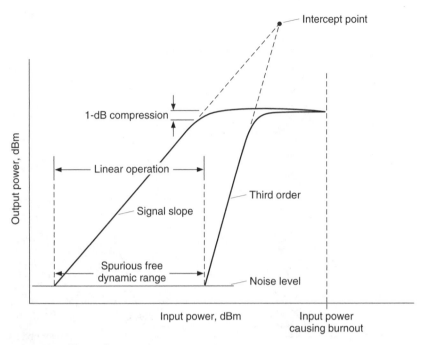

Figure 7.39 Dynamic range.

A large dynamic range for a receiver is a design priority and when comparing receiver to receiver, assuming they are operating in the same band and have the same bandwidth for receiving, the figure of merit (FOM) can be defined as in Eq. (7.7) to compare one receiver to another:

$$\text{Dynamic range FOM} = \text{input IP}_3 - \text{noise figure (NF)} \qquad (7.7)$$

If the output IP$_3$ is referenced for the receiver, the input IP$_3$ can be calculated once the gain of the device is known [Eq. (7.8)].

$$\text{Input IP}_3 = \text{output IP}_3 - \text{gain of device} \qquad (7.8)$$

The dynamic range FOM should always be referenced to the system input, and using this method a high dynamic range usually results in a positive FOM for the receiver.

Figure 7.40 shows how dynamic range is calculated.

7.5.4 SINAD

Signal to noise and distortion is called SINAD. SINAD is a ratio expressed in decibels which describes the receiver's performance in the presence of noise. SINAD is usually a reference value for determining a receiver's sensitivity, i.e., 12 dB SINAD. SINAD is a measure of the voltage ratio expressed in decibels at the audio output of the receiver.

Equation (7.9) is used to determine SINAD.

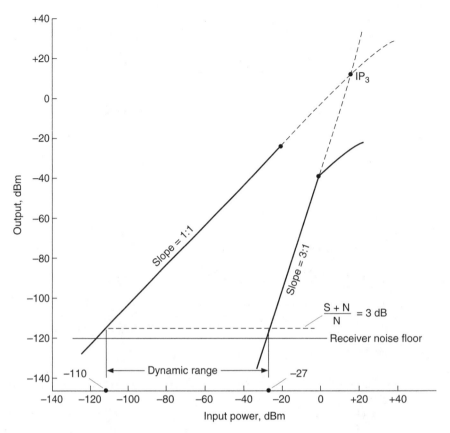

Figure 7.40 Dynamic range $= -110 - (-27) = 83$ dB.

$$\text{SINAD} = 20 \log_{10} \frac{\text{signal} + \text{noise} + \text{distortion}}{\text{noise} + \text{distortion}} \text{ dB} \qquad (7.9)$$

It can be measured in a few ways, and it is important to know the differences between them. One method, the normal method, is using C message weighting, and the other is not using C message weighting. The fundamental difference between the two is that C message weighting is a filter that is meant to pass voice only. C message filtering has a frequency range of between 300 and 3000 Hz. Therefore, when measuring without C message weighting for SINAD any signaling tones which are used in radio communication could distort the readings.

The methods of measurement involve injecting a 1-kHz signal with 2.9-kHz deviation and lowering the input level of the generator that is providing this signal until the noise is removed. Then a notch filter is used to remove the 1-kHz tone and the signal is attenuated more until noise becomes present. The difference in readings in terms of signal strength is the SINAD value either with or without C message filtering, depending on how you set up the test. S/N ratio is shown in Eq. (7.10).

$$\text{SNR (dB)} = 20 \log_{10} \text{(V signal rms/V noise rms)} \qquad (7.10)$$

It should be noted that SNR can be improved through a numerical operation called processor gains. In digitization processes the faster the signal is sampled the lower the noise floor becomes. The SNR does not improve; however, the noise energy is spread over a wider bandwidth (spectrum).

7.5.5   Distortion

Distortion simply means that unwanted signals appear at the output of any device in the RF path. A common place for distortion to occur is at the output of an amplifier; for this discussion it will be the preamplifier for the communication site. Distortion in a receive system generally takes on one or more forms, and the three most common forms are harmonic distortion, intermodulation distortion, and cross-modulation distortion. Whenever distortion occurs in the original signal some level of degradation occurs. Depending on the severity of the distortion, the communication content may or may not be disturbed.

Harmonic distortion occurs when the unwanted signals from multiple carriers are some integer multiple of the initial signal. For example, an amplifier boosts a signal at 835 MHz and at the same time generates a harmonic distortion starting at 1670 MHz for the second harmonic and 2505 MHz for the third harmonic. This has the negative effect of robbing the energy dedicated to the initial channel and creating unwanted signals in other parts of the spectrum.

Crossover distortion occurs when the amplitude-modulated signal from one transmitter is transferred to another carrier at the output of the device.

Intermodulation distortion, the most common form of distortion, is the product of several signals mixing together. The amount and levels of intermodulation distortion (IMD) are a direct result of the number of signals available to be mixed at any location. Intermodulation products for a second- and third-order mix for two signals are shown in Example 7.2.

Example 7.2

$$
\left.
\begin{array}{l}
A \ + B \\
A \ - B \\
B \ - A
\end{array}
\right\} \quad \text{second order}
$$

$$
\left.
\begin{array}{l}
2A + B \\
2A - B \\
2B + A \\
2B - A
\end{array}
\right\} \quad \text{third order}
$$

$\therefore$ if $A = 925$ MHz

$\qquad B = 870$ MHz

Order		MHz
$A + B$	925 + 870	1795
$A - B$	925 − 870	55
$B - A$	870 − 925	NA
$2A + B$	2(925) + 870	2720
$2A - B$	2(925) − 870	980
$2B + A$	2(870) + 925	2665
$2B - A$	2(870) − 925	815

Regardless of the type of distortion, the location for its occurrence can be at the audio or baseband prior to modulation, during demodulation, or somewhere at the RF level after modulation.

7.5.6 Quieting

Quieting is a measure of the receiver's ability to measure an unmodulated signal which will reduce the noise output at the receiver when it is unsquelched. The quieting level is referred to in terms of dBm. This method of measurement was and is the predecessor to SINAD measurements, since it did not really give a measure of the receiver's sensitivity.

The measurement method involved measuring the audio output of the receiver with no RF signal coming into it and then applying a carrier, unmodulated, to it until the audio output was reduced by 20 dB. Simply put, this is the amount of signal required to quiet the receiver from no signal to the level of an unmodulated carrier, which reduces the noise by 20 dB, as shown in Eq. (7.11).

$$20 \text{ dB}Q = kTB + NF + \text{IF}_{\text{SNR}} \tag{7.11}$$

where k = Boltzmann's constant, 1.38×10^{-23} J/K
T = temperature, K (290°)
B = bandwidth, Hz
NF = noise figure

7.5.7 Capture

FM receivers utilize capture effect where the strongest of several cochannel signals completely suppresses the weaker one. FM capture is experienced every day with a car radio when you receive signals from two different radio stations on the same channel. When one of the radio channels is dominant (captured), the other is not heard, but during that transition period both can be heard since neither signal has been captured by the receiver.

7.5.8 Noise

Noise for a communication system directly affects its overall performance. All receivers need to have a certain C/I or E_b/N_0 value to perform properly. If the overall noise that the receiver experiences or has to deal with increases, the desired signal needs to be increased in signal, without increasing the noise content, to ensure the proper ratio is maintained.

For the RF design engineer several components associated with the receive system comprise noise. The three items involve thermal, shot, and system noise. The latter can be reduced through proper frequency planning, power control, and appropriate use of selective filters along with isolation techniques. Both the thermal and shot noise comprise what is referred to as the *noise figure* for a receiver. Noise figure is one of the fundamental measures of a receiver's performance and should be measured at a predetermined location for the receiver itself. Noise figure for a receiver degrades, i.e., increases, with each successive stage in the receive path. A common point to measure noise figure is the audio output for a receiver, but with digital radios there is no audio output and the measurement point is then the IF output.

The noise figure of the receive system is directly related to the overall receiver sensitivity and is calculated as shown in Eqs. (7.12).

$$\text{Noise factor} = F_N = \frac{S_{in}/N_{in}}{S_{out}/N_{out}}$$

$$\text{Noise figure, dB} = 10 \log_{10} (F_N)$$

(7.12)

The noise figure for the system is normally set by the first amplifier in the receive path. The noise floor can be improved or attenuated by passive devices, but the SNR will not be improved unless the bandwidth is narrowed.

7.5.9 Thermal and shot noise

There are two components to noise, one thermal and the other shot. Thermal noise is a direct result of kinetic energy and is directly related to the temperature. Shot noise, on the other hand, is caused by the quantized and random nature of current flowing in a device. Both thermal and shot noise comprise noise that exists in a communication system.

Noise temperature has a direct relationship to operating frequency and the bandwidth of the receiver or signal to be detected. The relationship is shown in Eqs. (7.13).

$$kTB = (1.38 \times 10^{-23} \text{ J/K}) (290 \text{ K}) (115)$$
$$= 4.002 \times 10^{-21} \text{ J/s} = 4.002 \times 10^{-21} \text{W} \quad (7.13)$$

$$\text{Noise power per hertz} = 10 \log_{10} \left(\frac{4.002 \times 10^{-21} \text{ W}}{10^{-3} \text{ W}} \right) = -173.97 \text{ dB/Hz}$$

where k = Boltzmann's constant, 1.38×10^{-23} J/K
T = temperature, K (290°)
B = bandwidth, Hz

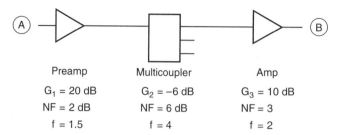

Preamp

$G_1 = 20$ dB

NF = 2 dB

f = 1.5

Multicoupler

$G_2 = -6$ dB

NF = 6 dB

f = 4

Amp

$G_3 = 10$ dB

NF = 3

f = 2

Figure 7.41   Noise representation.

What is of prime interest to the RF engineer is what effect each component in the receive system has on improving the overall sensitivity of the system. The first amplifier in the system sets what the noise figure for the system will be. Equation (7.14) determines the noise figure.

$$F = F_1 + \frac{F_2 - 1}{G_1} + \frac{F_1 - 1}{G_1 \cdot G_2} + \cdots \tag{7.14}$$

Example 7.3 shows how to calculate the noise figure for a system using specifications normally provided. Note that the decibel values are changed to reflect gain in terms of ratios. The noise figure for a passive device is the loss attributed to that device (Fig. 7.41).

Example 7.3

$$f_B = f_1 + \frac{f_2 - 1}{G_1} + \frac{f_3 - 1}{G_1 \cdot G_2}$$

$$= 1.5 + \frac{4 - 1}{100} + \frac{2 - 1}{(100)(.25)}$$

$$= 1.5 + \frac{3}{100} + \frac{1}{25}$$

$$= 1.57$$

$$NF_{A \to B} = 10 \log_{10} (f_B) = 1.958 \text{ dB}$$

The above example shows that the preamplifier, first active device in the chain, defines the overall noise figure for the system. As a rule of thumb, if the gain of the first amplifier is sufficient, the noise figure for the first stage can be used as a rough rule to determine the improvements in decibels for sensitivity.

Care must be taken in the determination of the gain for the first active stage in a receiver; too much gain can overdrive the later stages in the receive path and actually degrade the receiver's sensitivity. However, using Example 7.3, a quick relationship, shown in Example 7.4, can be determined between noise figure and receiver sensitivity.

Example 7.4

$$N = 10 \log_{10} kTB$$

where k = Boltzmann's constant, 1.38×10^{-23} J/K
T = temperature, K (290°)
B = bandwidth, Hz

Thermal noise = -174 dB/Hz (290 K)

Sensitivity (optimal) = -174 dB + $10 \log_{10}(B) + NF$

BW	30 kHz, dB	200 kHz, dB	1.25 MHz, dB
Bandwidth (dB) noise	44.77	53.01	60.96
$-174 + 10 \log_{10}(B)$	-129	121	113
NF	1.95	1.95	1.95
	127.05	119.05	111.05

The above calculations do not factor in the effects of the antenna noise, feedline loss, and required S/N or E_b/N_0 as part of the receiver sensitivity.

7.5.10 1-dB compression

The 1-dB compression point is a common reference term used to define the performance of a particular receiver, or rather amplifier in the receiver itself. When the 1-dB compression point is referenced for the entire receiver, however, the 1-dB compression point can also reference an individual amplifier, like the preamplifier.

The 1-dB compression point is where the power gain for the receiver is down 1 dB from the ideal gain. That is, if the input signal goes up by 2 dB and the output goes up by only 1 dB, this is the point where the 1-dB compression point occurs. Often the 1-dB compression point is referred to as blocking for the receiver. The blocking occurs in that the weaker signals are not amplified properly, leading to them being potentially blocked from being detected. The 1-dB compression point is part of the component for determining the receiver's overall dynamic range.

The 1-dB compression point is shown in Fig. 7.42. It can occur either as a direct result of trying to receive the desired signal, which is too hot, causing the overload condition, or through undesired signals overloading the receiver. Referring to Fig. 7.42, the 1-dB compression point can be directly affected by the receiver's overall gain setting.

7.5.11 Third-order intercept

The third-order intercept point (IP_3) is a figure of merit for a receiver. Specifically the IP_3 value directly influences and determines the receiver's dynamic range. The third-order intercept value, along with the second-order, determines the receiver's linearity.

It should be noted that the IP_3 value is a theoretical value which is achieved owing to extrapolation of the third-order curve. In addition the IP_3 value is

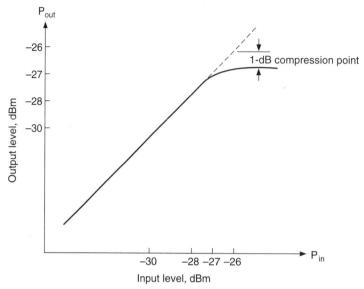

Figure 7.42   1-dB compression point.

frequency-dependent, and based on the selection of the test frequencies, a different IP_3 value may be achieved. However, most IP_3 tests are set up to produce a product that falls within the first IF of the receiver.

Basically the IP_3 value increases by 3 dB for every 1-dB increase for the desired signal. The slope of the IP_3 line is therefore 3:1, shown in Fig. 7.38. The IP_3 value is important for determining the receiver's performance since the presence of two larger signals can generate spurs caused by nonlinearities, or rather distortions, that can override the weaker desired signals.

7.5.12   Phase noise

Phase noise is a measure of the receiver's phase and frequency perturbations which are added to the initial input signal. The effect of phase noise is to distort the initial signal which either degrades the initial signal or degrades the receiver's sensitivity.

Phase noise is a specification of spectral purity, usually defined at –dBc/Hz, which is a certain offset from the center frequency. Phase noise is an important figure of merit, since it can lead to frequency translations occurring for out-of-band signals to fall within the receiver's pass band.

7.5.13   Desense

Any form of RF energy, whether manmade or natural, has the potential of adversely affecting the receiver. Any RF energy that adversely affects the receiver's ability is defined as RFI. Desense is technically a reduction in the receiver's overall sensitivity which is caused by RFI, either manmade or natural. Desense occurs when a very strong signal begins to overload the front end of the receiver and makes the detection of weaker signals more difficult.

All types of noise interference adversely affect the cell site's receiver performance, whether the desired signal is purely FM analog or TDMA. Noise interference can either be attributed to out-of-band emissions from nearby transmitters whose sidebands contribute to the effective noise floor of the cell site or it can come from in-band sources, cochannel, or IMD products.

The magnitude of the noise interference is determined by the amount of energy within the receiver pass band. Depending on the amount of noise interference, the effective receiver sensitivity can be degraded during increases in noise interference level because of reduced S/N or BER ratios.

When colocated with many transmitters the transmitter noise spectrum is noticeably close to the carrier at 60 dBc and gradually falls off in power as a function of effective bandwidth. However, at some power level the noise level will degrade receiver sensitivity. The noise interference power level caused by other transmitters can be reduced by increasing the physical separation between the offending transmitter(s) and the offended receiver(s). Another alternative is to place a narrower transmit filter on the transmitter or utilize a more selective receive filter to remove or seriously reduce the transmitter noise spectrum.

Whenever transmitter noise is present, there exists the potential for receiver desensitization. The cell site receiver can be desensitized by undesired signals, reducing the sensitivity of a receiver by reduction in the gain or an increase in noise level of certain amplifiers or mixers in the receive path of the cell site. It is possible, based on the receiver filter used, to pass out-of-band energy into the receive system and cause a reduction in receiver gain. The loss of gain in this manner is a form of desensitization.

Therefore, if a cell site is desensed by out-of-band emissions the installation of a more selective filter will make the site more sensitive. However, if the site is not being desensed by out-of-band emissions, the filter's introduction could show up as a reduction in effective noise floor for the cell site.

The basic function of any RF preamplifier is to increase the signal-to-noise ratio of the received signal. Since the RF preamplifier receives the desired signal at the lowest level of any receive stage in the cell site receive path, any noise or other disturbances introduced in this stage has a proportionally greater effect.

The performance of the cell site receiver with respect to weak signals depends on the performance of the preamplifier or rather the signal-to-noise ratio of its output. The key issue is that amplifiers do not discriminate between what is signal and what is noise within their pass band. In fact the preamplifier will amplify the desired signal plus any noise equally.

Any amplifier has a power budget, regardless of its gain and robustness. The preamplifier for the cell site will amplify the out-of-band emissions that made it past the cell site filters. Depending on the amount of out-of-band emissions allowed to pass into the receive path for the cell site, a decrease in overall receiver gain can be experienced.

The best method to overcome desense problems is to ensure that there is sufficient isolation between the potential offending transmitter and the receiver. Desense is and will continue to be more of a problem, especially in urban environments, as the use of wireless communications continues to expand. Additionally, receiver desense is a major problem experienced with handheld subscriber units.

References

1. Smith, Clint, and Curt Gervelis, *Cellular System Design and Optimization,* McGraw-Hill, New York, 1996.
2. Crescenzi, E. James, and Roger Fildes, "Miniature Tuner Serves Diverse IGNIT Applications," *Microwaves & RF,* 1995, pp. 69–79.
3. Breed, Gary, "Receiver Basics—Part 1: Performance Parameters," *RF Design,* February 1994, pp. 48–50.
4. Nezami, Mohamed, "Guidelines Dictate the Performance of MMDS Downconverters," *Microwaves & RF,* April 1996, pp. 58–72.
5. Schweber, Bill, "Converters Restructure Communication Architectures," *EDN,* August 1995, pp. 51–64.
6. Breed, Gary, "Receiver Basics—Part 2: Fundamental Receiver Architectures," *RF Design,* March 1994, pp. 84–89.
7. Brannon, Brad, "Using Wide Dynamic Range Converters for Wide Band Radios," *RF Design,* May 1995, pp. 50–65.
8. Herskovitz, Don, "Wide, Wider, Widest," *Microwave J.,* September 1995, pp. 26–40.
9. Gilbert, Barrie, and Rupert Baines, "Fundamentals of Active Mixers," *Applied Microwave & Wireless,* winter 1995, pp. 10–27.
10. Neuf, Donald, "Extended Dynamic Range Mixers," *Applied Microwave & Wireless,* winter 1996, pp. 24–39.
11. Oltman, Randy, "CDMA Noise Measurements," *Applied Microwave & Wireless,* winter 1996, pp. 40–48.
12. Pawlan, Jeffrey, "A Tutorial on Intermodulation Distortion: Part 2—Practical Steps for Accurate Computer Simulation," *RF Design,* March 1996, pp. 74–86.
13. IEEE Standard 184-1969, *IEEE Test Procedures for Frequency Modulated Mobile Communication Receivers,* IEEE, New York, New York.
14. Ciccarelli, Steven, "Specifying IF Filter Selectivity for FM Receivers," *Microwaves & RF,* March 1995, pp. 101–106.
15. Jacob, Gerald, "Why Special Signals Are Needed for Digital Communications Testing," *EE Evaluation Engineering,* November 1995, pp. 62–66.
16. Howard, Robert, "Receiver Recover SMSK- and MSK-Modulated Inputs," *Microwaves & RF,* January 1995, pp. 118–124.
17. Pawlan, Jeffrey, "A Tutorial on Intermodulation Distortion and Non-Linearity in RF Systems," *RF Design,* February 1996, pp. 71–74.
18. Watson, Robert, "Guidelines for Receiver Analysis," *Microwaves & RF,* December 1986, pp. 113–122.
19. Carlson, A. Bruce, *Communication Systems,* 2d ed., McGraw-Hill, New York, 1975.
20. American Radio Relay League, *The ARRL 1986 Handbook,* 63d ed., The American Radio Relay League, Newington, CT, 1986.
21. American Radio Relay League, *The ARRL Antenna Handbook,* 14th ed., The American Radio Relay League, Newington, CT, 1984.
22. AT&T, *Engineering and Operations in the Bell System,* 2d ed., AT&T Bell Laboratories, Murray Hill, NJ, 1983.
23. Brewster, *Telecommunications Technology,* Wiley, New York, 1986.
24. Carr, J. J., *Practical Antenna Handbook,* McGraw-Hill, 1989.
25. Code of Federal Regulations, CFR 47 Parts 1, 17, 22, 24, and 90.
26. DeRose, *The Wireless Data Handbook,* Quantum Publishing, Mendocino, CA, 1994.
27. Dixon, *Spread Spectrum Systems,* 2d ed., Wiley, New York, 1984.
28. Kaufman, M., and A. H Seidman, *Handbook of Electronics Calculations,* 2d ed., McGraw-Hill, New York, 1988.

29. Lee, W. C. Y., *Mobile Cellular Telecommunications Systems,* 2d ed., McGraw-Hill, New York, 1996.
30. Qualcom, *An Overview of the Application of Code Division Multiple Access (CDMA) to Digital Cellular Systems and Personal Cellular Networks,* Qualcom, San Diego, CA, May 21, 1992.
31. Sklonik, M. I., *Introduction to Radar Systems,* 2d ed., McGraw-Hill, New York, 1980.
32. Simo, "IS-95 Based SS-CDMA: Operational ISSUES," seminar, January 1995.
33. Stimson, *Introduction to Airborne Radar,* Hughes Aircraft Company, El Segundo, CA, 1983.
34. White, Duff, *Electromagnetic Interference and Compatibility,* Interference Control Technologies, Inc., Gainesville, GA, 1972.
35. Yarborough, *Electrical Engineering Reference Manual,* 5th ed., Professional Publications, Belmont, CA, 1990.
36. Qualcom, *CDMA Handbook,* Qualcom, San Diego, CA, 1996.
37. Fink, Donald, and Donald Christiansen, *Electronics Engineers Handbook,* 3d ed., McGraw-Hill, New York, 1989.
38. Schwartz, Bennett, and Stein, *Communication Systems and Technologies,* IEEE, New York, 1996.
39. Rappaport, *Wireless Communications Principles and Practices,* IEEE, 1996.
40. Fink, Beaty, *Standard Handbook for Electrical Engineers,* 13th ed., McGraw-Hill, New York, 1995.
41. Webb, Hanzo, *Modern Amplitude Modulations,* IEEE, 1994.
42. Steele, *Mobile Radio Communications,* IEEE, 1992.
43. Jakes, W. C., *Microwave Mobile Communications,* IEEE, New York, 1974.
44. Carlson, A. B., *Communications Systems,* 2d ed., McGraw-Hill, New York, 1975.
45. Johnson, R. C., and H. Jasik, *Antenna Engineering Handbook,* 2d ed., McGraw-Hill, New York, 1984.
46. Lathi, *Modern Digital and Analog Communication Systems,* CBS College Printing, New York, 1983.
47. MacDonald, "The Cellular Concept," *Bell System Technical J.,* vol. 58, no. 1, 1979.
48. Rappaport, *Cellular Radio & Personal Communications,* IEEE, New York, 1995.
49. *Reference Data for Radio Engineers,* 6th ed., Sams, 1983.
50. Mouly, Pautet, *The GSM System for Mobile Communications,* Mouly Pautet, 1992.
51. Williams, Taylor, *Electronic Filter Design Handbook,* 3d ed., McGraw-Hill, New York, 1995.

Transmitters

8.1 Introduction

This chapter covers transmitters, their types and applications. It also covers transmitter combining techniques. With the constant demand to extract all the possible performance from a communication site, understanding of some of the basic topics associated with transmitters is essential.

There are many types of transmitters (amplifiers) in a communication system. Most of the amplifiers are located in the receiver for the communication cell site. The fundamental difference between the amplifiers in the transmit portion and the receiver lies in the power they deliver to the desired load. The focus of this chapter is toward RF transmitters which are power-oriented.

Knowledge of RF transmitter types and different combining techniques enables an RF engineer to maximize the efficiency of a communication site through improving the amount of energy delivered to the antenna system or reducing the number of physical antennas at a site. The number of physical antennas available at a site may or may not be driven by economic reasons alone; local ordinances may have a more profound role in deciding the ultimate configuration for a communication site.

8.2 Transmitter System Building Block

The transmitter block diagrams for the three basic forms of radio communication are included next. The details for modulation are covered in Chap. 4. Figure 8.1 is a block diagram of an AM transmitter. The AM transmitter changes the amplitude of the carrier as a function of the information content.

Figure 8.2 is a brief block diagram of an FM transmitter. The FM transmitter modulates the information content by changing the frequency of the carrier as a function of the information content.

Figure 8.3 is a block diagram of a phase modulation (PM) transmitter. The PM transmitter places the information onto the carrier much as it is done with FM; however, the modulation is achieved through adjusting the phase of the information that rides on the carrier.

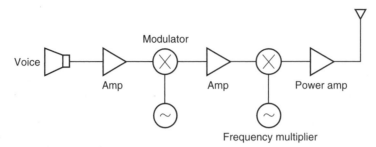

Figure 8.1 AM modulation.

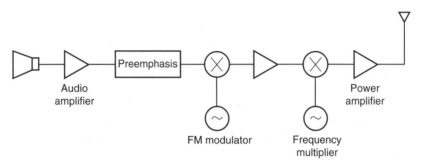

Figure 8.2 FM modulation.

8.3   Amplifier Types and Classes

There are numerous types and classifications for amplifiers. Each of the classes defined has advantages and disadvantages. The type of amplifier chosen needs to naturally meet the requirements set forth in the design objective, maximize output while consuming as little power as possible, and at the same time pass the desired modulation format with no distortion over its frequency and power operating range.

8.3.1   Class A

A class A amplifier is an amplifier that conducts during the entire 360° of the sine-wave input into the amplifier. Class A amplifiers offer high linearity with respect to the transfer function of the amplifier. The class A amplifier will remain linear as long as the amplifier does not saturate, properly dissipates its power, and operates within the defined linear range of the transfer function for the amplifier.

8.3.2   Class B

The class B amplifier is designed to conduct for only half, or rather 180°, of the sine-wave input. The class B amplifier is typically not used since it relies on a push-pull amplifier design. The push-pull amplifier design does not afford linearity across its defined operating range. The class B amplifier has the advantages of offering greater output power than the class A and is more energy-efficient. However, the class B amplifier is known to exhibit crossover distortion, which is why it is not a desired amplifier for RF.

8.3.3 Class AB

Class AB amplifiers are used in many cases where a single transmitter is needed. The class AB amplifier does not exhibit the amount of crossover distortion experienced in a class B amplifier and is not as linear as the class A amplifier. The class AB amplifier is finding more and more use in wireless applications.

8.3.4 Class C

Class C amplifiers are utilized throughout the wireless industry. The class C amplifier has the advantage of being more power-efficient than the A-, B-, or AB-type amplifiers. The class C amplifier does not conduct during the entire 360° of the sine wave but only a little less than 180°. The class C amplifier will distort the signal but frequency and phase modulation techniques do not need linearity for the amplification.

8.3.5 Feed-forward

The use of feed-forward amplifiers for communication sites wishing to amplify multiple RF channels at the same time has increased greatly. The feed-forward amplifier lends itself to excellent linearity and low-level combining of the RF signals for final amplification. However, the linear amplifier not only amplifies the desired signal but also generates intermodulation

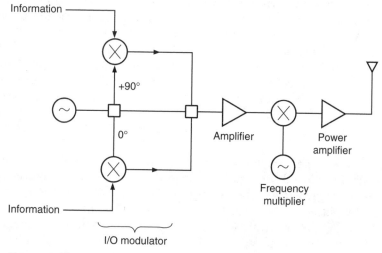

Figure 8.3 Phase modulation.

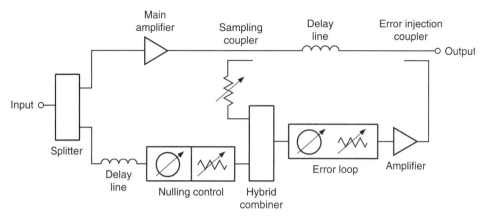

Figure 8.4   Feed-forward amplifier.

products at the same time. The feed-forward amplifier design relies on the ability of the system to cancel the unwanted signals and allow the desired signals to pass unimpeded.

A brief block diagram of a typical feed-forward amplifier is shown in Fig. 8.4. The chief advantage of utilizing a feed-forward amplifier is that this type of design reduces the amount of power required to achieve the same composite power as compared to conventional combining methods. The disadvantage of the feed-forward amplifier is the complexity.

8.4   Modulation

Transporting the information from one location to another in wireless communications involves transmitting the information content. The information content is sent via a carrier wave at a defined frequency, and the carrier wave, depending on the modulation format chosen, is either modulated or unmodulated. A more detailed description of modulation is included in Chap. 4.

If the amplification process used overmodulates the signal, distortion will occur. In addition as a result of *overmodulation* the generation of spurious harmonics accompanies this, and the transmitter sidebands increase in magnitude, creating the potential for adjacent channel interference.

8.5   Isolators

An isolator, when deployed in a communication site, has many functions. Specifically it is a device that conducts RF energy in one direction and rejects, or rather highly attenuates, energy in the opposite direction. The isolator is normally placed between the transmitter and the antenna system as a method of protecting the transmitter. Figure 8.5 shows a typical location in a communication system for an isolator. The isolator effectively isolates the transmitter from the antenna and minimizes the potential for intermodulation products to be generated. The isolator also acts as a balanced load for a transmitter, ensuring maximum power transfer to the antenna system.

An isolator is a three-port device that consists of a circulator and a load. The circulator is also sometimes referred to as a gyrator. The isolator is shown in Fig. 8.6.

Items to consider in the selection of an isolator include:

1. Power rating
2. Isolation requirements (dB)
3. Ambient temperature and heat for the application
4. Worst-case reflected power at the output of the isolator port

The isolator's construction consists of the circulator and the load, normally 50 Ω. The components that comprise a circulator are shown in Fig. 8.7. The circulator takes advantage of the magnetic field and the polarization that is created because of their placement in the device. On the center conductor shown in Fig. 8.7 are the relative port numbers referred to in Fig. 8.6.

Typically a single isolator has 20 dB of isolation capability, and to achieve, say, 40 dB of isolation it is necessary to gang two isolators together to achieve the required isolation (Fig. 8.8). Just how much isolation is required for a communication site is dependent upon a multitude of items. However, the isolation

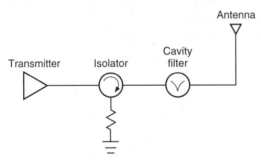

Figure 8.5 Typical isolator location.

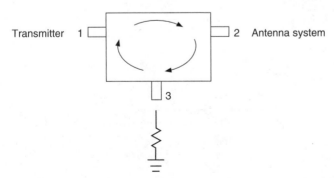

Figure 8.6 Isolator.

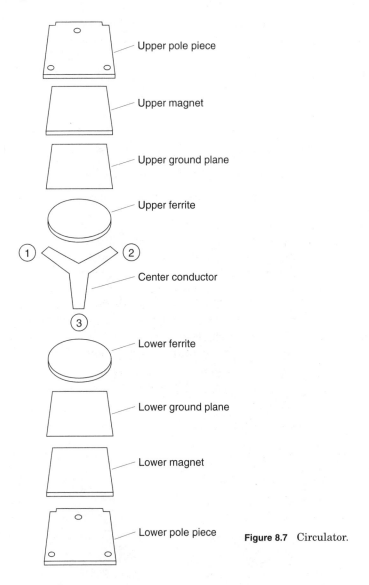

Upper pole piece

Upper magnet

Upper ground plane

Upper ferrite

Center conductor

Lower ferrite

Lower ground plane

Lower magnet

Lower pole piece

Figure 8.7   Circulator.

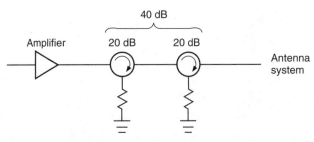

Figure 8.8   Cascaded isolators.

requirements are driven by frequency separation between the transmitters at a site. The closer the frequencies of the transmitters are to each other the greater the need for isolators. Typically in a cellular system a 40-dB isolator is at the output of each power amplifier right before the cavity or hybrid combiner input.

How much isolation is needed for a system can be calculated with the following method shown in Example 8.1. It should be noted that the spurious transmission value used is typical and can be exceeded depending on the manufacturer of the infrastructure equipment used.

Example 8.1   See Fig. 8.9 for a visual depiction of this method.

Transmitter A	+50 dBm
Path loss 1 (coupling)	−50 dB
Antenna gain B	+10 dBd
Line loss B	−1.5 dB
	+8.5 dBm power received at B_1

Spurious transmission (B)	−60 dBc
Line loss B	−1.5 dB
Antenna gain B	+10 dBd
Path loss 2	−50 dB
Antenna gain C	+11 dBd
Line loss C	−1.5 dB
	−83.5 dB IMD power received at C_1

Rx sensitivity	=	−116 dBm
IMD power	=	−83.5 dBm
		32.5 dB of isolation needed
		∴ 40 dB isolation

In Fig. 8.9 a total of 32.5 dB of isolation is computed as being needed between B and C. The isolation can be achieved through additional antenna isolation, i.e., moving them farther apart, or by the introduction of a cavity filter at B or by the placement of an isolator at B between the transmitter and the antenna system.

The example used shows that the placement of a 40-dB isolator in the transmit path for B would solve the potential problem. Also normally a transmit filter would accompany the installation of an isolator to ensure proper emission masks are met. Lastly the example is calculated only for one direction, and the other direction, i.e., IMD generated by A, should also be checked. Naturally with a multitude of transmitters located at a communication site the calculation of individual isolation requirements may prove daunting.

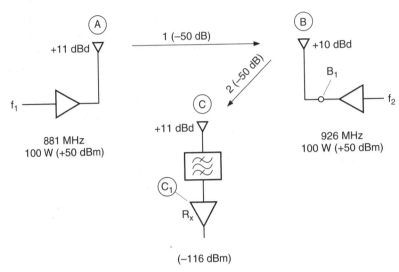

Figure 8.9 Calculating isolation for a cellular system.

8.6 Hybrid Combiners

Hybrid combiners are utilized in many locations in a communication system. The most common location for a hybrid combiner is in the transmit path of the radio system. A hybrid combiner's purpose is to combine different discrete frequencies from two, or multiple, paths onto one path (Fig. 8.10).

Often a hybrid combiner is used for combining signals when the isolation between frequency channels is not sufficient to ensure proper operation of the transmitter. The hybrid combiner also enables multiple signals to be combined at a low energy level for later amplification by a linear amplifier.

Several types of hybrid combiners are available, each with the same fundamental objective of combining two signals onto one port. Figures 8.11 and 8.12 below show the hybrid configuration of a hybrid ring combiner. The difference between the two ring combiners is simply the relationship between the ports. Fundamentally they work the same by taking advantage of the physical wavelength between the ports.

Another type of hybrid combiner is the Wilkinson hybrid combiner, shown in Fig. 8.13. The Wilkinson hybrid lends itself to PC board fabrication and is used extensively in wireless communication systems. The primary difference between the ring and the Wilkinson hybrids is the power-handling capabilities of both. The ring combiner affords a higher level of power that it can handle vs. the Wilkinson combiner. However, the Wilkinson combiner is rather simple to construct and lends itself to low-level power combining situations.

The typical isolation between ports is normally 40 dB. Therefore, the isolation between ports 1 and 4 is 40 dB, which enables closer channel spacing to occur without sacrificing excessive insertion losses due to reduced transmitter isolation. The insertion loss for a four-port, two in two out, hybrid combiner is a little greater than 3 dB. Also the hybrid can be used for receive as well as

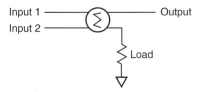

Figure 8.10   Hybrid combiner.

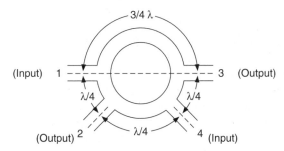

Figure 8.11   Hybrid ring combiner. No coupling between 1-4 and 2-3.

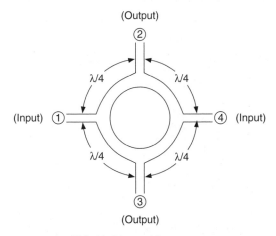

Figure 8.12   Hybrid ring combiner.

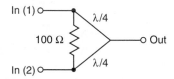

Figure 8.13   Wilkinson hybrid.

transmit paths. It should be noted that the hybrid combiner will alter the phase of the signal by 90°.

8.7   Cavities

Cavities are utilized in a communication site to provide isolation from the transmitter, ensure spectral purity, and facilitate combining multiple carriers onto a single antenna system. The arrangement of cavities, i.e., their physical location and frequency of operation, determines what their function is. Figure 8.14 is a diagram of one type of cavity. It is a cylindrical can that has a tuning rod which is either lowered or raised in the cavity itself. The physical placement of the tuning rod determines the exact frequency of operation the cavity will operate at. The cavity is effectively an LC tank circuit which resonates at a specified frequency. Therefore, the higher the cavity's Q the more selective it will be.

The coupling loops shown in Fig. 8.14 are used to couple the energy into and out of the cavity. Depending on the physical arrangement of the couplers, the cavity can act as either a band pass or band reject or notch filter.

In cellular communication systems the cavity filter is associated with the transmit portion of the cell site. In two-way operations the cavity filter can be used on the receive portion of the system as a method of obtaining high selectivity for a single channel.

8.8   Duplexers

A duplexer is a valuable piece of hardware at a communication site. The duplexer effectively allows a single antenna to be used by both the transmit

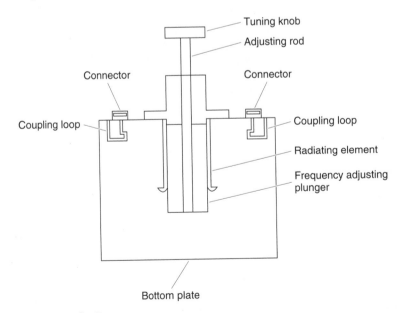

Figure 8.14   Cavity.

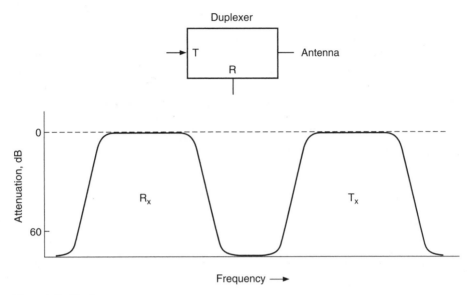

Figure 8.15 Duplexer.

and receive portions of a communication system. Most of the subscriber units that transmit and receive at the same time utilize a duplexer. Communication base stations often deploy a duplexer when the number of antennas need to be reduced for a variety of reasons.

Figure 8.15 is an example of a duplexer. The duplexer is a combination of a transmit and receive filter that are joined together into one package. It affords the communication system some level of isolation between the transmitter and receiver. The isolation is of course available only if there is sufficient frequency separation between the transmit frequency and the receive frequency.

A duplexer is sometimes referred to as a diplexer. However, there are devices that are triplexers and they allow more than one band of operation to utilize the same system. A triplexer can have two different transmit ports and one receive port for use.

8.9 Combining Techniques

In an ideal world there would be a separate antenna for every frequency transmitted and one for every frequency desired to be received. However, if this were the case in cellular and PCS environments the number of antennas required for a communication site would be ghastly. Therefore, maximizing the efficiency of a communication site for antenna real estate is a desirable objective. The maximization is achieved through reduction in the number of antennas required for a communication site, achieved through various transmit combining techniques. The following are some examples of transmit combining.

8.9.1 Cavity combining

Cavity combiners are used when there are multiple carriers that the operator wishes to combine into one antenna system. Some simple rules to follow are dependent upon the technology or rather modulation format and bandwidth used for the system. The rules generally refer to the minimum channel separation allowed for a group of channels to be combined. Table 8.1 illustrates some of the channel separation requirements that must be adhered to for certain specific technology platforms. Figure 8.16 illustrates combining channels with the use of cavity combiners.

TABLE 8.1   Frequency Separation Chart

	Separation, kHz	Channels
AMPS	630	21
TACS	600	24
GSM	600	3

Figure 8.16 shows that there is a physical separation between the cavities and the junction where they come together. The 0.25 wavelength relationship enables the other cavities to appear as shorts of very high impedance to each other. This impedance enables maximum power to be transferred to the antenna system.

The frequency separation for the channels is a function of the cavity's selectivity and the isolation that it can afford the combining system. As a general rule as the amount of isolation is reduced between the channels the insertion loss increases, making the combining method more inefficient.

8.9.2 Hybrid combiners

Hybrid combiners are used to avoid the use of a cavity combining method or when the frequency separation between the channels is not sufficient to

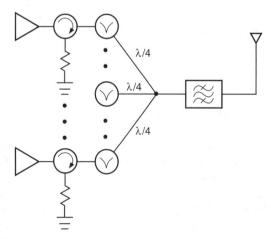

Figure 8.16 Cavity combining.

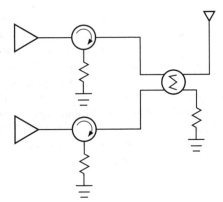

Figure 8.17 Hybrid combiner.

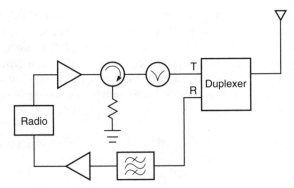

Figure 8.18 Duplexer.

ensure proper isolation and minimized insertion loss. The hybrid comes at a cost, however, in that it typically has a little more than 3 dB of insertion loss. An example of a hybrid combiner implementation is shown in Fig. 8.17.

8.9.3 Duplexer

The implementation of a duplexer in a communication system is normally done to reduce the physical number of antennas that are needed for a site. Sometimes a duplexer is used to resolve a particular short-term installation problem where a faulty antenna is discovered and a replacement is not directly available but service must be maintained.

Figure 8.18 is an example of a duplex configuration in a communication system. Note that the radio system can be one or multiple radios. If the system employs multiple radios, a cavity or hybrid combining method may be used to obtain the number of channels on one antenna.

8.9.4 Crossband couplers

One very interesting piece of equipment available to the RF engineer is a crossband coupler. The crossband coupler is used in locations where multiple bands are required to be combined into one system. The use of a crossband coupler is a method of combining multiple bands with minimal insertion loss.

Figure 8.19 shows two types of crossband couplers currently available on the market. The use of crossband couplers enables VHF and UHF systems to be combined into one. Obviously the antenna system utilized must be able to accommodate the entire frequency of operation for the system to make this a viable alternative. The crossband coupler has found direct use in multiple situations, and one in particular involves leaker feeder situations where multiple communication systems need to share the same transmit and receive platform. An example of this is shown in Fig. 8.20.

8.10   Filters

Filters are an essential component for a transmitter. Transmit filtering can be accomplished through a variety of methods. The fundamental concept, however, for filters involved with transmitters is to ensure that the energy outside of the band of interest is attenuated to a sufficient level as to not cause interference. What this means is that even though there are FCC transmission masks which must be adhered to it may become necessary to provide better filtering than that required to ensure your system will not degrade someone else's communication system.

The types and amount of filtering needed for a communication site are dependent upon a multitude of issues:

1. Operating band
2. Insertion loss
3. Power
4. Out-of-band emission suppression

A discussion of filters is covered in more detail in Chap. 6; however, some basic types of filters are used in the transmit path:

1. Cavity
2. Bandpass
3. Low pass
4. Harmonic
5. Duplexer

The location of the filter in the transmit path is shown in Fig. 8.21. It is not uncommon to deploy multiple filters in a transmit system.

8.11   Standing-Wave Ratio (SWR)

To ensure proper operation of the transmitter for the communication site, it is important to have maximum power delivered to the antenna system. To have

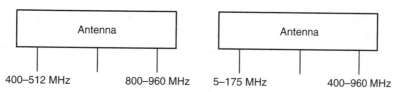

Figure 8.19 Crossband couplers.

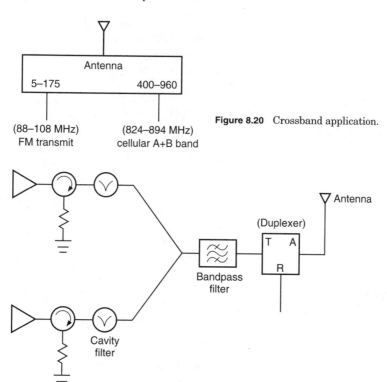

Figure 8.20 Crossband application.

Figure 8.21 Transmit filter systems.

the maximum amount of power delivered to the antenna system requires a low SWR for the system, meaning that no energy, or very little, is reflected back toward the transmitter.

Figure 8.22 shows the relationship between SWR and VSWR (voltage standing-wave ratio) and ISWR (current standing-wave ratio). In an ideal world there would be no reflection or SWR, since all the energy is delivered to the load.

$$\text{VSWR} = \frac{V_{\text{rms, max}}}{V_{\text{rms, min}}}$$

$$I_{\text{SWR}} = \frac{I_{\text{rms, max}}}{I_{\text{rms, min}}}$$

$$\text{SWR} = \text{VSWR} = \text{ISWR}$$

Figure 8.23 shows how to briefly determine what the VSWR is for a particular situation. The VSWR of 1.5:1 is a typical value referenced in the industry for determining the health and well-being of an antenna or antenna system.

$$V_{max} = 1.5$$

$$V_{min} = 1.0$$

$$VSWR = V_{rms,\,max} = \frac{1.5}{1.0} = 1.5$$

$$\therefore \; VSWR = 1.5:1$$

Example 8.2 shows the difference between a SWR of 1:1 and 1.5:1. Note that SWR is interchanged here with VSWR since they are directly related, as was shown in Fig. 8.22.

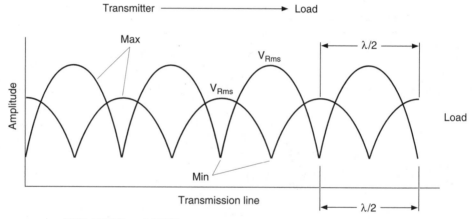

Figure 8.22   SWR, VSWR, and ISWR.

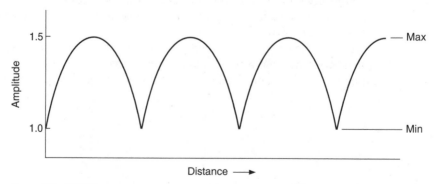

Figure 8.23   VSWR.

Example 8.2

$$\text{SWR} = \frac{Z_0}{R_{\text{load}}} \quad or \quad \text{SWR} = \frac{R_{\text{load}}}{Z_0}$$

If

$$Z_0 = R_{\text{load}} = 50 \ \Omega$$

$$\text{SWR} = \frac{Z_0}{R_{\text{load}}} = \frac{50}{50} = 1$$

$$\therefore \ \text{SWR} = 1:1$$

A SWR of 1:1 is a perfect match. However, if

$$Z_0 = 50 \ \Omega$$

$$R_L = 75 \ \Omega$$

$$\text{SWR} = \frac{R_L}{Z_0} = \frac{75}{50} = 1.5$$

$$\therefore \ \text{SWR} = 1.5:1$$

Knowing what the SWR is for a system is interesting, but it is not relevant unless it is related to its impact on the reflection coefficient for the system itself. The reflection coefficient increases as the SWR situation deteriorates. In other words, as the SWR increases, the amount of reflected energy increases, creating multiple problems for a communication system.

The relationship between the reflection coefficient and SWR is shown in Example 8.3.

Example 8.3

$$p = \frac{\text{SWR} - 1}{\text{SWR} + 1} = \text{reflection coefficient}$$

If

$$\text{SWR} = 1:1$$

$$p = \frac{1 - 1}{1 + 1} = 0$$

If

$$\text{SWR} = 1.5:1$$

$$p = \frac{1.5 - 1}{1.5 + 1} = \frac{0.5}{2.5} = 0.2$$

Now

$$\text{Percent reflected power} = p^2$$

and

$$\text{Return loss} = 10 \log_{10} (p^2)$$

Therefore

SWR	p^2	Return loss, dB
1:1	0	∞
1.5:1	0.04	-14
2:1	0.111	-9.5

The reflection coefficient is also used to determine how much energy is returned and thus is a determination of the return loss of a system. The reflection coefficient is directly proportional to the SWR for the system. However, the return loss is inversely proportional to the reflection coefficient. That is, as the reflection coefficient decreases the return loss increases, meaning that there is a better match between the transmitter and the antenna system.

Figure 8.24 and Table 8.2 show the relationship between SWR and return loss.

TABLE 8.2

SWR	Return loss
1.1	-26.9
1.2	-20.8
1.3	-17.7
1.4	-15.6
1.5	-14
1.6	-12.7
1.7	-11.7
1.8	-10.9
1.9	-10.8
2.0	-9.5
2.1	-9.0
2.2	-8.5
2.3	-8.1
2.4	-7.7
2.5	-7.4
2.6	-7.0
2.7	-6.8
2.8	-6.5
2.9	-6.2
3.0	-6.0

An important point to note is that if the SWR for a system is high, then either the PA for the channel of interest must put out more power to achieve the same ERP or the coverage, down link, will decrease as a direct result of the mismatch.

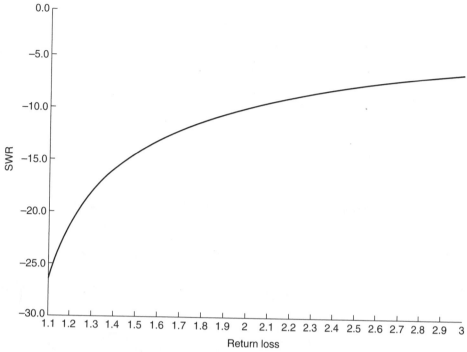

Figure 8.24   SWR vs. return loss.

8.12   ERP and EIRP

Effective radiated power (ERP) and effective isotropic radiated power (EIRP) are the two most common references used for determining the transmit power of a communication site. ERP and EIRP are directly related to each other, and a simple conversion can be achieved when one is known and the other is sought.

$$ERP = EIRP + 2.14 \text{ dB}$$

Example 8.4 shows how to calculate the ERP for the communication site.

Example 8.4   Figure 8.25 is calculated as shown below:

Transmitter	+42.75 dBm
Isolator	−0.25 dB
Filter	−1.0 dB
Feedline	−1.5 dB
Antenna gain	+10 dBd
	+50 dBm = 100 W ERP or 165 W EIRP

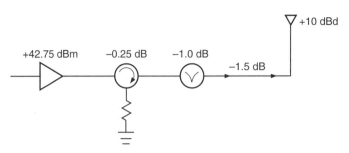

Figure 8.25

References

1. Smith, Clint, and Curt Gervelis, *Cellular System Design and Optimization,* McGraw-Hill, New York, 1996.
2. Buxton, Bob, "Measurement Methods Analyze Digital Modulation Signals," *Microwaves & RF,* July 1995, pp. 67–72.
3. Henderson, Brian, "How to Use Duplexers: Isolation Requirements," *Mobile Radio Technology,* July 1994, pp. 10–12.
4. Breed, Gary, "Classes of Power Amplification," *RF Design,* August 1993, pp. 80–81.
5. Vizmuller, Peter, "Gauge the Effects of Imbalance on Feedforward Amps," *Microwaves & RF,* December 1995, pp. 127–132.
6. Technical Papers, EMR Corp.
7. Eid Eid, Fadhel Ghannouchi, and Francois Beauregard, "Optimal Feedforward Linearization System Design," *Microwave J.,* November 1995, pp. 78–86.
8. Myer, Daniel, "A Multicarrier Feed-forward Amplifier Design," *Microwave J.,* October 1994, pp. 78–88.
9. Duff, William, and Donald White, *EMI Prediction and Analysis Techniques,* vol 5, p. 4.38, Don White Consultants, Inc.
10. American Radio Relay League, *The ARRL 1986 Handbook,* 63d ed., The American Radio Relay League, Newington, CT, 1986.
11. Brewster, *Telecommunications Technology,* Wiley, New York, 1986.
12. Kaufman, M., and A. H Seidman, *Handbook of Electronics Calculations,* 2d ed., McGraw-Hill, New York, 1988.
13. Lee, W. C. Y., *Mobile Cellular Telecommunications Systems,* 2d ed., McGraw-Hill, New York, 1996.
14. White, Duff, *Electromagnetic Interference and Compatibility,* Interference Control Technologies, Inc., Gainesville, GA, 1972.
15. Yarborough, *Electrical Engineering Reference Manual,* 5th ed., Professional Publications, Belmont, CA, 1990.
16. Fink, Donald, and Donald Christiansen, *Electronics Engineers Handbook,* 3d ed., McGraw-Hill, New York, 1989.
17. Fink, Beaty, *Standard Handbook for Electrical Engineers,* 13th ed., McGraw-Hill, New York, 1995.
18. Jakes, W. C., *Microwave Mobile Communications,* IEEE, New York, 1974.
19. Carlson, A. B., *Communications Systems,* 2d ed., McGraw-Hill, New York, 1975.
20. Lathi, *Modern Digital and Analog Communication Systems,* CBS College Printing, New York, 1983.
21. Rappaport, *Cellular Radio & Personal Communications,* IEEE, New York, 1995.
22. *Reference Data for Radio Engineers,* 6th ed., Sams, 1983.

Communication Sites

9.1 Introduction

Just what specifically is a communication site? Simply put, it is a physical location where there is radio equipment that is intended for receiving, transmitting, or both. There are almost an infinite number of different types of communication site configurations. For the RF engineer engaged in cellular and PCS communications the vast number of configurations that can be and are used is daunting.

This chapter discusses various issues associated with a wireless communication site. The focus of attention is directed toward the RF engineer and the issues associated with the design phase. The particulars associated with the operation and construction concerns that are an integral part of the communication sites design criteria are not covered here for a variety of reasons. However, several checklists included for reference can easily be used.

The chapter is focused on the types of cell sites, macro-, micro-, and picocells. In-building systems (rerads) as well as tunnel systems are covered. Additional topics include intermodulation, lightning protection, and grounding.

A typical cell site or rather communication site consists of the components in Fig. 9.1.

It cannot be overstressed that the installation decision for the antennas and equipment should be taken with great care. Not only must the engineering requirements be met but also operations concerns must be taken into account. This is of course provided that leasing and land use entitlement issues are met.

9.2 Macrocell

A macrocell is what most people have come to know or expect to see for a communication site for cellular operations. With the advent of PCS the need for macrocells was initially portrayed as being an item of the past; however, many PCS sites have the look, feel, and size of a macrocell.

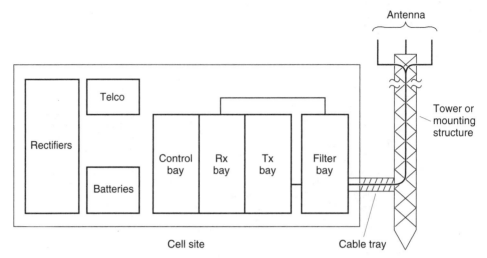

Figure 9.1   Communication site.

Multiple configurations are associated with each type of technology platform picked for the communication system. For instance, AMPS, TACS, GSM, CDMA, and NADC, to mention a few, can all be configured as either an omni, bidirectional, or three-sector cell depending on the application at hand.

What defines a macrocell is primarily the coverage area for which the cell is designed.

For the RF engineer the decision of using a macrocell is driven by multiple reasons. However, there are a few different perturbations with regard to cell sites that lead to interesting designs. As is often the case in real life in the city the amount of green field location is not large and in fact the desire to utilize an existing communication site receives much pressure. There are a multitude of reasons why an existing communication site should be used and also why it should not be used.

The reasons for utilizing an existing communication site lead to the issue of community affairs in that there is a strong public awareness of communication sites and the need to limit the number of towers that are erected. The utilization of existing structures also enables, hopefully, the time from inception to activation to be minimized. The other technical issue is that with power control being utilized by most wireless services, if colocating would minimize the potential for receiver overload due to mobile generated interference.

9.2.1   Omni

When the installation is for an omni site several methods can be used for antenna installations. The first is a simple installation on a monopole shown in Fig. 9.2. The transmit antenna is highest on the structure with the receive antennas located under the platform. The distance between the receive antennas should follow the separation requirements for the network, most likely following the equation that is listed on the figure itself. Care must be exercised when installing the receive antennas to ensure that their pattern is correct; i.e., the antenna is designed to be installed inverted. In addition the drain hole needs to be opened for the water to drain from the antennas. The location of the drain hole will of course be different for the transmit and receive antennas.

A few other variants to omni installations involve installing on a building and when the number of antennas is limited. Figure 9.3 is an example of an antenna installation that occurs on a building. Note that the location of antennas needs to meet the required setback rules. If the setback rules cannot be adhered to, it is possible to install the antennas near the edge of the roof; however, they may become visible to the public at this point or the landlord may

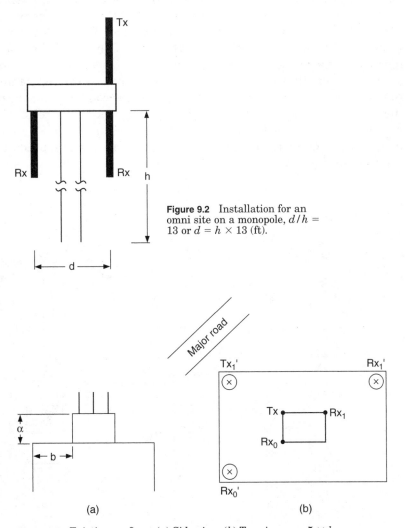

Figure 9.2 Installation for an omni site on a monopole, $d/h = 13$ or $d = h \times 13$ (ft).

Figure 9.3 Existing rooftop. (a) Side view. (b) Top view. $\alpha = 5 \times b$.

not wish this type of installation to take place, or the local ordinances may pro-
hibit this from occurring. Note that the placement of the receive antennas
should be such that if there is only one major road in the area for the cell to
cover then the horizontal diversity placement for the antennas should be such
that it is maximized in the direction toward the road.

In addition if the primary location is not achievable for mounting the antennas, moving them to the lower level is possible. However, based on the penthouse size significant blockage may occur in a direction. If a trade-off needs to
be made, the antennas should be structured such that the area of most concern is covered, as shown in Fig. 9.4.

Situations arise where the physical number of antennas needs to be reduced
for the reasons stated before or because of tower or structural loading issues,
to mention a few. The arrangement shown in Fig. 9.4 involves the elimination
of one of the antennas through use of a duplexer. The duplexer configuration
shown in Fig. 9.4 is arranged such that one of the receive branches will be
favored as the dominant branch (Rx_1), if conditions are the same at both
antennas. Note that using a duplexer also increases the risk of intermodulation problems. It is recommended that either a second duplexer be installed or
the leg that has the smallest feedline loss be the one that receives the duplexer
in order to balance the legs if possible. An obvious alternative would be to eliminate the site down to one antenna; however, this would eliminate the use of
diversity at the site as it is shown.

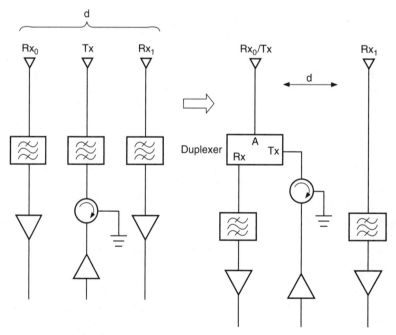

Figure 9.4   Omni duplex.

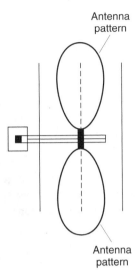

Antenna
pattern

Antenna
pattern

Figure 9.5   Directional site.

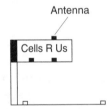

Antenna

Cells R Us

Figure 9.6   Billboard antenna applications.

9.2.2   Directional sites

Directional sites are often used for locations near the system borders or when coverage for a particular area is not desired. For instance, coverage may be desired only along a roadway and not anywhere else. If this is the case the use of a directional antenna site might be directly applicable. The example which follows is for a single-sector and two-sector cell site. Obviously a sector cell is a directional cell site but the configurations shown in the examples should be exceptions to the system design and not the norm.

Figure 9.5 is an example of single-sector cell site that is meant to cover only the highway. The antenna structure for this application is located on a billboard and is positioned so that coverage follows the roadway through use of a highly directional antenna (Fig. 9.6). The antenna system here uses two directional antennas that are phased together (Fig. 9.7), forming the scallop pattern. Only two antennas structures are shown, indicating that a duplexer is in operation. It is possible to forgo the use of one antenna with the elimination of possible diversity for fade margin protection.

Another application for a directional site can be where the edge of the system is, that is, the adjoinment to another carrier. The issue here is the amount of physical overlap you allow into their system. The amount of overlap is more

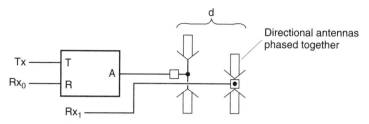

Figure 9.7 Antenna configuration.

a business issue than a technical issue. For this example it is assumed that some level of extension overlap is allowed but not enough to warrant an omni site at the desired power level. In this situation an omni configuration was assumed (Fig. 9.8).

The objective here is to maximize the coverage while not exceeding the contour extension and also cover the major highway shown in Fig. 9.8. A directional antenna was selected for the application that is about 180° in azimuth and oriented at 225°. The use of three, two, or one antennas for this application is dependent upon the site particulars.

9.2.3 Three sector

The three-sector cell site is one of the most popular cell site configurations utilized in the wireless industry, next to the omni cell. The three-sector cell has sectors that cover 120° each, thus having three sectors make a full circle. There are a multitude of combinations for transmit receive that can be used for establishing a three-sector cell site. However, for the following examples only a few of the more basic configurations are shown. Additional configurations can easily be derived from the material presented.

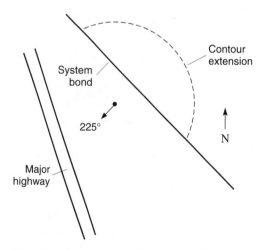

Figure 9.8 Contour extension.

Figure 9.9 is an example of a typical antenna arrangement that would be situated for a monopole application. The configuration in Fig. 9.9 has a separate antenna for each transmit and receive leg of the cell site. The general transmit and receive configuration is shown in Fig. 9.10. The particular transmitter and receiver multicoupler configurations are left off since they are directly dependent upon the infrastructure vendor chosen. If a second transmit antenna is needed it is placed next to the existing transmit antenna, between the receive antennas. It is suggested that if future considerations may require the need for an additional transmit antenna it be designed in at the beginning so as to avoid configuration problems in the future.

As is often the case it is desired to minimize the number of antennas for a site. Figure 9.11 is an example of a duplexed configuration for a three-sector cell site.

9.2.4 Six sector

The six-sector cell site is deployed in a situation where the $N = 4$ reuse pattern is deployed for frequency planning. The use of a six-sector cell site increases the number of available erlangs per square mile or kilometer.

A six-sector cell site antenna layout is shown in Fig. 9.12. The use of only one transmit antenna is shown in the drawing, but a second one could be added depending on the configuration requirements.

The six-sector configuration shown relies on horizontal space diversity and has a separate transmit and receive system. The brief transmit and receive configuration is shown in Fig. 9.13. The illustration in Fig. 9.14 is another example of a six-sector configuration. The difference between the six-sector configuration shown in Fig. 9.12 and that in Fig. 9.14 is that the number of antennas is drastically reduced and the use of angle diversity is relied upon instead of spatial diversity.

The drawing in Fig. 9.15 represents the extensive use of duplexers to achieve the six-sector configuration. Obviously there are other variants to implementing a six-sector cell site, but the antenna configurations shown are the more common.

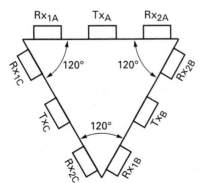

Figure 9.9 Three-sector antenna array.

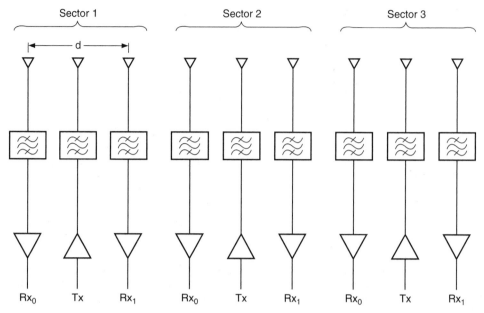

Figure 9.10   Three-sector Tx and Rx plumbing.

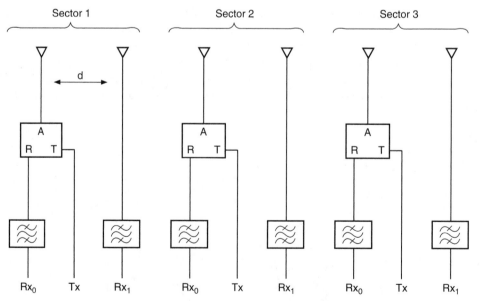

Figure 9.11   Three-sector duplex arrangement.

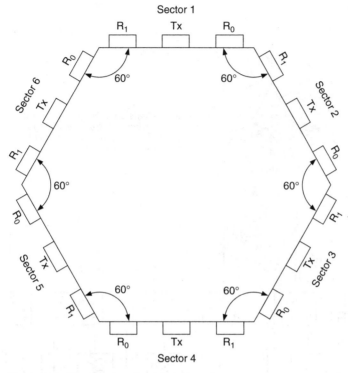

Figure 9.12 Six-sector configuration, top view.

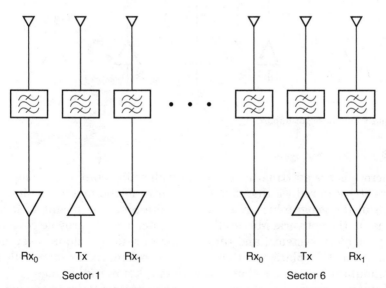

Figure 9.13 Six-sector configuration.

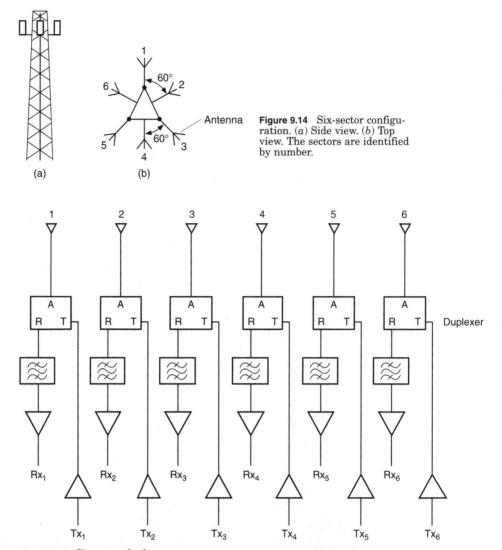

Figure 9.14 Six-sector configuration. (*a*) Side view. (*b*) Top view. The sectors are identified by number.

Figure 9.15 Six-sector duplex.

9.3 Microcells

Microcells are starting to become prevalent in cellular systems as the operators strive to reduce the physical footprint of the macrocell sites. The reduction in the geographic area each cell site covers is meant to facilitate more reuse in the network. Microcells are also deployed to provide coverage in buildings, subway systems, and tunnels and to resolve unique coverage problems. The technology platforms that tend to be referenced as microcells involve any communication system that is less than $\frac{1}{2}$ km in radius.

Currently several types of technology platforms fall into the general categorization called microcells:

1. Fiber-fed microcell

2. T1 microcell

3. Microwave microcell

4. High-power rerad

5. Low-power rerad

6. Bidirectional amplifier

The choice of which technology platform to utilize is driven by a variety of factors unique to that particular situation. One is the application which is being engineered for—capacity, coverage, or private wireless PBX. Another important factor in the technology platform decision is the configuration options available at that location for providing radio capacity. A third factor is driven by the overall cost of the solution for the network.

9.3.1 Fiber-fed microcell

The fiber-fed microcell can be and is being deployed in cellular systems. The choice of a fiber-fed microcell has many advantages and disadvantages associated with this technology platform. Currently two distinct types of fiber-fed microcells are available, analog and digital. Both of these microcells require the use of dark fiber to make them operate in a cellular network. Dark fiber is often not readily available in many of the areas where a microcell could be deployed and requires that the pair of fibers utilized for the microcell be dedicated only to that microcell. However, a fiber-fed microcell can be utilized with any of the cellular infrastructure vendors, making it vendor-transparent.

Figure 9.16 is a drawing of an analog fiber-fed microcell. The positive attributes to this microcell are that it provides easy transport of the RF modulated carrier and various technology platforms. The analog microcell will transport regular FM, TDMA, and CDMA signals over it because it utilizes AM modulation for transporting the RF information along the dark fiber. The microcells themselves are small in size and are excellent choices for establishing a centralized radio hub site allowing one cell site to feed many microcells.

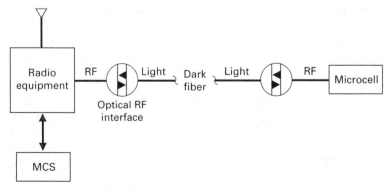

Figure 9.16 Analog fiber-fed microcell.

The negative attributes to the analog microcell tie into the requirement that it utilize dark fiber, the limited optical link budget, and the expense of the optical interface equipment. The optical interface equipment is needed to convert the RF modulated signal into a light format for transport along the dark fiber. At the other end of the fiber path the optical interface equipment is needed to convert the light format into the RF modulated signal again.

The other type of fiber-fed microcell is the digital fiber-fed microcell. The issues are primarily the same with the digital microcell except that it is able to transport the converted RF modulated signal over a greater distance than the analog- based unit. The digital microcell, however, may or may not be able to transport TDMA or CDMA signals based on the digitization process utilized. The digital microcell still reuses dark fiber to connect itself between the host and donor locations.

9.3.2 T1 microcell

The T1, or copper-fed microcell, has many potential applications in the wireless arena. The T1 fed microcell has the distinct advantage of providing full or near full functionality of a regular cell site in a small package. The T1 microcell generalized configuration is shown in Fig. 9.17. It has the distinct advantage of being able to utilize the readily available T1 facilities in a network. The obvious advantage of the utilization of the T1 facilities is the interconnect ease and lower cost with initial deployment. Utilizing the reduced footprint of the T1 microcell and its ability to capitalize on the available T1 facilities makes it suitable for a variety of applications.

However, the T1 microcell is presently infrastructure vendor-specific. The T1 microcell is part of cellular vendors' product offering and will not work with another vendors' equipment on its own. It is possible, however, to deploy a T1 microcell system in a network that utilizes another vendor's equipment through the use of intersystem protocols like IS-41.

9.3.3 Microwave microcell

Microwave-fed microcells are another type of microcell, although in some situations microwave is used to feed macrocells. The microwave microcell has

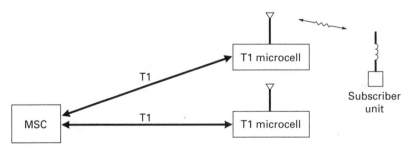

Figure 9.17 T1 microcell.

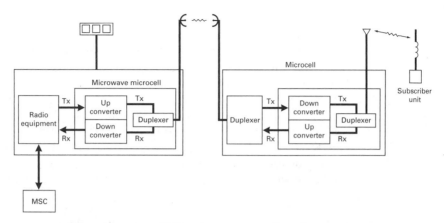

Figure 9.18 Microwave microcell. Tx = base transmit. Rx = base receive.

some very unique advantages associated with this type of technology. It is independent of backhaul and infrastructure vendors. It is also directly compatible with different technology platforms. A simple microwave microcell system drawing is shown in Fig. 9.18.

The microwave microcell utilizes microwave point-to-point communication to connect the feeding cell site with the remote locations. The advantage with utilizing a point-to-point method for connecting the microcell to the donor cell is the fact it doesn't require the use of landline facilities, T1 or fiber, to connect the donor cell and the microcell. The disadvantages of this type of microcell are frequency coordination of the microwave path, ensuring path clearance, and an additional antenna and feedline for the donor cell.

The microwave frequency coordination can place severe operational limitations on the utilization of this technology platform. Specifically the actual microwave path for the frequency of interest might not be available, requiring another operating band. The choice of the operating frequency band has a direct impact on the path length, i.e., the distance between the donor cell and the microcell. The operating frequency and the path length determine the path outages the system will experience because of weather conditions for the region.

The other issue with the microwave microcell is ensuring that there is sufficient path clearance for the path utilized. Depending on the microwave path required and the frequency of operation the donor of microcell microwave antennas might have to be placed at a significant elevation to ensure a reliable communication path.

9.3.4 Reradiators

Reradiators as the solution to a communication problem or situation have been used effectively by many operators. They take on three distinct forms, high-power, low-power translating, and nontranslating or bidirectional.

When utilizing a reradiator the only true objective can be coverage since this technology platform will not add capacity to the network. The reradiator, or repeater, redistributes capacity from one cell site to another area. The repeater can alter the capacity distribution of the network through the selection of the donor cell from which the capacity is drawn.

The primary negative attribute of a repeater is its ability to add interference into the network. The interference is added to the network as both mobile-to-base and base-to-mobile interference. The interference is a result of the repeater's extending the actual coverage of the donor cell beyond which it was designed to operate. Which type of reradiator to utilize depends on the design objective and the configuration options available for the situation.

High-power reradiator.   The high-power reradiator is one form of reradiator that is available for use in a network. The high-power reradiator (HP) is an extension of the donor cell utilizing different frequencies and is able to achieve transmit power levels comparable to that of a cell site itself. A simple diagram of a HP is shown in Fig. 9.19.

The high-power reradiator is often utilized when there is not sufficient isolation between the reradiator transmitter and receiver. Isolation is a requirement for low-power reradiators since they transmit the same frequency they receive, in both up- and down-link directions. As with all communication systems that are of a point-to-point nature it is imperative that a path clearance analysis be performed to ensure proper operation.

The HP is an extension of the donor cell site utilizing different frequencies to communicate with a subscriber unit. The diagram in Fig. 9.19 illustrates that the HP translates the donor cell sites transmit and receive frequencies by some value Z. The translated channel, however, is still within the cellular carrier's operating band and represents a specific cellular channel. The actual channels that HP translates the donor cell channels to is defined by the operator, which requires frequency coordination for interference protection inside the cellular network itself.

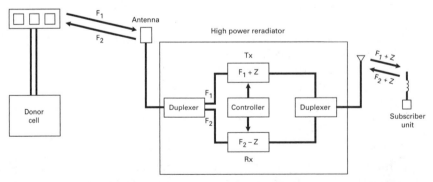

Figure 9.19   High-power reradiator. F_1 = donor cell, Tx frequency, F_2 = donor cell, Rx frequency.

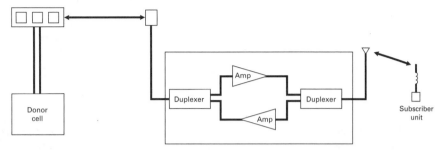

Figure 9.20   Low-power reradiator.

The HP also has its own cell site controller which has its own software. The controller software is utilized not only to control the translation of the radio channel frequency but also to perform a call-processing role, which is served by the HP when it tries to hand back the mobile to the system. The controller software instructs the mobile to retune from channel $f_1 + z$ to just f_1 in an attempt to transfer the mobile to the network for the continuation of the call. The HP tries to hand back the mobile because it has a limited footprint in the network.

Low-power reradiator. The low-power reradiator (LP) is very similar in configuration and purpose to the high-power reradiator except it has lower transmit power. The LP reradiates the signals it receives to improve radio coverage in a given area but does not translate the voice channels to another frequency. The LP, however, does translate the control, or setup, channel to another control channel frequency. Figure 9.20 shows a simplified block diagram of a low-power reradiator.

The LP configuration requires that close attention is paid to the transmit and receive isolation to prevent feedback problems with reradiating the same signal. The minimum isolation usually required for an LP is 70 dB, achieved through physical separation of the antennas themselves. However, at times an in-line attenuator is utilized in one branch of the antenna system to obtain the isolation required. The LP, as with all reradiators, requires that the proper path clearance is obtained between the donor and LP unit itself. The primary difference between the HP and LP reradiators is the isolation requirements, frequency translations, and output power.

Bidirectional amplifier. The bidirectional amplifier is very similar to the low-power reradiator. The block diagram used in Fig. 9.20 to represent a low-power reradiator is functionally the same for a bidirectional amplifier. The bidirectional amplifier, like the low-power reradiator, does not translate voice channel frequency and does not translate the control channel that it receives from the donor cell site. It has the same isolation and path clearance requirements as a low-power reradiator. It can be utilized for in-building applications either by itself or adjunct to microcell deployments. It offers a cost-effective method of providing coverage in an in-building environment.

9.4   Installation

Many issues with different elements of importance need to be factored into the installation of a cell, including the physical placement of the antennas themselves, which may or may not be in the design engineer's control.

9.4.1   Cable runs

Some of the physical installation issues that need to be factored into the design involve cable runs from the antenna system (each leg) to the base station equipment. While this may seem an obvious point, often the desired routing of the cables is not practical, making the real installation length much longer than desired. The additional cable run length when installation reality is factored in may have made the site nondesirable; however, if this situation occurs too far down the stream of the construction process it is too late to reject the site or make the appropriate design alterations to correct the situation.

9.4.2   Antenna mounting

Obviously the mounting of the antennas needs to be taken with extreme care. The following is a brief checklist to ensure that antenna mounting concerns are checked prior to acceptance of a cell site.

1. What are the number and type of antennas to be installed?
2. Maximum cable run allowed.
3. Identify and rank obstructions that would alter the desired coverage.
4. Rx antenna spacing adequate; diversity requirements met.
5. Isolation requirements met with other services.
6. Antenna AGL requirements met.
7. Antenna mounting parameters met.
8. Intermodulation analysis completed.
9. Path clearance analysis verified (if applicable).

This list is preliminary and can easily be altered based on the situation at hand. However, it should be modified to meet your particular system design requirements.

When installing on a tower the physical spacing (offset from the tower) must be selected so that the tower structure either enhances or does not alter the antenna pattern desired. In addition to the pattern issue, care must be taken in ensuring that no degradations caused the system because of unwanted energy from adjacent systems. It is suggested that an interference analysis be conducted for every site to ensure the proper isolation requirements are met.

9.4.3 Diversity spacing

The diversity spacing for the receive antennas is needed to ensure that the proper fade margin protection is designed into the system. Obviously the horizontal diversity spacing requirements listed next are a minimum requirement and more spacing can be placed between the receive antennas. However, there is a practical limit to the spacing between the receive antennas when the feedline length between the antennas becomes such that either the feedline loss exceeds the diversity advantage or the signals are completely decorrelated to eliminate any diversity combining gain possible [Eq. (9.1)].

$$\text{Diversity spacing } (f) = \frac{\text{AGL of antenna (ft)}}{11} \times \left(\frac{835}{f_0}\right) \qquad (9.1)$$

9.4.4 Roof mounting

When installing antennas on an existing roof or penthouse account must be taken of how high the antenna must be with respect to the roof surface. Obviously the ideal location is right at the roof edge. However, it may not be a viable installation design. When the antennas cannot be placed at the edge of the roof, a relationship between the distance from the edge of the roof and the antenna height exists. It is depicted in Fig. 9.21.

The above example assumes that there are no additional obstructions between the antenna and the roof edge. If there are obstructions, additional height may be needed. Examples of additional obstructions are HVAC units and window-cleaning apparatus. Remember that if severe downtilt is implemented into the design either at the present or in the future, the height requirements above the rooftop may need to be increased.

9.4.5 Wall mounting

For many building installations it may not be possible to install the antennas above the penthouse or other structures for the building. Often it is necessary to install the antennas on the penthouse or water tank of an existing building. The building architect has rarely factored in the potential installation of antennas at the onset of the building design. Therefore, as shown in Fig. 9.22 the building walls may meet one orientation needed for the system but rarely

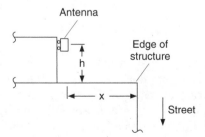

Figure 9.21 Roof mounting. $h = x/5$, where h = height of antenna above roof (ft) and x = distance from edge of roof (ft).

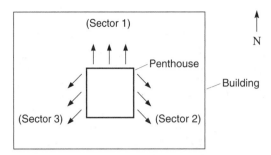

Figure 9.22 Three-sector building configuration.

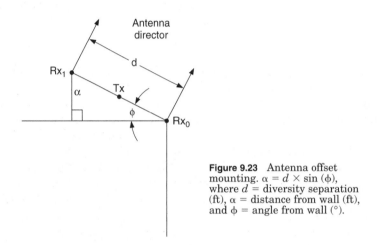

Figure 9.23 Antenna offset mounting. $\alpha = d \times \sin(\phi)$, where d = diversity separation (ft), α = distance from wall (ft), and ϕ = angle from wall (°).

all three, for a three-sector configuration. Therefore, it is necessary to determine what the offset from the wall of the building structure needs to be. Figure 9.23 illustrates the wall mounting offset required to ensure proper orientation for each sector. Obviously common sense must enter into the situation here when the inclusion of the offset brackets makes the site a metal monster. Trade-off can be made in the design when the orientation for each sector is within the design tolerance limits for sector orientation. The design tolerance should be within ±5° for a three-sector cell. Lastly the wall mounting offset must meet all the setback requirements for both antenna and local ordinances.

9.4.6   Towers

Numerous types of towers can and do exist in a wireless network. However, three basic types of towers are more common, self-supporting, guy wire, and monopole. The general configuration for each of the towers is shown in Figs. 9.24 to 9.26. The cheapest to construct is the guy tower followed by the monopole and then the self-supporting. Each has its advantages and disadvantages.

The guy tower requires a large amount of room for its guy wires shown in Fig. 9.25, and this can be either relaxed or increased depending on loading and height issues. The self-supporting tower will enable multiple carriers to operate at the facility while the monopole will also accommodate multiple users, but not as many as a self-supporting tower of similar height.

9.5   Path Clearance

Radio path clearance is an essential criterion for any point-to-point communication system. Many different types of point-to-point communication systems can be utilized in cellular communication. Some technology platforms that require a path clearance analysis involve point-to-point microwave and reradiators.

The path clearance analysis needs to be performed for every RF point-to-point communication link in the network. The path clearance analysis is not difficult to perform, and an example of a path clearance analysis is provided. The Fresnel zone is shown in Fig. 9.27. There are effectively an infinite number of Fresnel zones for any communication link. It is a function of the frequency of operation for the communication link. The primary energy of the propagation wave is contained within the first Fresnel zone. The Fresnel zone is important for the path clearance analysis since it determines the effect of the wave bending on the path above the Earth and the reflections caused by

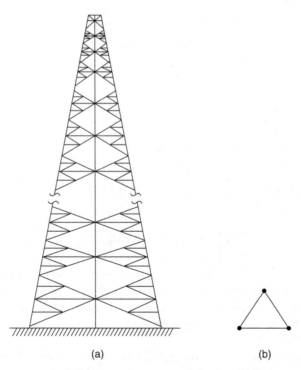

(a)   (b)

Figure 9.24   Self-supporting tower. (a) Side view. (b) Top view.

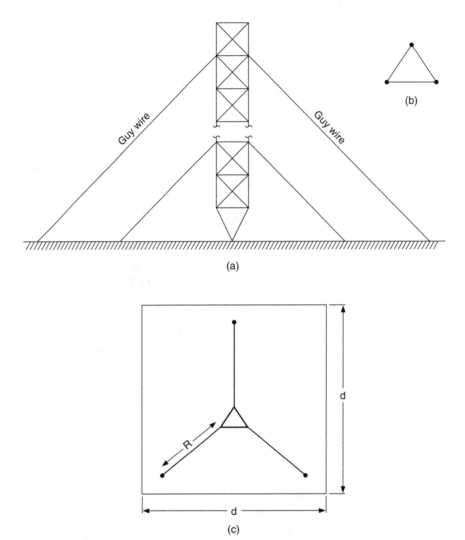

Figure 9.25 Guy wire tower. (*a*) Side view. (*b*) Top view. (*c*) R = height of tower × guy radius (%), $d = 2 \times R + 30$ (ft).

the Earth's surface itself. The odd-numbered Fresnel zones will reinforce the direct wave and the even Fresnel zones will cancel.

In a point-to-point communication system it is desirable to have at least a 0.6 first Fresnel zone clearance to achieve path attenuation approaching free-space loss between the two antennas. The clearance criteria apply to all sides of the radio beam, not just the top and bottom portions represented by the drawing in Fig. 9.27.

Environmental effects on the propagation path that have a direct influence on the point-to-point communication system include foliage, atmospheric moisture, terrain, and antenna height of the transmitter and receiver. The K factor utilized for point-to-point radio communication is 1.333 or $\frac{4}{3}$ Earth radius. It ties in the relationship between the earth's curvature and the atmospheric conditions that can bend the electromagnetic wave.

Example 9.1, for quickly determining the path clearance required for a point-to-point communication site follows.

Example 9.1   To determine the path clearance required for a point-to-point communication site refer to Fig. 9.28.

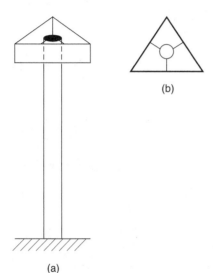

(a)

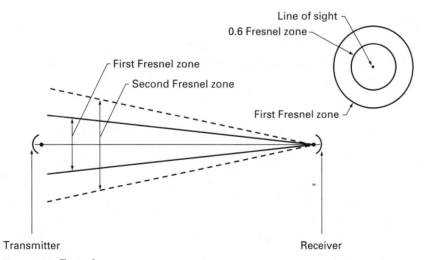

Figure 9.26   Monopole. (a) Side view. (b) Top view.

Figure 9.27   Fresnel zone.

First Fresnel zone:

$$R = 72 \sqrt{\frac{d_1 d_2}{D_r f}}$$

Earth curvature:

$$h = \frac{d_1 \cdot d_2}{1.5\,k}$$

with f = frequency in gigahertz, d_1, d_2, D_r in miles; x, h, h_T, h_R in feet, and where $d_1 = 1.6$ mi, $d_2 = 2.1$ mi, $D_T = 3.7$ mi, and $f = 0.88$ GHz (or 880 MHz).

$$R = 72 \sqrt{\frac{(1.6)(2.1)}{(3.7)(.88)}} = \sqrt{\frac{3.36}{3.256}} = 73.14 \text{ ft} \qquad R_1 = (0.6)\,R = 43.884$$

$$h = \frac{(1.6)(2.1)}{(1.5)(\frac{4}{3})} = 1.68 \text{ ft} \quad k = \frac{4}{3}$$

Assume that the transmitter, receiver, and obstruction have the same ASML.

Earth curvature	1.68
0.6 Fresnel zone	43.88
Obstruction height	100
	145.56 ft

The minimum Tx and Rx heights for the system are $h_T = h_R = 145.56$ ft.

9.6   In-building

Microcells have numerous applications for in-building applications. The applications include improving coverage for a convention center or large client, disaster recovery, or a wireless PBX, to mention a few. The propaga-

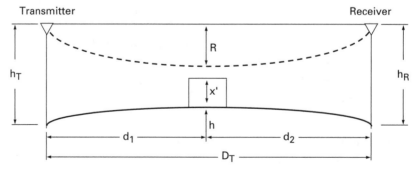

Figure 9.28 Path clearance.

tion of the radio-frequency energy, however, takes on unique characteristics in an in-building application as compared to an outdoor environment. The primary difference in propagation characteristics for in-building versus outdoor is fading, shadowing, and interference. The fading situation for in-building results in deeper and spatially closer fades when a system is deployed in an in-building application. Shadowing is also quite different in an in-building application owing to the lower antenna heights and excessive losses through floors, walls, and cubicles. The shadowing effects in an in-building application severally limit the effective coverage area to almost line-of-site (LOS) for cellular communications. The interference issue with in-building systems can actually benefit in-building applications since the interference is primarily noise-driven and not cochannel interference. The reason the in-building systems are primarily noise-driven is the attenuation experienced by external cell sites as they traverse into the buildings and various structures.

Some unique considerations must be taken into account regarding microcell system design inside a building:

1. Base-to-mobile power

2. Mobile-to-base power

3. Link budget

4. Coverage area

5. Antenna system type and placement

6. Frequency planning

The base-to-mobile power needs to be carefully considered to ensure that the desired coverage is met, deep fades are mitigated in the area of concern, the amplifier is not being over- or potentially underdriven, and mobile overload does not take place. The desired coverage that the in-building system is to provide might require several transmitters because of the limited output power available from the units themselves. For example, if the desired coverage area required 1W ERP to provide the desired result a 10-W amplifier would not be able to perform the task if you needed to deliver a total of 40 channels to that location, meaning only 25 mW of power per channel was really available. The power limitation can, and often does, make the limiting path in the communication system for an in-building system the forward link.

The forward link power problem is further complicated by the fact that portable and potential mobile units will be operating in very close proximity to the in-building systems antenna. If the forward energy is not properly set, a subscriber unit could easily go into gain compression, causing the radio to be desensitized.

The mobile-to-base power also needs to be factored into the in-building design. If the power windows (dynamic power control) are not set properly, imbalances could exist in the talkout to talkback path. Usually the reverse link in any in-building system is not the limiting factor, but the mobile-to-base path should be set so that there is a balanced path between the talkout and talkback paths.

Most in-building systems have the ability to utilize diversity receive but do not utilize it for a variety of reasons. The primary reason for not utilizing diversity receive in an in-building system is the need to place two distinct antenna systems in the same area.

The link budget for the communication system needs to be calculated in advance to ensure that both the forward and reverse links are set properly. The link budget analysis plays a very important role in determining where to place the antenna system, distributed or leaky feeder, and the amount of microcell systems required to meet the coverage area requirement.

The antenna system selected for the in-building application is directly related to the uniformity of the coverage and quality of the system. The antenna system, (no diversity) primarily provides LOS coverage to most of the areas desired in the defined coverage area. Based on the link budget requirements the antenna system can be either passive or active. The antenna system for an in-building system may take on the role of having passive and active components in different parts of the system to satisfy the design requirement.

Typically a passive antenna system is made up of a single or distributed antenna system or can also utilize a leaky coaxial system. A distributed antenna system can be used for delivering the service (Fig. 9.29). A leaky coaxial system could also be deployed within the same building to provide coverage for the elevator in the building.

The advantage a leaky coaxial system has over a distributed antenna is that it provides a more uniform coverage to the same area over a distributed antenna system. However, the leaky coaxial system does not lend itself to an aesthetic

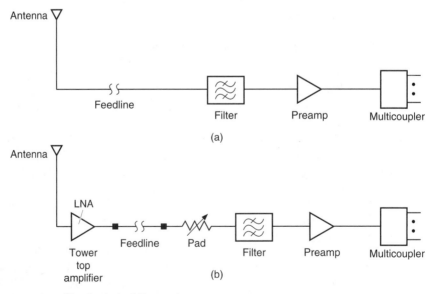

Figure 9.29 Passive in-building system.

installation in a building. The use of a distributed antenna system for providing coverage in an in-building system makes the communication system stealthy.

If the antenna system requires the use of active devices in the communication path, the level of complexity increases. It increases for active devices since they require ac or dc power and introduce another failure point in the communication system. However, the use of active devices in the in-building system can ultimately make the system work in a more cost-effective fashion. The most common active device in an in-building antenna system is a bidirectional amplifier.

The frequency planning for an indoor microcell system needs to be coordinated with the external cellular network. Most in-building systems are designed to facilitate handoffs between the in-building and external cellular system. If the in-building system is utilizing a microcell with its own dedicated channels assigned to it, it is imperative that the system be integrated into the macrocellular network.

9.7   Intermodulation

Intermodulation is the mixing of two or more signals that produce a third or fourth frequency which is undesired. All cell sites produce intermodulation since there is more than one channel at the site. However, the fact that intermodulation products are produced does not mean there is a problem.

Various intermodulation products are shown below for reference. The values used are simplistic in nature to facilitate the examples. In each of the examples $A = 880$ MHz, $B = 45$ MHz, $C = 931$ MHz, and D is the intermodulation product. The example listed below does not represent all the perturbations possible.

Second order:

$$A + B = D \text{ (925 MHz)}$$

$$A - B = D \text{ (835 MHz)}$$

Third order:

$$A + 2B = D \text{ (970 MHz)}$$

$$A - 2B = D \text{ (790 MHz)}$$

$$A + B + C = D \text{ (1856 MHz)}$$

$$A - B + C = D \text{ (1766 MHz)}$$

Fifth order:

$$2A - 2B - C = D \text{ (739 MHz)}$$

The various products that make up the mixing equation determine the order of the potential intermodulation. All too often when you conduct an intermodulation study for a cell site numerous potential problems are identified in the report. The key concept to remember is that the intermodulation report you are most likely looking at does not take into account power, modulation, or physical separation between the source and the victim, to mention a few. Therefore, the intermodulation report should be used as a prerequisite for any site visit so you have some potential candidates to investigate.

Intermodulation can also be caused by your own equipment through bad connectors, antennas, or faulty grounding systems. However, the majority of the intermodulation problems encountered were a result of a problem in the antenna system for the site and well within the control of the operator to fix.

Just how you go about isolating an intermodulation problem is part art and part science. I prefer the scientific approach since it is consistent and methodical in nature. The biggest step is identifying the actual problem and the rest of the steps will fall in line. Therefore, it is recommended that the following procedure be utilized for intermodulation site investigations.

Intermodulation Check Procedure

1. Determine if there are any colocated transmitters at this facility.

2. Collect information on each transmitter:
   Antenna types
   Emission type
   Transmit power
   Location of antennas
   Operator of equipment
   FCC license number

3. Conduct an intermodulation study report looking for hits in your own band or in another band based on the nature of the problem.

4. Allocate sufficient time to review the report.

5. Determine if there is a potential problem.

6. Formulate a hypothesis for the cause of the problem and engineer a solution.

Based on the actual problem encountered the resolution can take on many forms.

1. Is the problem identified feasible?

2. Can the problem be resolved through isolation alone?

3. Is the problem receiver overload-related?

If the intermodulation product is caused by the frequency assignment at the cell site, it will be necessary to alter the frequency plan for the site, but first remove the offending channels from service. If the intermodulation problem is due to receiver overload the situation can be resolved by placing a notch filter in the receive path if it is caused by a discrete frequency. If the overload is caused by cellular mobiles, using a notch filter will not resolve the situation. Instead mobile overload can be resolved by placing an attenuation in the receive path, prior to the first preamplifier, effectively reducing the sensitivity of the receiver.

9.8   Isolation

The amount of isolation needed for a communication system is dependent upon a multitude of issues.

1. Location of potential offending transmitter to receiver

2. Technology platform utilized

3. Receiver sensitivity

The methods that follow are based on the simple fact that there is no defective equipment or out-of-specification transmitters at the location in question. Please keep in mind that the isolation requirements may or may not be directly applicable to the communication facilities that are colocated with you. As often is the case, the offending transmitter is several buildings away.

Isolation can be achieved, once the offender(s) is identified, through antenna placement using both horizontal and vertical separation or through more selective filters. Example 9.2 shows how to determine the amount of isolation needed for a communication system.

Example 9.2

$$Tx = 852 \text{ MHz}$$

$$Rx = 849 \text{ MHz}$$

Tx power	+ 50 dBm
Rx 1-dB compression	$\underline{-27 \ \text{dBm}}$
	77 dB isolation needed

Filter attenuation of Tx in Rx band 20 dB

Isolation	77 dB
Filter attenuation	$\underline{-30 \text{ dB}}$
	47 dB isolation needed

47 dB isolation $\approx$ 10-ft vertical separation

$\therefore$ Filter attenuation	30 dB
Vertical isolation	$\underline{+ 50 \text{ dB}}$ at 10 ft
	80 dB isolation

$$80 \text{ dB} > 77 \text{ dB}$$

However, it is for cases where out-of-band emissions are the problem. When the problem is intermodulation-related, it is possible to obtain the necessary isolation to protect the receiver, if the mix is occurring at another location besides in the receive path itself, through simple path loss alone.

9.9   Tower Top Amplifiers

The use of tower top amplifiers has many applications in wireless communications. Until recently they were a device utilized in two-way communication and not in cellular or PCS. However, based on the situation it is possible to exploit the advantages of a tower top amplifier for use in cellular and PCS.

The primary purpose of the tower top amplifier is to improve receiver sensitivity through elimination of the feedline loss component. The TTA effectively establishes the noise figure and ultimate receiver sensitivity for the communication site. Care must be exercised, however, in ensuring that the installation of the TTA does not cause more problems than it was meant to solve. Specifically the TTA net gain must be such that it does not overdrive the front end of the cell site, therefore reducing the real dynamic range of the site and also its sensitivity. In addition the filtering used for the TTA needs to be studied since too wide a front end for the TTA can in fact cause it to go into compression when colocated with other transmitters, even the site's own transmitter. The reliability of the TTA in the event of lightning also needs to be factored into the analysis. In addition, the TTA should have an auto bypass feature in the event that the TTA fails.

The TTA net gain should be set so that it effectively removes the line loss component for the system and nothing more. False reports have been issued that indicate that an overall increase in gain from the TTA nets a similar gain in the cell's receive path. The latter is not true and usually comes about as a result of incorrect analysis.

Example 9.3 which shows how to set the net gain for a TTA situation is presented here. In the simple application here the site's coaxial cable run is excessive and it is desired to improve the site's performance by installation of a TTA.

Example 9.3   Feedline for site $\approx$ 250 ft

$$\text{Attenuation } \tfrac{7}{8} \text{ in at 2000 MHz} = 1.97 \text{ dB/ft}$$

$$\text{Feedline loss} = 4.925 \text{ dB}$$

$$\text{TTA net gain} = 4.925 \text{ dB} = 5 \text{ dB (reality)}$$

$$\text{TTA gain for linear operation} = 15 \text{ dB}$$

$$\therefore \quad 15 \text{ dB} \quad \text{gain}$$

$$- \quad \underline{5 \text{ dB}} \quad \text{net gain}$$

$$10 \text{ dB} \quad \text{pad}$$

If desire to use $\frac{1}{2}$-in cable instead,

$$\frac{1}{2} \text{ attenuation} = 3.45 \text{ dB/100 ft}$$

$$\text{Feedline } (\frac{1}{2}\text{-in}) = 8.625$$

$$\therefore \quad \begin{aligned} 15 \quad &\text{dB} \\ - \quad 8.625 \, &\text{dB} \\ \hline 6.375 \quad &= 7 \text{ dB pad} \end{aligned}$$

It is interesting to note that with the use of a TTA it is possible to relax the installation requirements for the site either in terms of overall cable length or by changing the cable size itself. For the example shown above a smaller cable size was picked, therefore reducing the cost and potential tower loading at the site.

9.10 Communication Site Checklist

The following is a brief summary of the major items that need to be checked prior to or during the commissioning of a communication site. The checklist that follows is generic and should be tailored for your particular application; i.e., add or remove parts where applicable. However, the list is an excellent first step in ensuring that everything is accounted for before the communication site goes commercial.

Site location issues

1. 24-hour access
2. Parking
3. Direction to site
4. Keys issued
5. Entry and access restrictions
6. Elevator operation hours
7. Copy of lease
8. Copy of building permits
9. Obtaining of loan releases
10. Certificate of occupancy

Utilities

1. Separate meter installed
2. Auxillary power (generator)
3. Rectifiers installed and balanced

4. Batteries installed

5. Batteries charged

6. Safety gear installed

7. Fan and venting supplied

Facilities

1. Copper or fiber

2. Power for fiber hookup (if applicable)

3. POTS lines for operations

4. Number of facilities identified by engineering

5. Spans shacked and baked

HVAC

1. Installation completed

2. HVAC tested

3. HVAC system accepted

Antenna system

1. FAA requirements met

2. Antennas mounted correctly

3. Antenna azimuth checked

4. Antenna plumbness checked

5. Antenna inclination verified

6. SWR check of antenna system

7. SWR record given to ops and engineering

8. Feedline connections sealed

9. Feedline grounds completed

Operations

1. User alarms defined

Engineering

1. Site parameters defined

2. Interference check completed

3. Installation MOP generated

4. FCC requirements document filled out

5. Drive test completed

6. Optimization completed

7. Performance package completed

Radio infrastructure

1. Bays installed

2. Equipment installed according to plans

3. Rx and Tx filters tested

4. Radio equipment ATP'd

5. Tx output measured and correct

6. Grounding complete

References

1. Smith, Clint, and Curt Gervelis, *Cellular System Design and Optimization,* McGraw-Hill, New York, 1996.
2. AT&T, *Engineering and Operations in the Bell System,* 2d ed., AT&T Bell Laboratories, Murray Hill, NJ, 1983.
3. Brewster, *Telecommunications Technology,* Wiley, New York, 1986.
4. Carr, J. J., *Practical Antenna Handbook,* McGraw-Hill, 1989.
5. Code of Federal Regulations, CFR 47 Parts 1,17,22,24, and 90.
6. DeRose, *The Wireless Data Handbook,* Quantum Publishing, Inc., Mendocino, CA, 1994.
7. Dixon, *Spread Spectrum Systems,* 2d ed., Wiley, New York, 1984.
8. Kaufman, M., and A. H. Seidman, *Handbook of Electronics Calculations,* 2d ed., McGraw-Hill, New York, 1988.
9. Lee, W. C. Y., *Mobile Cellular Telecommunications Systems,* 2d ed., McGraw-Hill, New York, 1996.
10. Qualcom, *An Overview of the Application of Code Division Multiple Access (CDMA) to Digital Cellular Systems and Personal Cellular Networks,* Qualcom, San Diego, CA, May 21, 1992.
11. Simo, "IS-95 Based SS-CDMA: Operational ISSUES," seminar, January 1995.
12. White, Duff, *Electromagnetic Interference and Compatibility,* Interference Control Technologies Inc., Gainesville, GA, 1972.
13. Yarborough, *Electrical Engineering Reference Manual,* 5th ed., Professional Publications, Inc., Belmont, CA, 1990.
14. Fink, Donald, and Donald Christiansen, *Electronics Engineers Handbook,* 3d ed., McGraw-Hill, New York, 1989.
15. Schwartz, Bennett, and Stein, *Communication Systems and Technologies,* IEEE, New York, 1996.
16. Rappaport, *Wireless Communications Principles and Practices,* IEEE, 1996.
17. Fink, Beaty, *Standard Handbook for Electrical Engineers,* 13th ed., McGraw-Hill, New York, 1995.
18. Webb, Hanzo, *Modern Amplitude Modulations,* IEEE, 1994.
19. Gibson, *The Mobile Communications Handbook,* IEEE, 1996.
20. Steele, *Mobile Radio Communications,* IEEE, 1992.
21. Jakes, W. C., *Microwave Mobile Communications,* IEEE, New York, 1974.
22. Carlson, A. B., *Communications Systems,* 2d ed., McGraw-Hill, New York, 1975.
23. Johnson, R. C., and H. Jasik, *Antenna Engineering Handbook,* 2d ed., McGraw-Hill, New York, 1984.
24. Lathi, *Modern Digital and Analog Communication Systems,* CBS College Printing, New York, 1983.
25. MacDonald, "The Cellular Concept," *Bell System Technical J.,* vol. 58, no. 1, 1979.
26. Rappaport, *Cellular Radio & Personal Communications,* IEEE, New York, 1995.
27. *Reference Data for Radio Engineers,* 6th ed., Sams, 1983.
28. Mouly, Pautet, *The GSM System for Mobile Communications,* Mouly Pautet, 1992.

10

System Design

10.1 Introduction

This chapter goes over the system design requirements for putting together or expanding an existing wireless system. Topics included involve the process that an engineer should utilize for designing a wireless system from the ground up or expanding an existing system. Issues covered include information needed to effectively design a network and some proposed output forms. Many technology and market-specific issues need to be addressed when putting together an RF design.

The RF system design planning for any network is exceptionally critical. Design-specific technology issues, both current and future, must be factored into the RF design. If not, the potential is there for the system to not meet the market requirements and thus become more costly in the long term.

The RF system design is important because it defines the how, what, when, and why aspects for the company's technical direction. The direction put forth from this plan is used to define most, if not all, of the projects that need to be completed over the report time frame by the technical community in the company, with regard to the radio environment.

The network design must be dovetailed into the RF system design to ensure that the network backbone is in place to support the RF design. This chapter does not address the network design issues, but these can be found in *Cellular System Design and Optimization.*[1] The network and RF design processes are linked in various points in the growth planning, and it is necessary to incorporate both aspects to ensure that a proper design is completed.

10.2 RF System Design Process

The RF system design process for a network is an ongoing process of refinements and adjustments based on a multitude of variables, most of which are not under the control of the engineering department. However, the RF system design cell site growth analysis can be used to help direct the limited resources of the company.

The final output of the RF system design is simply to identify the number of cell sites required for the network and their required on-air dates. The required on-air dates are most critical with an existing system, since they define when the cell site is specifically needed. Obviously the on-air date for a new system is the on-air date for the entire network and the sequence in which they are secured is not as important as ensuring they will all be available for system activation.

The other aspect of the RF system design is to factor into the design process the number of physical radios or rather traffic-handling capacity for each cell site and sector. Several excellent computer modeling systems exist for helping define the required channel elements for a cell site. The automated system incorporates the new and existing cell site traffic loads and adjusts the area accordingly for traffic distribution. However, the process presented in the next few sections assumes that no automated cell building program is in place. The process presented here can be used for analog, TDMA, and GSM but is not totally valid for CDMA since that process requires the use of a computer modeling program to account for various levels of external and internal noise.

The required process is as follows:

1. Marketing requirements

2. Methodology

3. Technology decision

4. Definition of cell site types

5. Establishing a link budget

6. Defining coverage requirements

7. Defining capacity requirements

8. Completing RF system design

9. Issuing search area

10. Site qualification test (SQT)

11. Site acceptance/site rejection

12. Land use entitlement process

13. Integration

14. Handing over to operations

Obviously many steps are incorporated into each of these various process milestones. But the process outlined is the basic process steps that need to be followed by RF engineering to design a system, regardless of the technology platform chosen.

At each of the steps of the process a brief design review takes place to ensure that the design is put together in accordance with the requirements for the network. If the design does not meet market requirements, the potential for failure is very high.

10.3 Methodology

The methodology for the RF system design needs to be established at the beginning of the design process. The establishment of the methodology utilized for the formulation of the design and ultimately the RF system design report is essential in the beginning stages to ensure the proper baseline assumptions are used and to prevent labor-intensive reworks and teeth gnashing.

Some of the issues that need to be identified at the beginning of study:

1. Time frames for the report to be based on

2. Subscriber growth projections (current and future by quarter)

3. Subscriber usage projection (current and forecast by quarter)

4. Subscriber types (mobile, portable, blend)

5. Design criteria (technology-specific issues)

6. Baseline system numbers for building on the growth study

7. Cell site construction expectations (ideal and with land use entitlement issues factored in)

8. New technology deployment and time frames

9. Budget constraints

10. Due date for report

11. Maximum and minimum off loading for cell sites when a new cell is added to the design

It is essential to have the time frames for the report established prior to the beginning of the system design and the report generation. The time frames will define the baseline, foundation, and how much of a future look the report will present. For a new system the on-air date obviously will be the starting point; however, this is only one of the milestones since it is when the system loading will officially begin. The design time frame needs to account for how long after system turn-on the design will be valid, assuming the marketing input and construction assumptions remain valid.

If the system already exists, the baseline data used will need to be decided upon. The baseline time frame is usually a particular month, and the design time frame that is selected will determine which set of data will be used for generating the report. The baseline is critical for an existing system, for if the wrong month is selected the data used for generating the design will alter the outcome of the report by either overprojecting or underprojecting the number of sites required.

The amount of time the report projection takes into account is also critical for the analysis. The decision to project 1 year, 2 years, 5 years, or even 10 years has a dramatic effect on the final outcome. In addition to the projection time frame it is important to establish the granularity of the reporting period, monthly, quarterly, 6 month, yearly, or some perturbation of them all.

The particular marketing plans also need to be factored into the report itself. They are the leading element in any RF system design. The basic input parameters needed for the RF design from the marketing department are listed next. The information is critical regardless of whether the system is a new or current system.

1. The projected subscriber growth for the system over the time frame for the study

2. The projected millierlangs* per subscriber expected at discrete time intervals for the study

3. The dilution rates for the subscriber usage over the time frame for the study

4. Types of subscriber equipment used in the network and percent distribution of CPE projections (that is, portable or mobile units in use and their percent distribution)

5. Special promotion plans over the time frame of the study like local calling or free weekend use

6. The projected number of mobile data users over the time frame of the study

7. Top 10 customer complaint areas in the network requiring coverage improvements

8. Identification of key coverage areas in the network that need to be included for system turn-on

A multitude of other items are needed from the marketing department for determining RF system design. However, the basic eight topics listed above obtained from the marketing department will be enough to adequately start the RF system design.

The establishment of the design criteria used for the report is another key criterion when putting together the plan. It is recommended that the design criteria used for the study be signed off by the director of engineering to ensure that nothing is missed.

The items in the RF design criteria need to include the following as a minimum.

1. Marketing input

2. RF spectrum available for the RF plan

---

*The erlang is a dimensionless unit since the numerator and denominator are of the same unit value and thus cancel out. In the wireless industry, an erlang is simply defined as 3600 seconds of usage in a 1-hour time period. Thus, if you have two radios and each is occupied for 1 hour, 3600 seconds, this represents 2 erlangs of usage.

3. Type of grade of service table to be used for plan, that is, erlang B P02

4. Minimum and maximum off-loading factors for new cells

5. Coverage requirements

6. Identification of coverage sites

7. BH peak traffic, 10-day high average for month (existing system)

8. Infrastructure equipment constraints

9. Digital and analog radio growth

10. New technology considerations

11. Cell site configurations used for new cells

12. Baseline system numbers

13. Cell site deployment considerations

The above list should be used as the foundation for establishing the design criteria for the RF system design plan.

The baseline system numbers should be listed in the design criteria for the growth plan. The baseline numbers involve defining the time period for the growth projection. In particular the decision to, say, use June data instead of July data has a significant impact in the plan as well as using peak versus an average value for traffic.

The construction aspects of the system also need to be factored into the report itself. They pertain to the proposed new cell sites that are or will shortly be under construction. Other construction aspects involve the possibility of actually building what is requested in the specified time frame dictated in the report. The realistic time frame for construction is the reality dose that needs to be placed in the report saying that although 100 sites are needed in the next 3 months only 20 will actually be available for operation.

The construction aspects of the design may force a redesign to take place to accommodate the RF coverage or capacity requirements. Specifically if a coverage site is needed for an area within 6 months but the land use entitlement process may take upward of 18 months, an alternative design will be sought. The alternative design may or may not be feasible, and therefore no solution may be available for the situation at hand other than to have a coverage hole for a defined period of time.

New technology deployments will also need to be factored into the plan. They might be converting from a TDMA system to a CDMA system, or a change in the vocoder utilized. Other technology deployment issues that need to be factored in might be the cell site infrastructure equipment where a new vendor's equipment is planned to be deployed. Also involved with technology deployments could be the introduction of a new switch that has improved processor capacity for key trigger points which change the requirements for future growth.

The budget constraints imposed on any operator must be folded into the RF plan. Failure to incorporate budget constraints in the plan will only make it unrealistic and squander everyone's time needlessly. Often if the plan calls for more sites than are budgeted for, as is the case most times, it is imperative to help establish a ranking order for which sites are truly needed versus which are nice to have.

The last part of the preparation for the report is finding out when the report itself is due, the level of detail required, and who is assigned to work on the report itself. All too often a plan will be put together but for various reasons the actual due date is often allowed to slip. The negative impact of this ties directly to providing some guidance to the other departments which need the RF design to complete their work, since the RF design is usually the gating project for their efforts. If the RF design is taking longer than planned, it is advisable that an interim report be issued to allow various departments, primarily real estate, to begin their efforts.

10.4   Technology Decision

The technology decision is a critical issue in putting together an RF design. Specifically the issues associated with an analog system are different from those for a digital system. The technology chosen will have a profound impact on how the system is designed.

For instance, the design criteria for an analog system differ from those of a digital system. More specifically the issues associated with which digital system is chosen impact the design decisions made throughout the process. For instance, in cellular the digital technology chosen impacts the amount of spectrum allocated for digital use. The initial deployment of digital may require only that a few analog channels are converted to digital use or upward of 1.8 MHz of spectrum are allocated to the digital. The obvious issue here is initial capacity reduction in a network to satisfy the long-term growth for the network.

PCS, however, has several issues that impact the method used for determining system growth issues. The most obvious is the spectrum requirements that the technology requires for initial and future implementation. This simple concept has a major impact on initial deployment because of microwave clearance issues alone, in the United States. GSM requires less spectrum per GSM carrier than say a CDMA system, 200 kHz vs. 1.25 MHz. However, TDMA (IS-136) requires less spectrum for initial deployment than GSM because of its bandwidth requirements, 30 kHz vs. 200 kHz. In all the above examples guard bands were not factored into the comparison.

If the decision is to utilize multiple technology platforms within the network for a given market, the various compatibility issues need to be factored into the design process. For example, in AMPS if CDPD is deployed, the choice of a dedicated channel set or frequency hopping needs to be factored into the design. In addition, if GSM and CDMA are meant to coexist in a market, decisions must be made as to which portion of the allocated spectrum needs to be used by which technology platform. The obvious point is that the technology decision not only will impact the services that may be provided by the operating company to the subscriber but will also have a profound impact on how the network is laid out.

10.5   Link Budget

This section documents the link budget assumptions made for the RF design. The link budget is defined in Chap. 3. However, the link budget decided upon for the system design needs to account for not only the physical issues defined in setting up the link budget but also marketing issues.

The objective of defining the link budget is to arrive at the size of the cell sites needed for the network design. Specifically the D/R ratio chosen for the system determines the radius of the site and also the distance between the cells themselves. The link budget also factors in the local issues that drive the financial requirements for the market.

A sample link budget is included in Fig. 10.1. The link budget is meant to accommodate multiple technology platforms and traverse over a wide range of radio spectrum issues. The link budget format presented here can and should be used when putting together the design of a system. Some issues covered in the link budget may or may not pertain to the particular potential cell site or technology platform utilized.

10.6   Cell Sites

The type of cell site or sites that are chosen for the design of the network will have a profound impact on the overall success or failure of the network design. For instance, if the goal is to design a low-cost system for the operating company, then not only the number of cell sites need to be factored into the design but also their impact to recurring costs. The recurring costs of the system involve leases, facilities, and other maintenance issues, which also include engineering efforts.

The use of smaller cell sites with lower antenna footprints is one way of reducing the lease costs for a system. However, if this is used for an initial system deployment, the sheer number of cell sites required to cover the same geographic area will increase. On the other hand, if taller sites are chosen as the design of choice, they would need to potentially be larger in physical size in both equipment and antenna real estate. There is of course a blend or middle ground that could be used with large and small cell site designs that would potentially meet the requirements for the network.

The different types of cell sites are included next to illustrate the number of different types of cell sites that can simultaneously exist in a network. The proposed listing that follows is not all-inclusive, and there are many perturbations to those shown next.

1. Omni

2. Directional

3. Three-sector

System: Example 1   Date _____
Frequency of operation
   Tx
   Rx

Down-link path

Base station parameters
   Tx PA output power   dBm
   Tx combiner loss   dB
   Tx duplexer loss   dB
   Jumper and connector loss   dB
   Lightning arrester loss   dB
   Feedline loss   dB
   Jumper and connector loss   dB
   Tower top amp gain   dB
   Antenna gain   dBd

   Cell ERP   0 dBm

Environmental margins
   Tx diversity gain   dB
   Fading margin   dB
   Environmental attenuation   dB
   Cell overlap   dB

   Environmental margin   0 dB

Subscriber unit parameters
   Antenna gain   dBd
   Rx diversity gain   dB
   Antenna cable loss   dB
   C/I or E_b/N_0 dB
   Rx sensitivity   dBm

   Subscriber unit effective sensitivity   0 dBm

Maximum down-link path loss allowed
   Base station ERP   0 dBm
   Subscriber unit effective sensitivity   0 dBm

   Maximum down-link path loss   0 dB

Uplink path

Subscriber unit parameters
   Tx PA output   dBm
   Cable and jumper loss   dB
   Antenna gain   dBd

   Subscriber unit ERP   0 dBm

Environmental margins
   Fading margin   dB
   Environmental attenuation   dB
   Cell overlap   dB

   Environmental margin   0 dB

Figure 10.1 Link budget.

```
Base station parameters
    Rx antenna gain                         dBd
    Tower top amp net gain                  dB
    Jumper and connector loss               dB
    Feedline loss                           dB
    Lightning arrester loss                 dB
    Jumper and connector loss               dB
    Duplexer loss                           dB
    Receive configuration loss              dB
    Rx diversity gain                       dB
    C/I or Eb/N0                            dB
    Rx sensitivity                          dBm

    Base station effective sensitivity      0 dBm

Maximum up-link path loss allowed
    Subscriber unit ERP                     0 dBm
    Base station effective sensitivity      0 dBm

    Maximum up-link path loss               0 dB
```

Figure 10.1   (*Continued*)

4. Six-sector

5. Microcell

6. Picocell

Utilizing the link budget, a rough estimate of the number of cell sites required for network design can be established. The rough estimate for cell sites needs to factor in capacity requirements for the network. At this stage of the design process the particular offloading adjustments for each cell site do not need to be accounted for. This is where an estimate of the type and number of cells needed for the design is identified. If the initial cell site count is above what the business case indicates, an interactive step needs to take place where the assumptions used for the initial cell site count need to be adjusted and recalculated.

10.7   Coverage Requirements

One important step in the RF system design process is to determine all the coverage requirements needed for the network. They will obviously be different for a new system than for an existing one. For a new system the coverage requirements will be driven by marketing in that they will define and rank order the areas needed for initial system deployment based on the market they intend to target.

However, for an existing system the defining of coverage requirements is not as straightforward as that for a new system. For an existing system the coverage requirements need to be more granularly defined in order to allocate the resources of the company. The coverage requirements needed for the network

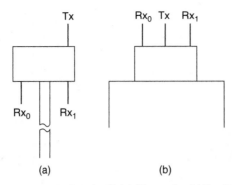

Figure 10.2   Omni cell. (*a*) Monopole. (*b*) Rooftop.

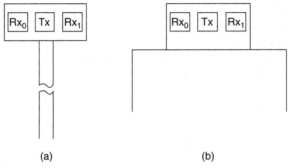

Figure 10.3   Directional (single-sector). (*a*) Monopole. (*b*) Rooftop (panel antenna).

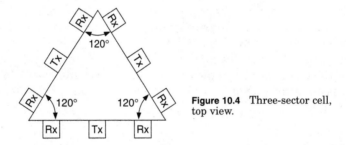

Figure 10.4   Three-sector cell, top view.

will be rank ordered in the final process to ensure that all the proper input parameters are taken into account.

RF coverage identification process for new system

1. Coverage requirements are defined by marketing and sales.

2. Using the design criteria of RF engineering, determine how many sites will be needed to satisfy the design goals.

3. Using the list of cell sites identified in step 2, rank order them according to the point system methodology.

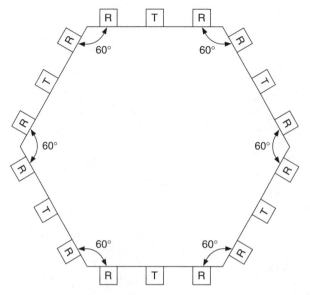

Figure 10.5 Six-sector cell.

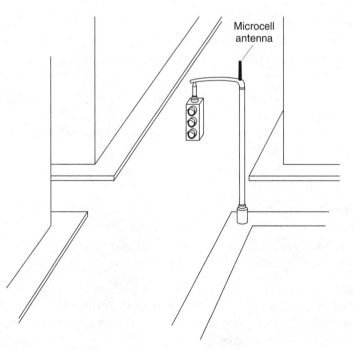

Figure 10.6 Microcell.

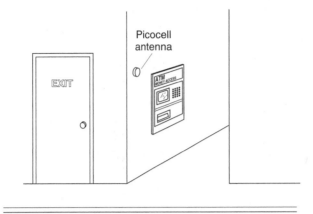

Figure 10.7   Picocell.

RF coverage identification process for existing system

1. Coverage requirements identified by
   Marketing and sales
   System performance
   Operations
   Customer care
   RF engineering

2. Generate a propagation plot of the system, or subregions, that reflects the current.

3. Generate a propagation plot of the system, or subregions, that reflects the current and known future cell sites that are under construction.

4. Utilizing physical field measurements, generate a plot of the system, or subregions.

5. Using plots from parts 2 and 3, compare this against areas identified in part 1 for correlation.

6. Using the field measurement plots, compare it against parts 1 and 2 for correlation.

7. Using the design criteria used by RF engineering, determine how many sites will be needed to satisfy the design goals.

8. Using the list of cell sites identified in step 7, rank order them according to the following point system methodology.

The point system will involve five key parameters all ranked from a scale of 1 to 5 based on its severity. Each of the five categories receives a value which is then multiplied by each field to arrive at a ranking. The ranking methodology uses the following key fields:

1. Coverage

2. Erlang potential

3. Customer care problems

4. Marketing and sales needs

5. System performance requirements

To help clarify the ranking methodology, a simple table (Table 10.1) with values added will assist in the explanation. Table 10.1 shows that for an existing system cell 102 has a higher weight then 101 in the identification of prioritization. The ranking methodology should be applied to all the potential cell sites in a network. The rationale behind establishing a ranking system folds into the budgetary issues of having a method in place to determine where to cut or add sites to the build program.

TABLE 10.1

Cell	Coverage	Erlang potential	Customer care	Marketing	System performance	Total
101	3	1	2	4	2	48
102	2	4	3	1	4	96

10.8   Capacity Cell Sites Required

The next step in the RF system design is to determine the sites required for capacity relief. Obviously if the system is new, the potential for capacity relief is minimal; however, it would be a pleasant surprise. The process shown applies directly to AMPS but can easily be converted for use in any of the technology platforms used.

The sites required for capacity relief involve a process of determining which sites and/or sectors expend their current capacity and require some level of relief. The time frame is very important in that you do not want to deploy equipment or cell sites too much in advance of capacity exhaustion. Deploying too late, however, would have potentially negative customer satisfaction impact. The capacity relief can come from a variety of options:

1. Radio additions

2. Parameter adjustments

3. Antenna system alteration

4. New cells

More capacity relief methods are available for redistributing the traffic loads of cell sites. The first step in identifying the capacity cell sites required for a network involves use of a spreadsheet to determine where the problems are anticipated. The recommended spreadsheet format is shown in Table 10.2.

TABLE 10.2   RF Growth Plan

System X
Date _____
RF growth plan
Quarter of interest (4Q95)

	Baseline					
Cells	Erlangs	Channels	Channels	Offload	Acquired	Adjusted channels
1A	5.4	17	21	(0.4)	0	15
2A	4.0	12	16	0.0	0	16
101	0	0	0	0.6	3.1	12
XXX	*a*	*b*	*c*	*d*	*e*	*f*

[a] This is the baseline erlang value used to begin the chart. The baseline erlang value is arrived at based on the data collected in the design criteria. For example, if the baseline is for the average of the 10 busiest days in June, the erlang value for this site's contribution is then deposited here.

[b] This is the number of physical channels currently available at the site during the baseline.

[c] This value is the number of channels needed for the site based on traffic projections arrived at in the design criteria. The channels needed are pulled from a look-up table, usually erlang B, which cross-references capacity to a physical number of channels. An example of the erlang B chart is in Table 10.1 The value that is used for the look-up is arrived at using this equation:

$$\text{System erlangs projected} = (\text{millierlang/S}) \times (\text{total system subscribers for quarter})$$

$$\text{Cell site system capacity*} = (\text{erlangs for sector})/(\text{total system erlangs})$$

The part marked with an asterisk uses the data from the previous quarter. The value will not be 100 percent correct but is more than sufficient to use for planning purposes.

$$\text{Cell site projected erlangs} = (\text{system erlangs projected}) \times (\text{cell site system capacity})$$

The cell site projected erlang value is then used in the look-up table for determining how many radios are needed for the site during this quarter.

[d] The offload value is used for shedding traffic to another cell site. The methodology used for arriving at how to shed traffic is discussed later. However, if it is determined that you can shed 40 percent of the site's projected traffic to another cell site, a value of 0.4 is entered.

[e] This is the acquired traffic the site receives from adjacent cell sites as a result of adjustments made to them. The acquired traffic portion should equal the total amount of traffic offloaded in section *d*. The values arrived at in this section pertain to how the traffic is distributed in the offloading cell site. For example, cell 5 may offload cell 3 by 40 percent, 0.4, but the traffic will not all go onto one sector, usually. Therefore, you must determine how this is distributed around the site. Continuing this example, cell 5 will have cell 3's traffic distributed, with sector 1 getting 70 percent and sector 2 getting 30 percent. I strongly suggest keeping meticulous records about how much each site offloads the other and the distribution percentages used.

[f] The adjusted channel is the resultant of offloading and acquired traffic to the site. The formula to use in this example is

$$\text{Adjusted erlangs} = (\text{cell site projected erlangs}) \times (\text{offload value} + \text{acquired traffic})$$

The adjusted erlang value is then used with the same erlang look-up table used for *c,* and the RF channels needed are then listed.

Obviously when looking at the traffic table a recursive feedback loop process is really implemented. The recursive feedback loop involves additional offloading and acquired erlang values until the proper design criteria are met.

Regarding the acquired traffic for a new site, it is suggested that a minimum value be added to the overall value for every sector added. The added erlang value is used to ensure that a minimum of one radio channel is put in place for every new cell. However, doing this will inflate the overall system erlang value beyond the projection for that quarter only. As you can see, this methodology can be easily deployed for a variety of technology platforms.

10.8.1   RF traffic offloading

The offloading of traffic to another cell site can be accomplished through an elaborate computer simulation method or educated engineering guesses. The elaborate computer modeling capability is the ultimate desired method since it eliminates some guesswork involved in arriving at the offloading values. However, the computer model method is only as good as the model itself.

If you are using the computer model method, the offloading and acquired values should be readily extracted from the algorithm output and input into the spreadsheet either manually or automated.

If you are not fortunate to have a sophisticated computer modeling method, the following method is recommended. It has been used repeatedly with a high level of success.

Referring to Fig. 10.8, a four-cell subsystem is listed for ease of illustration. The situation for this case will not involve the need for the site involving coverage-related issues. The example here involves only the introduction of a new cell site for the sole purpose of providing traffic relief to adjacent cell sites.

Traffic projections were done using the spreadsheet method listed above and it was found that sites 3 and 4 required some form of traffic shedding. It was determined that the only available method to ensure sufficient traffic shedding for both sites was the introduction of cell site 5 shown in Fig. 10.9.

Referring to Fig. 10.9, cell site 5's placement in the network can ensure that it will offload several of the adjacent cell sites by the percentages listed. The offloading values were determined through an interactive process involving the following data:

1. Current coverage of zone by the existing cell sites

2. Determination of voice and control channel dominant server areas, by cell and sector

3. Evaluation of the cell site configurations in the zone

4. Incorporation of cell 5 into data provided for parts 1, 2, and 3

5. Comparison of propagation plots to actual field measurement data

6. Establishment of offloading and acquired percentages

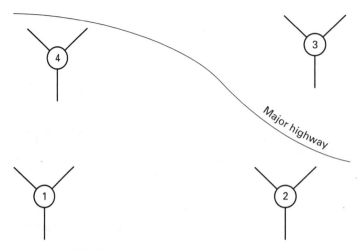

Figure 10.8 Offload.

The offloaded traffic is then assigned to the sectors of cell 5 as shown in Fig. 10.10. The values represented in Figs. 10.9 and 10.10 are then inputted into the tables of the growth plan to ensure that the desired effect is accomplished.

10.8.2 Radio growth

The radio growth portion of the RF system design plan involves analysis of the individual sectors for every cell site currently in the network and proposed. The radio growth projections are then compiled and fed into the process. The objective is to identify how to meet the system capacity requirements for a localized area without the introduction of a new cell site.

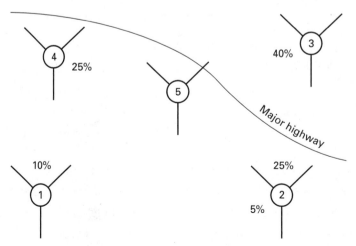

Figure 10.9 Offload cell.

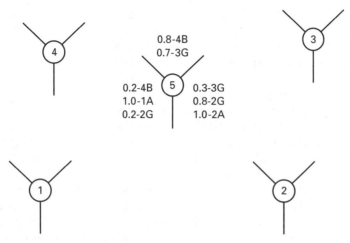

Figure 10.10 Offload percentages.

The capacity requirements for a cell or sector of a cell may be satisfied though the introduction of radio channels. The introduction of the radio channels can also defer the building of a new cell site through properly exploiting the existing infrastructure of a system. However, as with building a cell site for capacity, the introduction of radios into the network has the negative impact of increasing the capital and recurring costs with a network. However, the recurring costs are not as great as that of a new cell site. Ideally the introduction of new radios into the network should be done in such a fashion as to have them arrive just in time for deployment.

10.9 RF Design Implementation Guidelines

No true RF engineering can take place without some form of RF design guidelines, whether formal or informal. However, with the level of complexity rising every day in wireless communication systems, the lack of a clear definitive set of design guidelines is fraught with potential disaster. While this concept seems straightforward and simple, many wireless engineering departments when pushed have a difficult time defining what exactly their design guidelines are.

The actual format (method of how it is conducted) should be structured to facilitate ease, documentation, and minimization for formal meetings. For most design reviews a formal overhead presentation is not required. Instead a sit-down with the manager of the department is in most cases the level of review needed. The important point is that another qualified member of the engineering staff reviews the material to prevent common or simple mistakes. Ensuring a design review process does not eliminate the chances of mistakes. Design reviews ensure that when mistakes do take place the how, why, and when issues needed to expedite the restoration process are already in place.

It is highly recommended that department RF design guidelines are reasonably documented and updated on a predetermined basis, yearly at a minimum.

The use of design guidelines will facilitate the design review process and establish a clear set of directions for the engineering department to follow. The RF design guidelines will also ensure a consistent approach to designing and operating the capital infrastructure that has been or will be put into place within the network.

The actual design guidelines utilized by the RF engineers need to be well documented and distributed. The design guidelines, however, do not need to consist of voluminous amounts of data. They should consist of a few pages of information that can be used as a quick reference sheet by engineering. The design guideline sheet has to be based on the system design goals and objectives set forth in the RF system design.

The actual content of the design guideline can and will vary from operator to operator. However, it is essential that a list of design guidelines be put together and distributed. Publication and distribution of RF design guidelines will ensure a minimum level of RF design specifications in the network. A proposed guideline is shown in Table 10.3. It is for an AMPS system. The guideline can easily be crafted to reflect the particular design guidelines utilized for the market where it will be applied.

10.10   Cell Site Design

While this is not necessarily the first step in any design process, it is one of the most important for the RF engineering department. The reason the cell site design is critical for the RF engineering department lies in the fact that it is where the bulk of the capital is spent. The cell site design guidelines listed below can be utilized directly or modified to meet your own particular requirements.

The use of a defined set of criteria will help facilitate the cell site build program by improving interdepartmental coordination and will provide the proper documentation for any new engineer to review and understand the entire process with ease. Often when a new engineer comes onto a project all the previous work done by the last engineer is reinvented primarily because of a lack of documentation and/or design guidelines.

The cell site design process takes on many facets, and each company's internal processes are different. However, no matter what the internal process, the following items are needed as a minimum:

1. Search area
2. Site qualification test (SQT)
3. Site acceptance
4. Site rejection
5. FCC guidelines
6. FAA guidelines
7. Planning and zoning board
8. EMF compliance

TABLE 10.3   RF Design Guidelines

System name
Date

	RSSI, dBm	ERP, W	Cell area, km/sq	Antenna type
Urban	−80	16	3.14	12 dBd 90H/14E
Suburban	−85	40	19.5	12 dBd 90H/14E
Rural	−90	100	78.5	10 dBd 110H/18E

Voice channel C/I	17 dB (90th percentile)
Frequency reuse	$N = 7$
Maximum no. channels per sector	19

Antenna system
  Sector cell orientation   0,120,240
  Antenna height   100 ft or 30 m
  Antenna pass band   825–894 MHz
  Antenna feedline loss   2 dB
  Antenna system return loss   20–25 dB
  Diversity spacing $d = h/11$ (d = receive antenna spacing, h = antenna AGL)
  Receive antennas per sector   2
  Transmit antennas per sector   1
  Roof height offset $h = x/5$ (h = height of antenna from roof, x = distance from roof edge)

Performance criteria
  Lost call rate   <2%
  Attempt failure   <2%
  RF blocking   1%><2%

BER/FER   1%

10.11   Search Area

The definition of a search area and the information content provided is a critical first step in the cell site design process. The search area request is a key source document used by the real estate acquisition department of the company. The selection and form of the material presented should not be taken lightly, since more times than not RF engineers rely heavily upon the real estate group to find a suitable location for the communication facility to exist. If the search area definition is not done properly in the initial phase, it should not be a surprise when the selection of candidate properties is poor.

The search area issued needs to follow the design objectives for the area following RF system design objectives. The search area should be put together by the RF engineer responsible for the site design. The final paper needs to be reviewed and signed by the appropriate reviewing process, usually the department manager, to ensure that there is a check and balance in the process. The specifications for the search area document need to meet not only the RF engineering department requirements but also the real estate and construction group needs. Therefore, the form in Fig. 10.11 needs

RF Engineering

Search area code:_____   Capital funding code:_____

Target-on-air date:_____

Search area type: (<u>capacity, coverage, frequency plan, competitive, new technology</u>)

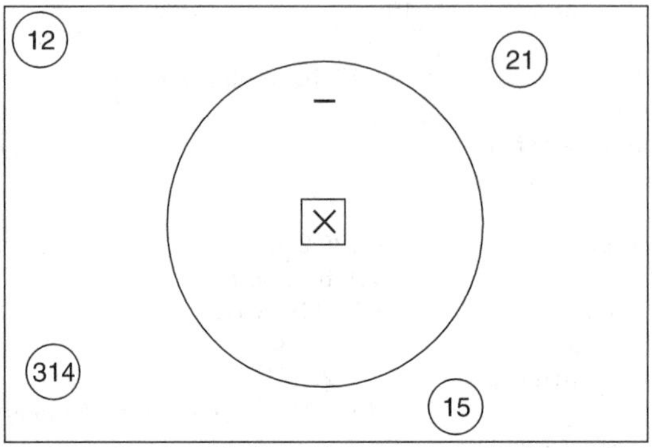

Search area map

Cell site configuration: (<u>omni, 3 sector, 6 sector, other</u>)

Type of infrastructure:_____

Physical size of equipment room:_____ (ft^2)

Antenna info:

   1. Number of antennas:_____

   2. Type of antennas: (<u>attach manufacturer's specification</u>)

   3. Antenna height

      AGL:_____

      AMSL:_____

Maximum cable length:_____

Comments: _____

           _____

           _____

           _____

| **Search area request** |
| Document: _____   Date:_____ |
| Design engineer:_____ |
| Reviewed by:_____ |
| Revision |

Figure 10.11   Search area request.

to be approved by the various groups but issued by the RF engineering department. It is imperative that the search area request issued under go a design review prior to its issuance. The proposed format is shown in Fig. 10.11.

Referring to Fig. 10.11, the following comments need to be made. The map included with the search area request needs to include as much information as practical for the real estate acquisition group to help locate the proper sites. The map used in this form will minimize the number of dud sites presented to RF engineering for consideration in the system design. The search area request map needs to include area-specific information, which varies from location to location. The information content for the map is different if the proposed site is in a very rural area or a dense urban environment. The variations in the map format are largely dependent upon design criteria for the site. If the search area ring is very defined, as is the case with mature systems, it is imperative that the adjacent existing sites and search areas are identified on the map itself. The rationale behind including adjacent sites and the current search areas is that it will better define options available to the real estate acquisition group.

A propagation plot for the search area request form also needs to be generated and included with the search area folder maintained by RF engineering. A propagation plot will define the search area and coverage rings put forth in the map provided. The propagation plots will be used as part of the site acceptance procedure listed later. The propagation plots are one of the steps taken to ensure the proposed site meets its desired objective.

On the search area request form the search area code should be identified along with its capital funding number. The capital funding number and the search area code can and should be the same.

The on-air target date is meant to identify when the site is to be placed in commercial service. The on-air date should match the dates put forth in the RF system design plan. Its purpose is to help prioritize the internal resources of the company by helping define the importance of the site.

The search area type is meant to notify internal and external departments of the actual purpose of the site. There are functionally five main groupings for cell site types: capacity, coverage, frequency planning, competitive, and new technology. The type of search area this site represents is important for everyone involved to understand.

The design objective for the site is a critical issue with the search area request form. Stating what the design objective is for the site will assist in the site acceptance design reviews. Stating the design objective will also assist other engineers in determining the rationale behind why the site is needed. The design objective should match the goals set forth in the RF system design.

The cell site configuration and infrastructure type is meant to assist the real estate acquisition department in determining the parameters for the physical site. It is important to have them defined in advance, since this directly impacts the equipment space and power requirements. The floor space requirements obviously are different if the site is a microcell or a macrocell.

The total number of antennas and their type are also important to define in advance for the site search. If a site requires 3 versus 16 antennas, the discussion the real estate acquisition group takes with the potential landlord is different.

The rationale behind listing the above ground level (AGL) value, and above mean sea level (AMSL), will help define the site location options for the real estate acquisition group. The AGL will help structure the search for suitable properties that fit the design parameters. Sometimes sites are available that are 10 m tall, have a willing landlord, and permit accessibility. However, the design specification calls for a 30-m antenna height, disqualifying the site for consideration and resource expenditure. If a site needs to be 30 m tall, (about 100 ft), its AMSL is important to define because it might be possible to have the antennas at the 30-m height but the location of the property is in a gully, requiring a 50-m-tall antenna installation. The comment section of the form in Fig. 10.11 is meant to provide an area for the designer to specify any particulars desired for the site.

On the search area request form itself is a section for documentation control which is meant to track who issued the search area request, what revision it is, and the dates associated with it. Also on the document control portion of the form is a section for the design reviewer to sign off on the request. It cannot be overstressed that the information on the search area request form will largely define the success or failure of the property search. All the search area requests need to undergo a design review.

10.12   Site Qualification Test (SQT)

The site qualification test (SQT) is an integral part of any RF system design. Some form of transmitter or site qualification test should be conducted at every site. The fundamental reason behind requiring a test is to assure the site is a viable candidate before a large amount of company capital is spent on building. This test also makes sure the site will operate well within the network. The financial implications associated with accepting or rejecting a transmitter necessitate a few thousand dollars expended in the front end of the build process. If a site is accepted that will not perform its intended mission statement, additional capital will be needed to accomplish it.

Based on the volume of sites required within a specified time frame, it may not be possible to physically test every cell site candidate. Therefore, it is essential that a goal be defined of how many sites should be physically tested, i.e., 75 percent. This will help establish the risk factors associated with the build-out of the network.

Regardless of whether a site is to be physically tested or evaluated through use of a computer simulation, several stages need to be done in this process. It is very important that the SQT be performed properly, since this will determine whether over \$500,000 to \$1 million will ultimately be spent on the potential facility.

It is strongly recommended that the RF engineer responsible for the final site design visit the location prior to any SQT. This site visit will facilitate several items. First the engineer will now have a better idea of the potential usefulness of the site and its ability to be built, and can provide more accurate instructions to the testing team.

It is strongly recommended that the RF engineer does not design the test on the fly by instructing the testing team on that day where to place the transmitter and which routes to drive. The desired approach is to have the engineer determine where to place the transmitter, either which part of the tower, the rooftop, or the location for the crane. The RF engineer then puts together a test plan, identifying the location of the transmitter antenna, the ERP, drive routes, and any particular desired variations. The test plan is then submitted to the manager of the department for approval and is passed to the SQT team.

The rationale behind putting this step in the process is the capital dollars needed to be spent and the system interoperability issues. The SQT is a very critical step in the process and needs to be well thought out in advance. Improperly defining the test routes or transmitter locations can lead to a poor site being accepted or a potentially good site not getting accepted. The driving issue here as always is that a well-planned test will save time and money many times over.

The proposed form to use for the site qualification test (SQT) process is shown in Fig. 10.12. The format of the SQT form needs to directly match the input requirements of the RF engineers, the SQT measurement team, and the group responsible for postprocessing the data.

It is imperative that the testing plan be reviewed and signed off as part of the design review process. The actual test conducted, antenna placement, physical routes traversed, and postprocessing criteria directly determine the viability of the location by completing the criteria set forth, because a poorly designed test plan can cause the failure of a site's acceptance even though it may meet the needs set forth in the growth plan. Of course, the opposite situation can also occur, when it meets the design criteria set forth in the growth plan. The test plan can also set the groundwork to approve a site even though the introduction of the site could negatively impact the network once it goes commercial.

The SQT needs to include a sketch of the test location defining where to place the actual test transmitter antenna. For example, when using a crane to place the transmitter antenna it is important to define where the crane should be parked and the test antenna height actually used. If the test location is not properly defined in advance, an error could occur in the test transmitter placement significant enough to pass or fail the SQT requirements. Figure 10.13 has several example diagrams of test transmitter antenna locations.

Site Qualification Test

Date:_____
Search area code:_____   Capital funding code:_____
Test site address: _____

_____

_____

Site contact: Name:_____
                   Phone #:_____
Requested test date:_____
Test parameters:
   1. 7.5-minute map: (name or # of map(s) that encompass SQT
   2. Test antenna: (make, model #)
   3. Test antenna height
      AGL:_____
      AMSL:_____
   4. Test ERP:_____watts
   5. Antenna orientation:_____
   6. Test frequency/channel:_____
   7. Clearance/coordination:_____
Test implementation:
   1. Antenna mounting info: (roof, tower, crane, water tank, misc.)
   2. Rigger required: (Y/N)
   3. Antenna test location sketch attached: (Y/N)
   4. Test routes attached: (Y/N)
   5. SQT team leader:_____
Postprocessing:
   1. Map scale:_____
   2. Color code:_____
   3. Data reduction method:_____
Test equipment:
   1. SQT test equipment cal date_____
   2. SQT transmitter cal date_____

SQT	
Document #:_____	Date:_____
Design engineer:_____	Reviewed by:_____

Figure 10.12   Site qualification test.

Figure 10.14 is an example drive test route for a SQT. The RF engineer working on the SQT must make certain that the drive test route defined matches the design criteria for their specified site. The individual line items of the SQT form shown in Fig. 10.14 are self-explanatory. The 7.5-minute map portion is meant to help specify the actual grid location of the test which can be used as

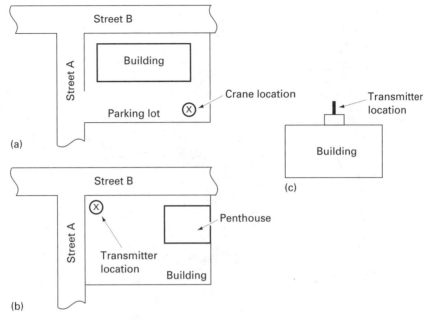

Figure 10.13 SQT transmitter location drawings.

part of the archiving process for SQTs. The test antenna type, height, and ERP obviously should be defined in advance to ensure the testing is conducted in accordance with the design specifications. The calibration requests on the form are meant to ensure that the equipment used is within calibration. It is important to always check the calibration for any test equipment used to be certain it is within specification and thus provides reliable measurements.

10.13 Site Acceptance (SA)

Once a site has been tested for its potential use in the network, it is determined to be either acceptable or not acceptable. For this section the assumption will be the site is acceptable for use by the RF engineering department as a communication facility. It is imperative that the desires of the RF engineering department be properly communicated to all the departments within the company in a timely fashion. The method of communication can be done verbally at first, based on time constraints, but a level of documentation must follow that will ensure that the design objectives are properly communicated.

The form listed below is meant as a general guide. It might need to be modified based on particular requirements. Before the SA is released it is imperative that it go through the design review process to ensure that nothing is overlooked. The SA will be used to communicate the RF engineering's intention for the site and will be a key source document used by real estate, construction, operations, and the various subgroups within engineering itself. The SA will also need to be sourced with a document control number to

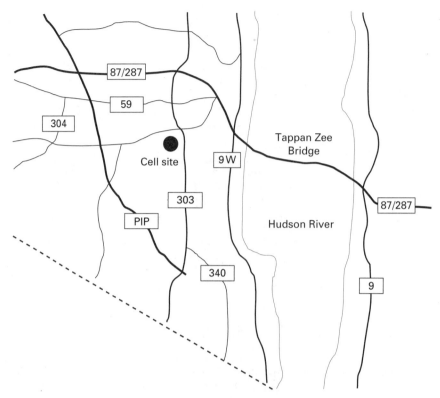

Figure 10.14 Drive route.

ensure that changes in personnel during the project's life are as transparent as possible.

The proposed site acceptance form is included in Fig. 10.15. The SA form shown can be easily expanded to ensure that all the relevant information required within the organization is provided. Whatever the form or method ultimately utilized it is important to include the information listed in the figure as the minimum requirements.

Most of the information included in Fig. 10.15 is self-explanatory. It is imperative that like all the other steps in the design process a design review and sign-off takes place, establishing a formal paper flow. The SA should include a copy of the predicted propagation used to generate the search area request and also a copy of the actual SQT plot utilized to approve the site for RF engineering.

The SA should also include a copy of the proposed antenna installation configuration used by equipment engineering and construction to evaluate the feasibility of the proposed installation. The antenna configuration drawing is also used as part of the lease exhibit information. A copy of the SA and the supporting documents should be stored in a secure central filing location so that all the information pertaining to this location and search is stored in one location and not distributed among many people's cubes.

RF Engineering

Search area code:_____   Capital funding code:_____
SAF document #:_____
Site address:_____
_____
_____

Latitude:_____   AGL:_____
Longitude:_____   AMSL:_____

Regulatory issues:
    1. FAA analysis attached: (Y/N)
    2. FAA lighting/marking required: (Y/N)
    3. FCC contour extension required: (Y/N)

Site-specific information:
    1. Antenna configuration attached: (Y/N)
    2. Radio equipment location defined: (Y/N)
    3. Radio equipment location sketch attached: (Y/N)
    4. Radio equipment type: _____
    5. Antenna structure: (roof, tower monopole, water tank)
    6. Equipment room: (prefab, interior fitup, exterior)
    7. Approximate cable length:_____

Type and quantity of antenna

Sector	Type	Quantity	Orientation	ERP

    8. Existing transmitters on structure: (Y/N)
    9. If (yes), state frequency, ERP, call signal, and physical location for each

Qualification information:
    1. SQT document #:_____
    2. Design propagation plot attached: (Y/N)
    3. SQT drive test plot attached: (Y/N)
    4. Site type
        Coverage (Y/N)
        Capacity (Y/N)
    5. IMD study completed: (Y/N)
    6. Site particular comments:_____
    _____
    _____

SAF
Document #:_____   Date:_____
Design engineer:_____   Reviewed by:____

Figure 10.15 Site acceptance form.

10.14   Site Rejection

In the unfortunate event that a potential site has been tested and is determined not to be suitable for potential use in the network, a site rejection form needs to be filled out. This may seem trivial until there is a change of personnel and the site is tested again at a later date. The site rejection form serves several purposes. The first is that it formally lets the real estate acquisition team know that the site is not acceptable for engineering and they need to pursue an alternative location. The second is that this process identifies why the site did not qualify as a potential communication site. The third ties into future use where the SQT data is stored and can be used for future system designs when the site might be more favorable for the network.

It is recommended that the site rejection process include a design review with a sign-off by the manager. This is to ensure that the reasons for rejecting the site are truly valid and the issues are properly communicated. The form proposed in the SR needs to be distributed to the same parties that the SA would be sent to (Fig. 10.16).

The fact that a site does not meet the design criteria specified at this time in the network design does not mean it will always be unsuitable. Therefore, it is imperative that the SQT information collected for this site be stored in the search area master file. This will assist efforts in later design issues that could involve capacity or relocation of existing sites to reduce lease costs.

10.15   Site Activation

The activation of a cell site into the network is exciting. It is at this point that the determination is made for how effective the design of the cell site is in

---

RF Engineering

Search area code: _____

The (name of test location) was tested on (date of test) and did not meet the design criteria for the search area defined.

The test location did not meet the design criteria for the following reasons (state reasons).

_____

_____

_____

_____

RF engineer: _____

Engineering manager: _____

Figure 10.16   Site rejection form.

resolving the problem area. There are numerous steps after the site acceptance process previously listed which need to take place. The degree of involvement with each of these steps is largely dependent upon the company resources available and the interaction required between the engineering and construction departments. As a minimum these two groups should perform site visits together involving the group responsible for the cell site architectural drawings and overall design of the site structure. Regardless of the interaction between the groups, when it comes to "show time" it is imperative to have a plan of action to implement.

10.16   FCC Guidelines

The FCC guidelines are unique for each type of radio service deployed. They should be known by the design engineer and the technical management of the company. All the FCC guidelines affecting the radio community are contained in various parts of the Code of Federal Regulations Title 47, CFR 47. Which section of CFR 47 applies to the particular system at hand depends on the license that the operator is utilizing to provide service. Each different part of CFR 47 dictates what can and cannot be used for the particular service. However, with the continuous flow of regulator changes that take place it is strongly advisable to obtain the most recent CFR 47 and check the FCC's web page for dockets. For example, CFR 47 part 22 applies to cellular, part 24 to PCS, and part 90 to SMR operators. With the rewrite of part 22 of the Code of Federal Regulations Title 47, CFR 47, and part 24 the paperwork required by the FCC is significantly reduced. However, the reduced paperwork the government requires does not eliminate the fundamental issue of ensuring that the system remains within FCC compliance rules. Most current operators no longer need to file with the FCC when they activate or modify a cell site in the network provided it does not change the outer contour of the network itself.

The potential problem with not being required to file for everything you do can lead to a potential compliance issue in the future. It is highly recommended that an internal filing process with sign-offs be put in place following the same process, with the exception of public filings, for filing a contour extension. The objective with following this methodology is to ensure that the operator remains in compliance and that in the event of an audit the proper paperwork for the system is in place for every site.

The FCC filings should be stored not only in the legal department but also with the site-specific files for each site. The master control number for each of the sites should be the key source reference number used followed by a date for a definer. The completion of the FCC form needs to include a sign-off process to, again, ensure that the overall process is being adhered to. It is also suggested that the proper paperwork is completed to ensure that no site is activated or modified without completing the FCC process.

The legal compliance issue is not the only reason why you should adhere to producing an internal filing for every site added or modified. In many communication companies the source documents specifying the current con-figuration of a site tend to be scattered about the engineering department in an unorganized manner. The purpose of following a procedure is to pre-vent this from occurring and to enable new and existing engineering and operations personnel to determine, without visiting the site, its major design attributes. Following the process will ultimately expedite the design or troubleshooting of the site.

Figure 10.17 covers items to be documented and the level of detail required. The FCC form 600 should be used as the reference document, with particulars filled out for a new site. However, the information in Fig. 10.17 is the minimum you should have available. Failure to remain in compliance can and will result in fines to the company posing both internal and external problems.

10.17   FAA Guidelines

FAA compliance is mandatory for all the sites within a system. Although you do not need to file every site with the FAA, it is necessary to ensure that every

Date:					
Cell Name:					
Activation/Change Date:					
Site Revision Number: 001					
Latitude		Longitude			
ASML AGL					
Tallest Point on Structure:					
Type of Cell:					
Antenna	Sector	Orientation	Cantination	Antenna Type	ERP
0					
1					
2					
3					
401 Contour Required (Y/N) If Yes Attach Contour					
FAA Study					
Lighting/Marking Required (Y/N)					
Colocated Transmitters (Y/N)					
Call Sign					
ERP					
EMF Power Budget Attached (Y/N)					

Figure 10.17   Site information form.

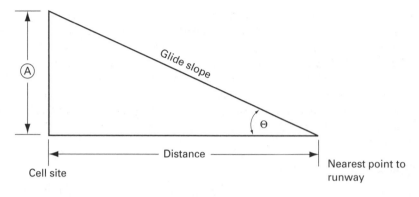

NOTE:

A = tallest point on structure

Glide slope = distance (feet)
                    factor

Factor = 100   if runway is > 3200 ft and within 20,000 ft of cell site
              50   if runway is < 3200 ft and within 10,000 ft of cell site
              25   if runway is a heliport and within 5000 ft of cell site

Figure 10.18   Glide slope calculation.

site is within compliance. The files for ensuring compliance are listed in CFR 47, part 17. The verification of whether the site is within FAA compliance should be covered during the design review process. If a site does not conform within the FAA guidelines, a potential redesign might be in order to ensure FAA compliance. For example , in one actual case the lowering of the height of the antenna on a rooftop by 2 ft made the site FAA compatible.

The overall key elements that need to be followed for compliance are:

1. Height
2. Glide slope
3. Alarming
4. Marking and lighting

Verification of the height and glide slope calculations is needed for every site. It is recommended that every site have the FAA compliance checked and included in the master site reference document. If there is no documented record for a site regarding FAA compliance, it is strongly recommended that this be done immediately. The time and effort required to check FAA compliance could be done within a week for a several hundred cell system.

An example of a glide slope calculation is shown in Fig. 10.18. The actual glide slope calculation needs to be included with the site acceptance form to ensure that the process is done.

A sample of when the FAA must be notified of an impending communication site is shown in Figs. 10.19 and 10.20. The figures are meant as a guideline to follow when installing antennas on an existing building. Note that in the illustrations below it is assumed that the sites referenced do not violate glide slope requirements. If the structure is in the glide slope path, it is prudent to register the site. Lastly it is stressed that the communication facility if requiring registration receive the proper clearance and/or requirements prior to the erection of the facility that forced the registration.

10.18   Planning and Zoning Board

Preparing for a zoning or planning board should be part of the design review process. Not only is the presentation important, but it might be possible with a modification to the original site design to eliminate this process entirely. In many actual cases a modification to the site design would have eliminated the need to request a variance from the town and thus prevented massive delays in the cell site build program. While this step seems obvious, checking local ordinances and incorporating them into the design process is rare and more times than not a "forgotten child" in the design process. It is recommended that the local ordinances for the site be included with the site source documents.

When it comes time to present the case of why the site is needed to the local planning and often zoning boards, a well-rehearsed presentation is needed. It is recommended that the program is rehearsed prior to the meeting night, as is most times the case, to ensure that everyone knows what each one is going to say and when.

Engineering's role in the process tends to focus on why the site is needed and health and safety issues associated with electromagnetic fields (EMF). The items that should be presented or prepared for should include as a minimum:

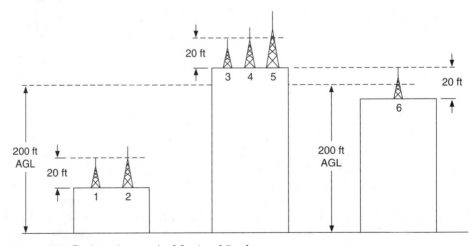

Figure 10.19   Registration required for 4 and 5 only.

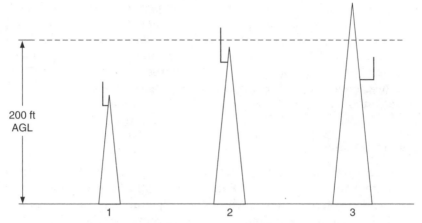

Figure 10.20 Registration required for 2 and 3. Note registration for 3 is needed regardless of where antenna(s) are installed.

1. Description of why the site is needed
2. How the site will improve the network
3. Drawing of what the site will look like
4. Views from local residences
5. EMF compliance chart
6. EMF information sheets and handouts for the audience

Before the meeting it is essential that the local concerns be identified in advance so they can be specifically addressed before or at the meeting. It is also recommended that the presentation be focused at the public and the board members, not just the board members. As in every case it is imperative that all the issues needed to launch a successful appeal for a negative ruling by the council be covered. The overall preparation for the meeting is essential, since the comments made by the company employees or consultants are a matter of public record and will be used solely for the appeal process.

10.19 EMF Compliance

EMF compliance needs to be factored into the design process and continued operation of the communication facility. The use of an EMF budget is strongly recommended to ensure personnel safety and government compliance. A simple source for the EMF compliance issue should be the company's EMF policy.

An EMF power budget (Fig. 10.21) should be incorporated into the master source documents for the site and stored on the site itself, identifying the transmitters used, power, who calculated the numbers, and when it was last done. As a regular part of the preventive maintenance process the site should be checked for compliance and changes to the fundamental budget calculation.

The method for calculating the compliance issue is included in the IEEE C95.1-1991 specification with measurement techniques included in C95.3. Both cellular and PCS utilize the same C95.1-1991 standard.

Figure 10.21 is a sample EMF power budget. It should be signed off by the manager for the department and shared with the operations department of the company. An EMF budget needs to be completed for every cell site in operation and also for those proposed.

10.20   Frequency Planning

Frequency planning, or rather frequency management, is an integral part of the system design. Frequency planning is a critical function for all wireless communication systems. How much frequency planning there is in a network is largely determined by the technology platform chosen by the operator. There are many variants to frequency planning, ranging from coordination of a single transmit channel to orchestrating the manipulation of hundreds of radio channels. Within the cellular and PCS arenas the amount of frequency planning can range from segmentation of the available spectrum to defining the different PN short codes for CDMA.

All too often it is the frequency planner who sets the direction for the performance of a communication system. The frequency planning process needs to be rigorously checked on a continuous basis to always refine the system. As a minimum for frequency planning, the designs, no matter how minor they seem, need to be passed through a design review process.

Frequency planning as a general rule is more of an art form than a defined science. C/I ratios, E_b/N_0, and BER rates are just some of the issues which help define the frequency planning criteria. In addition the choice of an $N = 3$, $N = 4$, $N = 7$, $N = 9$, $N = 12$, or $N = 19$ pattern is more clinical. The reuse pattern is primarily driven on the technology platform chosen and the infrastructure configuration used. However, how you go about defining what the frequency plan design trade-offs are for an area is an art. Some engineers are better artists than others, but the fundamental issue of controlling interference and how well it is done falls on the design review process.

Numerous technical books and articles, some of which are listed at the end of this chapter, go over how to frequency plan a network from a theoretical standpoint. It is very important to understand the fundamental principles of frequency planning in order to design a frequency management plan for any network. Failure to adhere to a defined frequency design guideline will limit the system's expansion capability.

The rationale behind defining the RF design process as an art is the multitude of perturbations available for any given frequency management plan. Several methods are available for use in defining the frequency management of a network. The method chosen by the mobile carrier needs to be factored into the frequency management scheme capacity requirements, capital outlays, and adjacent market integration issues, to mention a few. Obviously the

Cell:
Date:

Sector 1:

Number of channels	19
ERP/channel	100 W
Total power	1900 W

Colocated transmitters:

Paging (931.875 MHz)	
ERP	1000 W

Data points:

Location 1	Distance	Total power	Power density	Max for band	% budget
Cell site	25 ft	1900 W	0.260527768	0.586666667	44%
Paging	20 ft	1000 W	0.214249809	0.62125	34%
				Total	**79%**

Location 2	Distance	Total power	Power density	Max for band	% budget
Cell site	100 ft	1900 W	0.016282986	0.586666667	3%
Paging	110 ft	1000 W	0.007082638	0.62125	1%
				Total	**4%**

Figure 10.21 EMF power budget.

method used for the frequency plan also has to ensure that the best possible *C*/*I* ratio is obtained for both cochannel and adjacent channel RF interference.

The use of a grid is essential for initial planning, but when "sprinkled with reality" the notion is academic because the site acquisition process tends to drive the system configuration and not the other way around. It is dealing with the irregularities of the site coverage, traffic loading, and configurations that requires continued maintenance of the network frequency plan.

The specific channels available for cellular systems to operate with are defined based on the license held, either A or B band. In the United States an operator can operate only in either the A or B blocks and has available for its use the spectrum range shown in Fig. 10.22. The spectrum chart shown is not contiguous, meaning other wireless operators are using parts of the spectrum adjacent to cellular receive and transmit.

The spectrum allocated for PCS operation in the United States is shown in Fig. 10.23. The various frequency planning methodologies are driven by the technology platform chosen. The specific frequency planning methods are covered in a later chapter because of the number of possible variants.

10.21   Site Activation

The philosophy of site turn-on (activation) varies from company to company. Some of the philosophies are driven by engineering, others by financial objectives. Several philosophies are used in the wireless industry. The first is that when a site is finished being constructed it should be activated into the network. The second is that the site's depreciation should be minimized or maximized, depending on the accounting method employed by the company. In the third the site or sites are not activated until the implementation plan put forth by engineering dictates the timing of the new cells. A fourth site activation method involves a combination of philosophies two and three.

The first philosophy of cell site activation, turn it on immediately, has an emotional and upper management appeal. The appeal is that the site is being constructed to resolve some system problem and the sooner it is put into service the sooner the problem the site is designed for will be resolved. While this simple philosophy has its direct merit, it also has a few key drawbacks which if done incorrectly can create more system problems than it was intended to fix.

Specifically the drawback involves timing and coordination of the engineering plan to bring the site into the network. If site A requires handoff changes and is activated when the last ATP function is completed, the possi-

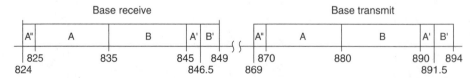

Figure 10.22 U.S. cellular spectrum.

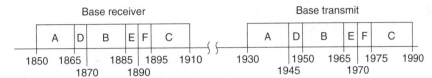

Figure 10.23 U.S. PCS spectrum.

bility of the handoff changes being implemented at this time is low. One alternative to this is to have the topology changes done in advance of the site's activation, but if it is done incorrectly handoff problems and possible lost calls could result. Another situation could occur where there are too many handoff candidates in the topology tables of the sites, complicating the frequency plan for the area. Another major problem with this activation philosophy is the coordination of resources. As in most cases a site needs some level of inner- and intersystem coordination; if the site is activated at a seemingly random time there is no guarantee that all the required coordination has been completed.

An additional disadvantage with this method is the queuing of post-turn-off resources for the system troubleshooting phase of a site turn-on. The post-turn-on efforts, for their full potential, require coordination in terms of timing. If a site is to be turned on anywhere within a 3-day window based on implementation problems, it is difficult to ensure that post-turn-on testing will begin right after turn-on. If you want to ensure that post-turn-on testing begins right after turn-on of the cell, however, additional opportunity costs will be associated with this effort, since resources will be significantly mismanaged.

Still another disadvantage with this effort is the most important aspect, the customer impact. By turning on when finishing the implementation process, this will most likely occur during the day. It is strongly advised that any site or major system action be conducted in the system maintenance window. The simple objective here is to minimize any negative impact the subscriber might experience and try to allow enough turnaround time for the engineering and operations teams to correct any problems before the subscribers find them.

The second philosophy of activating the sites to maximize or minimize depreciation costs is largely driven by the financial requirements of the company. In maximizing depreciation costs the operator usually scrambles to activate as many sites as possible by the end of the fiscal year, usually the calendar year. The objective here is to maximize the potential depreciation expense the company will have in any fiscal year.

The minimization of depreciation philosophy involves attempts to defer the depreciation of the new cell site into the next fiscal year. What is typically done here is that a site is prepared for activation into the network but will not be turned on until the next fiscal year. The objective here is to minimize the amount of expense reported by the subsidiary to its parent company. Usually this philosophy is "turn it on now." It involves a plan issued by engineering which matches the financial goals of the company.

The third philosophy for site activation is where no new cell site, or system change, is done without a plan being issued and approved by engineering. The objective here is to ensure that the introduction of the new cell into the network has been sufficiently thought out, resources are planned and staged, and all the coordination required has been or will be done in concert with the activation. This philosophy is essential to ensure that new cells are introduced into a mature system gracefully. For this process usually more than one cell site is activated at the same time, if logistically possible. This philosophy enables a maximization of post-turn-on resources to focus on the issues at hand and also ensures the minimization of system alterations required. The map in Fig. 10.24 shows how combining several cell sites into a single turn-on for the system will facilitate maximizing resources and minimize the number of changes required to the network.

The largest problem associated with trying to utilize this philosophy is getting upper management approval. The opposition occurs when a site may have to wait several weeks for activation because of configuration changes needed in the network to ensure its smooth transition. The issue of having a site ready for service and not activating it immediately is the hardest obstacle to overcome when presenting the case. However, several key advantages need to be stressed with this philosophy which might or might not be apparent:

1. Reduced system problems
   Interference
   Handoffs
   Parameter settings

Figure 10.24 Group activation.

2. Coordinated efforts among all departments

3. Design reviews of integration plan

4. Inner- and intersystem coordination more fluid

5. Pre-turn-on testing conducted

6. Minimized negative customer impact through activation in maintenance window

There are obviously more positive attributes; however, the key issues are reduced system problems, pre-turn-on testing, and minimization of negative customer impacting issues as a result of following a plan. It is essential that for every cell site brought into a network a plan is generated for its introduction and the plan is then carried out.

The design reviews necessary for a new site activation into the network need to be conducted by several parties. There are several levels of design reviews for this process. The first level of design reviews involves the RF engineer and the performance engineer discussing the activation plans and reviewing the plan of action put forth. The second level of design reviews involves having the manager of the RF engineering group sign off on the implementation design with full concurrence with the performance manager. The third level of design review involves reviewing the plan with the director for engineering and operations personnel to ensure that all the pieces are in place and that something has not been left out, like who will do the actual work.

After the design reviews are completed, the MOP for the activation is released. It should be noted that during the design phases the MOP should have been crafted and all the parties involved informed of their roles. A sample MOP is listed below for comparison; obviously the exact MOP for the situation is different and needs to be individually crafted.

Method of Procedure for New Cell Site Integration

Preactivation Process
Date

X-X-XX	New cell sites to be activated defined
X-X-XX	Project leader(s) defined and timetables specified as well as the scope of work associated with the project
X-X-XX	Phase 1 design review (frequency planning only and RF engineer for site)
X-X-XX	Phase 2 design review (all engineering)
X-X-XX	Phase 3 design review (operations and engineering)
X-X-XX	Phase 4 design review (adjacent markets if applicable)
X-X-XX	Frequency assignment and handoff topology sheets given to operations
X-X-XX	New cell site integration procedure meeting
X-X-XX	Performance evaluation test completed
X-X-XX	Executive decision to proceed with new cell site integration
X-X-XX	Adjacent markets contacted and informed of decision
X-X-XX	Secure post cell site activation war room area
X-X-XX	Briefing meeting with drive test teams
X-X-XX	MIS support group confirms readiness for postprocessing efforts
X-X-XX	Customer care and sales notified of impending actions

New Cell Site Activation Process (begins X-X-XX at time XXXX)

X-X-XX   Operations informs Key personnel of new cell site activation results

Operations personnel conduct brief post-turn-on test to ensure call processing is working on every channel and that handoff and handins are occurring with the new cell site

Operations manager notifies key personnel of testing results

Post-Turn-On Process (begins X-X-XX at time XXXX)

Voice mail message left from engineering indicating status of new cell sites (time)

Begin post-turn-on drive testing phase 1 (time)

Database check takes place

Statistics analysis takes place

Voice mail message left from RF engineering indicating status of post-retune effort (time)

Phase 2 of post-turn-on drive testing begins

Commit decision made with directors for new cell site (time)

Phase 3 of post-turn-on drive testing begins

X-X-XX

Continue drive testing areas affected

Statistics analysis

Conduct post-turn-on analysis and corrections where required

X-X-XX

Post-turn-on closure report produced

New site files updated and all relevant information about the site transferred to performance engineering

It is essential to always include a backout procedure for cell site activation in the event of a major disaster. The escalation procedure should be defined in the MOP and the decision to go or not go needs to be at the director level, usually the engineering director or the operations director.

After the MOP is released and the design reviews are completed it is essential that the potential new cell be visited by the RF and performance engineers at various stages of the construction period. However, prior to activation it is essential that a pre-turn-on (PTO) take place. The PTO is meant to ensure that the site is configured and installed properly so that when the site is activated into the network the basic integrity of the site is known. The PTO procedure that should be followed is listed later in this chapter.

Internal coordination involving a new site or sites being introduced into the network is essential. The MOP listed above focuses on voice mail notifications to many groups inside and outside the company. However, it is essential that the activation of new cells and major system activities be announced to other departments in the company to inform them of the positive efforts being put forth by engineering and operations.

The primary groups to ensure that some level of notification takes place are

1. Sales

2. Marketing

3. Customer service

4. Operations, real estate, and engineering

5. Corporate communications

6. Legal and regulatory

Primarily the entire company needs to be notified of the positive events that take place. One of the most effective methods is through the company's internal voice mail system associated with their extensions. However, not everyone will have an individual extension, and a voice mail account should therefore be available.

To ensure that all the people are notified of the new site activation into the network, a series of communications can be accomplished. One method is to issue an electronic mail message to all the employees notifying them of the new sites and any particulars about the intended improvements, if any, to the network. Another method is to slay the trees by issuing a memo to everyone in the company declaring the activation of the sites and the improvements that have arrived.

External coordination for new sites is as essential as internal coordination. Specifically the neighboring systems should and need to know when you are bringing new sites into the network and other major activities. The reason behind this effort is that your actions may have an unintended consequence, either positive or negative, which they need to know. In the same light, by providing your neighboring systems with new site activation information the same level of communication can be reciprocal.

After the site(s) are activated into the network it is essential that post-turn-on testing begins immediately. There has never been a site activated into a network, that I am aware of, which did not have some type of problem with it. Therefore, it is essential that the efforts put forth in this stage of the site activation process receive as much attention as the design phases did.

The key parameters or factors which need to be checked as part of the post-turn-on activities are:

Site configuration checks

Metrics analysis

Drive test analysis

1. *Site configurations from the switch's point of view.* The objective here is to check all the cell site parameters for the site as reported by the switch to those intended for the initial design. What you are looking for here is a possible entry mistake or even a design mistake made during the design process. Usually a fat finger mistake is found in this process or an entry is left out. It is imperative that the neighbor cell sites also be checked in this stage of the process.

2. *Metrics analysis.* The objective with this part of the post-turn-on activities is to help identify and isolate for problem resolution problems reported in the network by the system statistics. This process requires continued attention to detail and an overall view of the network at the same time. The metrics that you should focus on involve the following items:

Lost calls
Blocking
Usage
Access failures
BER/FER
Soft handoff percent
Customer complaints
Usage/RF loss
Handoff failures
RF call completion ratio
Radios out of service
Cell site span outage
Reported field problems called in by technicians or the drive test team

The statistics monitored should be the primary site(s) activated and their neighboring cells. The issue here is to not only look at the sites being brought into service but also to ensure that their introduction did not negatively impact the system.

The actual numbers to use for comparison need to be at least one week's prior data for benchmarking, if possible. In addition the numbers used for the new cell should be compared against the design objective to ensure that the site is meeting the stated design objectives.

3. *Drive testing.* The post-turn-on drive test data analysis needs to take place here. This effort usually begins at the specified time after turn-on, usually early in the morning or late at night, depending on the activation schedule. The drive tests are broken down into three main categories.

Phase 1 of the driving involves focusing on areas where there is the highest probability of experiencing a system design problem. The identification of these areas can be through prior experience, C/I plots, or SWAG.

Phase 2 of the drive testing involves targeting the rest of the areas involved with the site activation activities, usually the remaining class 1 and class 2 roads not already driven.

Phase 3 of the drive testing involves driving areas that were uncovered as problems in phases 1 and 2. This level of testing either verifies that the problem identified previously is still in existence or that the change introduced into the network did its job.

4. *New site performance report.* The last stage in the new site activation process is the issuance of the new site performance report. This will include all the key design documents associated with the new site. They should be stored in a central location instead of a collection of people's cubes. The information contained in the report is critical for the next stage of the site's life, which involves ongoing performance and maintenance issues.

To ensure that poor designs do not continue in the network, it is essential that the new site meets or exceeds the performance goals set forth for the network. If the site does not meet the requirements set forth, it should remain in the design phase and not the ongoing system operation phase. The concept of not letting the design group pass system problems over to another group is essential if your goal is to improve the network.

The new site performance report needs to include the following items in it as the minimum set of criteria.

1. Search area request form
2. Site acceptance report
3. New cell site integration MOP
4. Cell site configuration drawing
5. Frequency plan for site
6. Handoff and cell site parameters
7. System performance report indicating the following parameters one week after site activation:
   a. Lost calls
   b. Blocking
   c Access failures
   d. Customer complaints
   e. Usage/RF loss
   f. BER/FER
   g. Soft handoff percent
   h. Handoff failures
   i. RF call completion ratio
   j. Radios out of service
   k. Cell site span outage
   l. Technician trouble reports
8. FCC site information
9. FAA clearance analysis
10. EMF power budget
11. Copy of lease
12. Copy of any special planning or zoning board requirements for the site

The new cell site performance report is an essential step in the continued process for system improvements. Only once a site is performing at its predetermined performance criteria should the site transition from the design phase to the maintenance phase.

10.22   RF System Design Report

The following is an example of a RF system plan that you can use as a boiler plate for putting together your own report. The report is based on one that deals with an existing system. When crafting a particular report that is applied to a new system, issues associated with the new system can easily be crafted into the report. These are included in Chap. 11.

RF Design for System X

1. *Executive summary.*   The attached RF system design study is a continuation of a series of quarterly reports which define the growth requirements for

the system. The results from the analysis conducted indicate that for the next year a total of 75 macrocells are required to support the expected expansion of the network. In addition to the cell site expansion requirements a new switch is required and is proposed to be colocated with the existing MTSO location 3.

2. *Introduction.*   This study was performed as part of the ongoing effort to determine the RF system requirements necessary to support the overall subscriber growth projections, performance improvements, and marketing-driven initiatives. It is the latest in a series of quarterly update reports which define the capital and expense requirements of system X. This report spans the 1996 to 1998 time frame broken down on a quarterly basis.

The particular components analyzed in the study include cell sites, mobile switching center (MSC), facility requirements, and all the other auxiliary systems which comprise the system X market. The network interconnect requirements were included in this study. This document is structured in a format that is representative of the steps or building blocks in which it was developed.

3. *Design criteria*

Customer growth rate	45%	1996–1997
	40%	1997–1998
Baseline date	June 1995	
Traffic		
Cell site	Average of 10 busiest days per sector/month	
CPU	Average of 10 busiest days/month	
Subscribers	75,000	
Cell sites	170	
Radios	5810	
	Switches	2
Dilution	5%	1996–1997
	10%	1997–1998
Max radios/sector	19	($N = 7$)

1. The cells identified in the construction status chart, projected to be operational within the next 6 months, were used as part of the assumptions.
2. Coverage requirement sites identified by marketing and system performance were also included in the analysis.
3. Usage and subscriber dilutions throughout the network would remain in the same proportion for the entire study period.

4. *Cell site analysis*

Analysis. The growth projections identified in Section 2 were applied to the baseline sector data and used to determine the radio additions required for each quarter during the study period. Once the radio limit (19) was reached on a particular sector, a new site (search area) would be generated to offload the overloaded sector and support future growth. In addition to the trigger site offloading, all adjacent cells that would be affected by the new site were analyzed and deloaded. After the capacity sites were identified, the strategic coverage sites were added and their impact to the adjacent cell was factored in. The radio requirements for each of the sites were then determined.

Link budget. The attached is the link budget(s) and their associated assumptions that were used for the RF system design.

Coverage maps. Attached are propagation plots showing the various stages of the RF design. The first plot is that of the system as it exists today. The second plot shows only the cells that are projected to be added into the network as defined by the RF system design. The third plot is a composite plot indicating the system configuration at the end of the current design phase which this report encompasses.

5. *Summary of requirements.* As of June 30, 1996, the system consisted of 170 cell sites. Twenty-one additional sites are planned for the next 6 months. To accommodate the projected growth in usage, an additional 34 capacity sites would be required over the next year. The portable coverage requirements for 1997 trigger an additional 20 coverage sites.

The results of the analysis included: (1) the quantity and locations of new sites, and (2) the quantity of radios per sector and total required for each quarter during the study period. The associated chart with the various breakdowns is included for reference.

System X									
	4Q95	1Q96	2Q96	3Q96	4Q96	1Q97	2Q97	3Q97	4Q97
Area 1	1	3	3	5	7	4	3	2	3
Area 2	1	4	4	5	3	2	2	3	1
Area 3	2	5	4	2	4	2	1	5	1
Area 4	2	3	1	3	2	4	2	1	5

6. *Network capital requirements.* The capital requirements necessary to support the network and RF growth plan depicted in this study are summarized in the table below.

	4Q96	1Q97	2Q97	3Q97	Total
Cell site (new)	$1,269.0	$7,321.0	$ 1030	$5219	$ 8,250
Cell site (growth/mod)	$ 660.0	$ 740.0	$ 960.0	$ 420.0	$ 2,780
Total	$1,929.0	$8,061.0	$1,990.0	$5639.0	$11,030

10.23   Presentation

When presenting the material to upper management and your fellow engineers, it might be beneficial to craft two versions of the presentation. One presentation should be prepared that will be given to the technical departments as a whole. The other presentation is meant for upper management to view. Both presentations should involve a combination of visual aids and handouts. The visual aids should consist of several free-standing charts depicting the current and future network configuration. Additional visual aids include the use of overhead slides describing the key attributes of the plan. The handouts distributed should reflect the exact same information that is shown in the overhead projections.

The engineering presentation should take about 2 to 4 hours to present, depending on whether it is a quarterly update or a yearly plan. The material presented should be of sufficient detail to ensure that all the departments within the technical organization understand the general implications of the information in the report. It is recommended that a member of each department, usually a member on the growth plan, provide an option of their group's projects planned over the report period. The discussion of the various subplans should also include a projected start and end date for each of the topics discussed.

The upper management report is very similar in nature to the presentation that is given to the technical groups of the company. The primary difference in the presentations is the time frame and emphasis of material. The upper management presentation needs to last about 1 hour as the extreme upper limit. The use of visual aids is the proper method to use for this discussion.It is recommended that the technical aspects take on a very high level approach and not focus on details, unless instructed to do so. The presentation should include a few charts showing the growth trends of the network, critical triggers for the network, new cell sites expected, and cash requirements. The props used for the discussion with technical departments should also be presented as background material.

References

1. Smith, Clint, and Curt Gervelis, *Cellular System Design and Optimization,* McGraw-Hill, New York, 1996.
2. *Antenna Structure Registration,* FCC, Washington, D.C., May 1996.
3. AT&T, *Engineering and Operations in the Bell System,* 2d ed., AT&T Bell Laboratories, Murray Hill, NJ, 1983.
4. Code of Federal Regulations, CFR 47, Parts 1, 17, 22, 24, and 90.
5. Kaufman, M., and A. H. Seidman, *Handbook of Electronics Calculations,* 2d ed., McGraw-Hill, New York, 1988.
6. Lee, W. C. Y., *Mobile Cellular Telecommunications Systems,* 2d ed., McGraw-Hill, New York, 1996.
7. White, Duff, *Electromagnetic Interference and Compatibility,* Interference Control Technologies, Gainesville, GA, 1972.
8. Yarborough, *Electrical Engineering Reference Manual,* 5th ed., Professional Publications, Belmont, CA, 1990.
9. Fink, Donald, and Donald Christiansen, *Electronics Engineers Handbook,* 3d ed., McGraw-Hill, New York, 1989.
10. Fink, Beaty, *Standard Handbook for Electrical Engineers,* 13th ed., McGraw-Hill, New York, 1995.
11. Jakes, W. C., *Microwave Mobile Communications,* IEEE, New York, 1974.
12. Johnson, R. C., and H. Jasik, *Antenna Engineering Handbook,* 2d ed., McGraw-Hill, New York, 1984.
13. Lathi, *Modern Digital and Analog Communication Systems,* CBS College Printing, New York, 1983.
14. MacDonald, "The Cellular Concept," *Bell System Technical J.,* vol. 58, no. 1, 1979.
15. *Reference Data for Radio Engineers,* 6th ed., Sams, 1983.

System Design Guidelines

11.1 Introduction

This chapter provides a sample guideline for designing wireless communication systems utilizing different technologies. The intention is not to provide a comparison of one technology versus another, since that is a decision that is ultimately left to the operating company. However, with the plethora of technologies that exist in the marketplace right now it is very important to refrain from obtaining a religion regarding a technology platform.

The format of this chapter is to first provide the foundation for putting together a system. It can be used for an existing or new system. The approach here can easily be tailored to the specifics of the market at hand. As more systems become operational in the same market and demands for service change, the design guidelines here will need to reflect those alterations. However, a few steps must be taken regardless of whether the system is a new system, a retrofit system, or simply the continued expansion of the network.

The first stage of the system design is to follow the system design methodology.

1. Define the objective.
2. Identify the variables.
3. Isolate system components.
4. Design system.
5. Present report (design review).
6. Communicate design to respective departments.
7. Implement design.

The process for designing a system involves the general flow outlined below. Obviously there are many variants to what is discussed below, but the general gist is no matter what the process flow your organization or situation warrants the system design methodology should be followed. However, the process is slightly different for a new system than it is for an existing system. The issues, however, that are common to either a new or existing system involve the following items:

1. Define marketing requirements.
2. Methodology.
3. Technology decision.
4. Define the types of cell sites.
5. Establish a link budget.
6. Define coverage requirements.
7. Define capacity requirements.
8. Complete RF system design.
9. Issue search area.
10. Site qualification test (SQT).
11. Site acceptance/site rejection.
12. Land use entitlement process.
13. Integration.
14. Hand over to operations.

The primary difference between an existing system and a new system is that the existing system obviously has some traffic loading and performance parameters associated with the system that are based on actual data, not estimates.

The three fundamental variants to system designs for a wireless communication system involve:

1. Existing system expansion
2. New system design
3. Introduction of new technology platform to existing system

The focus of this chapter is on the establishment of new systems. The establishment of new communication systems can also incorporate the introduction of new technology platforms into an existing market. The new technology platform introductions are in fact a new type of system design.

New system

1. Obtain marketing plan and objectives.
2. Establish system coverage area.
3. Establish system on-air projections.
4. Establish technology platform decisions.
5. Determine maximum radius per cell (link budget).

6. Establish environmental corrections.

7. Determine desired signal level.

8. Establish the maximum number of cells to cover area.

9. Generate coverage propagation plot for system.

10. Determine subscriber usage.

11. Determine usage/per square kilometer.

12. Determine maximum number of cells for capacity.

13. Determine if system is capacity- or coverage-driven.

14. Establish total number of cells required for coverage and capacity.

15. Generate coverage plot incorporating coverage and capacity cell sites (if different).

16. Reevaluate results and make assumption corrections.

17. Determine revised (if applicable) number of cells required for coverage and capacity.

18. Check number of sites against budget objective; if too many, reevaluate design.

19. Using known database of sites overlay on system design and check of matches or close match (,0.2R).

20. Adjust system design using site-specific parameters from known database matches.

21. Generate propagation and usage plots for system design.

22. Evaluate design objective with time frame and budgetary constraints and readjust if necessary.

23. Issue search rings.

Existing system

1. Obtain marketing plan.

2. Identify coverage problem areas.

3. Establish technology platform decisions (CDPD, etc.).

4. Determine maximum radius per cell (link budget).

5. Establish environmental corrections.

6. Determine desired signal level.

7. Establish the maximum number of cells to cover area(s).

8. Generate coverage propagation plot for system and areas showing before and after coverage.

9. Determine subscriber usage.

10. Allocate percentage of system usage to each cell.

11. Determine maximum number of cells for capacity (technology-dependent).

12. Establish which cells need capacity relief.

13. Determine new cells needed for capacity relief.

14. Establish total number of cells required for coverage and capacity.

15. Generate coverage plot incorporating coverage and capacity cell sites (if different).

16. Reevaluate results and make assumption corrections.

17. Determine revised (if applicable) number of cells required for coverage and capacity.

18. Check number of sites against budget objective; if too many sites, reevaluate design.

19. Using known database of sites overlay on system design and check of matches or close match ($<0.2R$).

20. Adjust system design using site-specific parameters from known database matches.

21. Generate propagation and usage plots for system design.

22. Evaluate design objective with time frame and budgetary constraints and readjust if necessary.

23. Issue search rings.

New technology platform introduction to existing system

1. Obtain marketing plan.

2. Establish technology platform introduction timetable.

3. Determine new technology implementation trade-offs.

4. Determine new technology implementation methodology.

5. Identify coverage problem areas.

6. Determine maximum radius per cell (link budget for each technology platform).

7. Establish environmental corrections.

8. Determine desired signal level (for each technology platform).

9. Establish the maximum number of cells to cover area(s).

10. Generate coverage propagation plot for system and areas showing before and after coverage.

11. Determine subscriber usage.

12. Allocate percentage system usage to each cell.

13. Adjust cell maximum capacity by spectrum reallocation method (if applicable).

14. Determine maximum number of cells for capacity (technology-dependent).

15 Establish which cells need capacity relief.

16. Determine new cells needed for capacity relief.

17. Establish total number of cells required for coverage and capacity.

18. Generate coverage plot incorporating coverage and capacity cell sites (if different).

19. Reevaluate results and make assumption corrections.

20. Determine revised (if applicable) number of cells required for coverage and capacity.

21. Check number of sites against budget objective; if too many, reevaluate design.

22. Using known database of sites overlay on system design and check of matches or close match ($<0.2R$).

23. Adjust system design using site-specific parameters from known database matches.

24. Generate propagation and usage plots for system design.

25. Evaluate design objective with time frame and budgetary constraints and readjust if necessary.

26. Issue search rings.

Figure 11.1 is an example of a general system coverage requirement to design a system for the area shown. Figure 11.2 is an example of receiving marketing information regarding areas requiring coverage and capacity.

Sometimes the data comes in the form of percentage of traffic or subscribers per area. Using the percentage of traffic method, Table 11.1 is constructed showing that for system busy hour the design is to support a total of 3500 erlangs distributed in the fashion indicated. The system erlang number that was used to compute the rest of the table often comes from the information about the subscriber population as a whole or segmented into different target audiences. If the data is an aggregate of the whole, the total system erlang number is found by simply multiplying the number of subscribers by their anticipated busy hour usage (i.e., 0.7 merlang per subscriber). See Table 11.2.

The next sections in this chapter focus on different technology platforms and some of the design issues associated with each.

11.2 AMPS

The design of an AMPS system from the ground up is almost a nonissue for most of the world since this technology has been in use for some time now. In some applications, however, in developing areas the potential still exists for

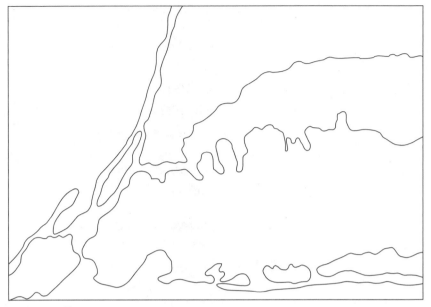

Figure 11.1 System area.

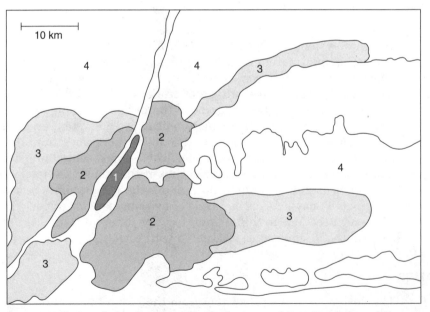

Figure 11.2 System design. Requirements for the geographic regions 1 through 4 are listed in Table 11.1.

TABLE 11.1

Region	Area, km^2	Traffic, %	Erlangs	Erlangs/km^2
1	100	20	700	7.0
2	1,500	40	1,400	0.933
3	2,500	30	1,050	0.420
4	5,900	10	350	0.0593
Total	10,000	100	3,500	0.35

500k subscribers at 7 merlangs/subscriber = 3500 erlangs, system busy hour.

TABLE 11.2

Region	Area, km^2	No. of subscribers	Erlangs	Erlangs/km^2
1	100	100,000	700	7.0
2	1,500	200,000	1,400	0.933
3	2,500	150,000	1,050	0.420
4	5,900	50,000	350	0.0593
Total	10,000	500,000	3,500	0.35

0.7 merlang/subscriber.

the design of an AMPS system. The effort here can also be used to help redesign an existing analog network. The process needed for an AMPS design involves a multitude of issues, most of which are covered in the system design methodology and guidelines listed previously. The addition of cell sites into an existing network is covered in Chap. 10.

11.2.1 Frequency

The AMPS system is meant to operate with 30-kHz-wide channels utilizing frequency modulation. The spectrum that AMPS occupies is shown in Fig. 11.3. The spectrum chart indicates that there are two operators, A- and B-band, and the respective spectrum is shown. The channel allocations for cellular are defined by the D/R ratio, which is a function of the reuse ratio. The more directional the site the smaller the D/R ratio can be. Therefore, if the cell had directivity that was selective enough to focus only on the subscriber utilizing the particular channel, the D/R ratio could be even lower than that of a $N = 4$ reuse pattern while maintaining the same or better C/I ratio.

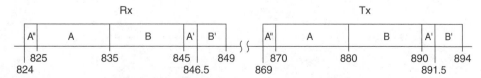

Figure 11.3 AMPS spectrum. SAT = 0, 5970 Hz; 1, 6000 Hz; 2, 6030 Hz. Setup channels: A band, 313–333; B band, 334–354. DCC = 0, 1, 2, 3.

Many different frequency allocation schemes exist today. Some of the more popular reuse schemes utilize either a $N = 12$, $N = 7$, or $N = 4$ pattern as the fundamental point that can be used to grow or change from (Figs. 11.4 to 11.6). The reuse pattern chosen for a system should factor into it the potential for growth in the network.

Specifically the growth issues, besides subscriber growth, involve deployment of alternative technology platforms on top of the existing system. Other alternative technology platforms which directly support analog involve the design of micro- and picocells in the network.

The common channel assignment charts are included in Tables 11.3 to 11.5.

11.2.2   Link budget

The link budget for the AMPS system is shown in Fig. 11.7.

11.2.3   Required cells

During the initial design phase of any system it is imperative that the determination be made regarding whether the system or parts of the proposed system are coverage- or capacity-driven. The capacity and coverage requirements should be driven from the marketing plan.

The system that will be first looked at is shown in Fig. 11.2. The requirements for the system in terms of geographic area to cover and capacity are listed in Table 11.1.

The next step is determining how many cell sites are required to meet the objectives. (See Example 11.1.) The two types of cells that will be used for this example are omni and sector (three-sector).

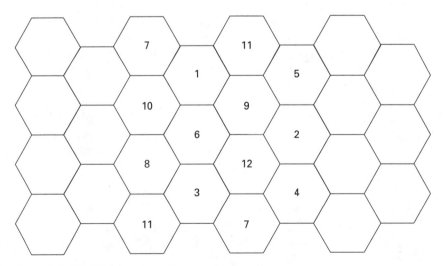

Figure 11.4 $N = 12$ frequency grid.

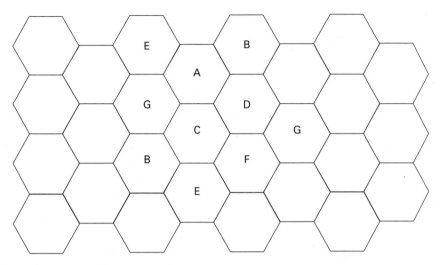

Figure 11.5 $N = 7$ frequency grid.

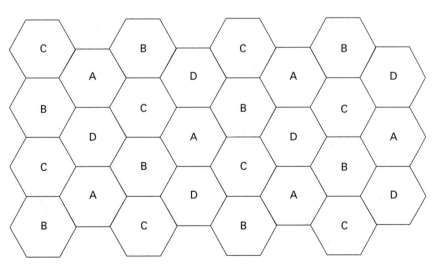

Figure 11.6 $N = 4$ frequency grid.

TABLE 11.3   FCC Channel Chart for $N=7$

Channel group:	A1	B1	C1	D1	E1	F1	G1	A2	B2	C2
			Wireline B-band channels							
Control channel:	334	335	336	337	338	339	340	341	324	343
	335	356	357	358	359	360	361	362	345	364
	376	377	378	379	380	381	382	383	366	385
	397	398	399	400	401	402	403	404	387	406
	418	419	420	421	422	423	424	425	408	427
	439	440	441	442	443	444	445	446	429	448
	460	461	462	463	464	465	466	467	450	469
	481	482	483	484	485	486	487	488	471	490
	502	503	504	505	506	507	508	509	492	511
	523	524	525	526	527	528	529	530	513	532
	544	545	546	547	548	549	550	551	534	553
	565	566	567	568	569	570	571	572	555	574
	586	587	588	589	590	591	592	593	576	595
	607	608	609	610	611	612	613	614	597	616
	628	629	630	631	632	633	634	635	618	637
	649	650	651	652	653	654	655	656	639	658
	717	718	719	720	721	722	723	724	725	726
	738	739	740	741	742	743	744	745	746	747
	759	760	761	762	763	764	765	766	767	768
	780	781	782	783	784	785	786	787	788	789
			Nonwireline A-band channels							
Control channel:	333	332	331	330	329	328	327	326	325	324
	312	311	310	309	308	307	306	305	304	303
	291	290	289	288	287	286	285	284	283	282
	270	269	268	267	266	265	264	263	262	261
	249	248	247	246	245	244	243	242	241	240
	228	227	226	225	224	223	222	221	220	219
	207	206	205	204	203	202	201	200	199	198
	186	185	184	183	182	181	180	179	178	177
	165	164	163	162	161	160	159	158	157	156
	144	143	142	141	140	139	138	137	136	135
	123	122	121	120	119	118	117	116	115	114
	102	101	100	99	98	97	96	95	94	93
	81	80	79	78	77	76	75	74	73	72
	60	59	58	57	56	55	54	53	52	51
	39	38	37	36	35	34	33	32	31	30
	18	17	16	15	14	13	12	11	10	9
	1020	1019	1018	1017	1016	1015	1014	1013	1012	1011
	725	724	723	722	721	720	719	718	717	716
	704	703	702	701	700	699	698	697	696	695
	683	682	681	680	679	678	677	676	675	674

D2	E2	F2	G2	A3	B3	C3	D3	E3	F3	G3
Wireline B-band channels										
344	345	346	347	348	349	350	351	352	353	354
365	366	367	368	369	370	371	372	373	374	375
386	387	388	389	390	391	392	393	394	395	396
407	408	409	410	411	412	413	414	415	416	417
428	429	430	431	432	433	434	435	436	437	438
449	450	451	452	453	454	455	456	457	458	459
470	471	472	473	474	475	476	477	478	479	480
491	492	493	494	495	496	497	498	499	500	501
512	513	514	515	516	517	518	519	520	521	522
533	534	535	536	537	538	539	540	541	542	543
554	555	556	557	558	559	560	561	562	563	564
575	576	577	578	579	580	581	582	583	584	585
596	597	598	599	600	601	602	603	604	605	606
617	618	619	620	621	622	623	624	625	626	627
638	639	640	641	642	643	644	645	646	647	648
659	660	661	662	663	664	665	666			
727	728	729	730	731	732	733	734	735	736	737
748	749	750	751	752	753	754	755	756	757	758
769	770	771	772	773	774	775	776	777	778	779
790	791	792	793	794	795	796	797	798	799	
Nonwireline A-band channels										
323	322	321	320	319	318	317	316	315	314	313
302	301	300	299	298	297	296	295	294	293	292
281	280	279	278	277	276	275	274	273	272	271
260	259	258	257	256	255	254	253	252	251	250
239	238	237	236	235	234	233	232	231	230	229
218	217	216	215	214	213	212	211	210	209	208
197	196	195	194	193	192	191	190	189	188	187
176	175	174	173	172	171	170	169	168	167	166
155	154	153	152	151	150	149	148	147	146	145
134	133	132	131	130	129	128	127	126	125	124
113	112	111	110	109	108	107	106	105	104	103
92	91	90	89	88	87	86	85	84	83	82
71	70	69	68	67	66	65	64	63	62	61
50	49	48	47	46	45	44	43	42	41	40
29	28	27	26	25	24	23	22	21	20	19
8	7	6	5	4	3	2	1			
								1023	1022	1021
1010	1009	1008	1007	1006	1005	1004	1003	1002	1001	1000
715	714	713	712	711	710	709	708	707	706	705
694	693	692	691	690	689	688	687	686	685	684
673	672	671	670	669	668	667				

TABLE 11.4   FCC Channel Chart for $N = 4$

Channel group:	A1	B1	C1	D1	A2	B2	C2	D2	A3	B3
				Wireline B-band channels						
Control channel:	334	335	336	337	338	339	340	341	324	343
		356			359	360	361	362	345	364
		377			380	381	382	383	366	385
		398			401	402	403	404	387	406
		419			422	423	424	425	408	427
		440			443	444	445	446	429	448
	460	461	462	463		465			450	469
	481	482	483	484		486			471	490
	502	503	504	505		507			492	511
	523	524	525	526	527	528	529	530	531	532
	544	545	546	547	548	549	550	551		553
	565	566	567	568	569	570	571	572		574
	586	587	588	589	590	591	592	593		595
	607	608	609	610	611	612	613	614	615	616
	628	629	630	631	632	633	634	635	636	637
	649	650	651	652	653	654	655	656	657	658
	717	718	719	720	721	722	723	724	725	726
	741	742	743	744	745	746	747	748	749	750
	765	766	767	768	769	770	771	772	773	774
	789	790	791	792	793	794	795	796	797	798
				Nonwireline A-band channels						
Control channel:	333	332	331	330	329	328	327	326	325	324
	312	311	310	309	308	307	306	305	304	303
		290			287	286	285	284	283	282
		269			266	265	264	263	262	261
		248			245	244	243	242	241	240
		227			224	223	222	221	220	219
		206			203	202	201	200	199	198
	186	185	184	183		181			178	177
	165	164	163	162		160			157	156
	144	143	142	141		139			136	135
	123	122	121	120	119	118	117	116		114
	102	101	100	99	98	97	96	95		93
	81	80	79	78	77	76	75	74		72
	60	59	58	57	56	55	54	53	52	51
	39	38	37	36	35	34	33	32	31	30
	18	17	16	15	14	13	12	11	10	9
	1020	1019	1018	1017	1016	1015	1014	1013	1012	1011
	999	998	997	996			994			991
	711	710	709	708	707	706	705	704	703	702
	690	689	688	687	686	685	684	683	682	681

C3	D3	A4	B4	C4	D4	A5	B5	C5	D5	B6	A6	C6	D6
				Wireline B-band channels									
344	345	346	347	348	349	350	351	352	353	354			
365	366	367	368	369	370	371	372	373	374	375	355	357	358
386	387	388	389	390	391	392	393	394	395	396	376	378	379
407	408	409	410	411	412	413	414	415	416	417	397	399	400
428	429	430	431	432	433	434	435	436	437	438	418	420	421
449	450	451	452	453	454	455	456	457	458	459	439	441	442
470	471	472	473	474	475	476	477	478	479	480	509	488	467
491	492	493	494	495	496	497	498	499	500	501	485	487	488
512	513	514	515	516	517	518	519	520	521	522	506	508	509
533	534	535	536	537	538	539	540	541	542	543	552	554	555
		556	557	558	559	560	561	562	563	564	573	575	576
		577	578	579	580	581	582	583	584	585	594	596	597
		598	599	600	601	602	603	604	605	606	623	625	626
617	618	619	620	621	622		624			627	644	646	647
638	639	640	641	642	643		645			648	665		
659	660	661	662	663	664		666						
727	728	729	730	731	732	733	734	735	736	737	738	739	740
751	752	753	754	755	756	757	758	759	760	761	762	763	764
775	776	777	778	779	780	781	782	783	784	785	786	787	788
799													
				Nonwireline A-band channels									
323	322	321	320	319	318	317	316	315	314	313			
302	301	300	299	298	297	296	295	294	293	292			
281	280	279	278	277	276	275	274	273	272	271	291	289	288
260	259	258	257	256	255	254	253	252	251	250	270	268	267
239	238	237	236	235	234	233	232	231	230	229	249	247	246
218	217	216	215	214	213	212	211	210	209	208	228	226	225
197	196	195	194	193	192	191	190	189	188	187	207	205	204
176	175	174	173	172	171	170	169	168	167	166	182	180	179
155	154	153	152	151	150	149	148	147	146	145	161	159	158
134	133	132	131	130	129	128	127	126	125	124	140	138	137
		111	110	109	108	107	106	105	104	103	115	113	112
		90	89	88	87	86	85	84	83	82	94	92	91
		69	68	67	66	65	64	63	62	61	73	71	70
50	49	48	47	46	45		43			40	44	42	41
29	28	27	26	25	24		22			19	23	21	20
8	7	6	5	4	3		1				2		
							1023	1022	1021				
1010	1009	1008	1007	1006	1005	1004	1003	1002	1001	1000	995	993	992
					716	715	714	713	712				
701	700	699	698	697	696	695	694	693	692	691			
680	679	678	677	676	675	674	673	672	671	670			

TABLE 11.5   FCC Channel Chart for *N* = 4 with 24 Channel Spacing

Channel group:	A1	B1	C1	D1	A2	B2	C2	D2	A3	B3
				Wireline B-band channels						
Control channel:	334	335	336	337	338	339	340	341	342	343
	355	356	357	358	359	360	361	362	363	364
	379	380	381	382	383	384	385	386	387	388
	403	404	405	406	407	408	409	410	411	412
	427	428	429	430	431	432	433	434	435	436
	451	452	453	454	455	456	457	458	459	460
	475	476	477	478	479	480	481	482	483	484
	499	500	501	502	503	504	505	506	507	508
	523	524	525	526	527	528	529	530	531	532
	547	548	549	550	551	552	553	554	555	556
	571	572	573	574	575	576	577	578	579	580
	595	596	597	598	599	600	601	602	603	604
	619	620	621	622	623	624	625	626	627	628
	643	644	645	646	647	648	649	650	651	652
	717	718	719	720	721	722	723	724	725	726
	741	742	743	744	745	746	747	748	749	750
	765	766	767	768	769	770	771	772	773	774
	789	790	791	792	793	794	795	796	797	798
				Nonwireline A-band channels						
Control channel:	333	332	331	330	329	328	327	326	325	324
	312	311	310	309	308	307	306	305	304	303
	288	287	286	285	284	283	282	281	280	279
	264	263	262	261	260	259	258	257	256	255
	240	239	238	237	236	235	234	233	232	231
	216	215	214	213	212	211	210	209	208	207
	192	191	190	189	188	187	186	185	184	183
	168	167	166	165	164	163	162	161	160	159
	144	143	142	141	140	139	138	137	136	135
	120	119	118	117	116	115	114	113	112	111
	96	95	94	93	92	91	90	89	88	87
	72	71	70	69	68	67	66	65	64	63
	48	47	46	45	44	43	42	41	40	39
	24	23	22	21	20	19	18	17	16	15
	1023	1022	1021	1020	1019	1018	1017	1016	1015	1014
	999	998	997	996	995	994	993	992	991	716
	701	700	699	698	697	696	695	694	693	692
	677	676	675	674	673	672	671	670	669	668

C3	D3	A4	B4	C4	D4	A5	B5	C5	D5	B6	A6	C6	D6
Wireline B-band channels													
344	345	346	347	348	349	350	351	352	353	354			
365	366	367	368	369	370	371	372	373	374	375	376	377	378
389	390	391	392	393	394	395	396	397	398	399	400	401	402
413	414	415	416	417	418	419	420	421	422	423	424	425	426
437	438	439	440	441	442	443	444	445	446	447	448	449	450
461	462	463	464	465	466	467	468	469	470	471	472	473	474
485	486	487	488	489	490	491	492	493	494	495	496	497	498
509	510	511	512	513	514	515	516	517	518	519	520	521	522
533	534	535	536	537	538	539	540	541	542	543	544	545	546
557	558	559	560	561	562	563	564	565	566	567	568	569	570
581	582	583	584	585	586	587	588	589	590	591	592	593	594
605	606	607	608	609	610	611	612	613	614	615	616	617	618
629	630	631	632	633	634	635	636	637	638	639	640	641	642
653	654	655	656	657	658	659	660	661	662	663	664	665	666
727	728	729	730	731	732	733	734	735	736	737	738	739	740
751	752	753	754	755	756	757	758	759	760	761	762	763	764
775	776	777	778	779	780	781	782	783	784	785	786	787	788
799													
Nonwireline A-band channels													
323	322	321	320	319	318	317	316	315	314	313			
302	301	300	299	298	297	296	295	294	293	292	291	290	289
278	277	276	275	274	273	272	271	270	269	268	267	266	265
254	253	252	251	250	249	248	247	246	245	244	243	242	241
230	229	228	227	226	225	224	223	222	221	220	219	218	217
206	205	204	203	202	201	200	199	198	197	196	195	194	193
182	181	180	179	178	177	176	175	174	173	172	171	170	169
158	157	156	155	154	153	152	151	150	149	148	147	146	145
134	133	132	131	130	129	128	127	126	125	124	123	122	121
110	109	108	107	106	105	104	103	102	101	100	99	98	97
86	85	84	83	82	81	80	79	78	77	76	75	74	73
62	61	60	59	58	57	56	55	54	53	52	51	50	49
38	37	36	35	34	33	32	31	30	29	28	27	26	25
14	13	12	11	10	9	8	7	6	5	4	3	2	1
1013	1012	1011	1010	1009	1008	1007	1006	1005	1004	1003	1002	1001	1000
715	714	713	712	711	710	709	708	707	706	705	704	703	702
691	690	689	688	687	686	685	684	683	682	681	680	679	678
667													

System: Example 1		Date _____

System: Example 1   Date _____

Frequency of operation
    Tx
    Rx

Down-link path

Base station parameters		
Tx PA output power	42.7	dBm
Tx combiner loss	1.5	dB
Tx duplexer loss	0	dB
Jumper and connector loss	0.25	dB
Lightning arrester loss	0.2	dB
Feedline loss	1.5	dB
Jumper and connector loss	0.25	dB
Tower top amp gain	0	dB
Antenna gain	11	dBd
Cell ERP	50	dBm

Environmental margins		
Tx diversity gain	0	dB
Fading margin	8	dB
Environmental attenuation	15	dB
Cell overlap	5	dB
Environmental margin	−28	dB

Subscriber unit parameters		
Antenna gain	3	dBd
Rx diversity gain	0	dB
Antenna cable loss	2	dB
C/I or E_b/N_0	18	dB
Rx sensitivity	−116	dBm
Subscriber unit effective sensitivity	−99	dBm

Maximum down-link path loss allowed		
Base station ERP	50	dBm
Subscriber unit effective sensitivity	−99	dBm
Maximum down-link path loss	149	dB

Up-link path

Subscriber unit parameters		
Tx PA output	28	dBm
Cable and jumper loss	0	dB
Antenna gain	0	dBd
Subscriber unit ERP	28	dBm

Environmental margins		
Fading margin	8	dB
Environmental attenuation	15	dB
Cell overlap	5	dB
Environmental margin	28	dB

Figure 11.7 AMPS link budget.

```
Base station parameters
    Rx antenna gain                         11      dBd
    Tower top amp net gain                   0      dB
    Jumper and connector loss               0.25    dB
    Feedline loss                           1.5     dB
    Lightning arrester loss                 0.2     dB
    Jumper and connector loss               0.25    dB
    Duplexer loss                            0      dB
    Receive configuration loss               0      dB
    Rx diversity gain                        3      dB
    C/I or E_b/N_0                          18      dB
    Rx sensitivity                        −116      dBm

    Base station effective sensitivity     109.8    dBm

Maximum up-link path loss allowed
    Subscriber unit ERP                     28      dBm
    Base station effective sensitivity     109.8    dBm

    Maximum up-link path loss              137.8    dB
```

Figure 11.7   (*Continued*)

Example 11.1   To determine capacity:
Cell erlang

1. Omni = 12.3 erlangs/cell
2. Sector erlang × sector gain = 12.3 erlangs/sector × 2.64 = 32.47 erlangs/cell

Now determine the number of capacity cells for region 1

$$N_{capacity} = \frac{\text{capacity region 1}}{\text{capacity cell}} = \frac{700}{12.3} = 57 \text{ cells} \qquad \text{omni}$$

Or,

$$\frac{700}{32.47} = 22 \text{ cells} \qquad \text{sector}$$

Region 1:

$$PL = 121 + 36 \log_{10}(d) + \text{overlap}$$

$$137.8 = 121 + 36 \log_{10}(d) + 4 \text{ dB}$$

$$\therefore d = 2.2675 \text{ km}$$

$$A_{\text{cell region 1}} = \pi r^2 = 16.15 \text{ km}^2$$

Now to determine the number of cells, this equates to region 1 for coverage only.

$$N_{cell} = \frac{\text{area region 1}}{\text{area cell}} = \frac{100 \text{ km}^2}{16.15} \doteq 7 \text{ cells}$$

Therefore, 22 sector cells are needed because region 1 is the capacity driven.

Region 2:

$$PL = 121 + 36 \log_{10}(d) + \text{overlap} - 5 \text{ dB}$$

$$\therefore d = 3.12 \text{ km}$$

$$A_{\text{cell region 2}} = \pi r^2 = 30.62 \text{ km}^2$$

$$N_{\text{cell}} = \frac{\text{area region 2}}{\text{area cell}} = \frac{1500 \text{ km}^2}{30.62} = 49 \text{ cells}$$

$$N_{\text{capacity}} = \frac{\text{capacity region 2}}{\text{capacity cell}} = \frac{1400}{12.3} = 114 \text{ cells} \qquad \text{omni}$$

Or,

$$N = \frac{1400}{32.47} = 43 \text{ cells} \qquad \text{sector}$$

Region 3:

$$PL = 121 + 36 \log_{10}(d) + 4 - 10 \text{ dB}$$

$$\therefore d = 4.298 \text{ km}$$

$$A_{\text{cell region 3}} = 58.05 \text{ km}^2$$

$$N_{\text{cell}} = \frac{\text{area region 2}}{\text{area cell}} = \frac{2500}{58.05} = 43 \text{ cells}$$

$$N_{\text{capacity}} = \frac{1050}{12.3} = 86 \text{ cells} \qquad \text{omni}$$

Or,

$$N_{\text{capacity}} = \frac{1050}{32.47} = 33 \text{ cells} \qquad \text{sector}$$

Therefore, 43 sectors are needed for region 3.

Region 4:

$$PL = 121 + 36 \log_{10}(d) + 4 - 15 \text{ dB}$$

$$\therefore d = 5.918 \text{ km}$$

$$A_{\text{cell region 4}} = 110.05 \text{ km}^2$$

$$N_{\text{cell}} = \frac{5900}{110.05} = 54 \text{ cells}$$

$$N_{\text{capacity}} = \frac{350}{12.3} = 28.45 \qquad \text{omni}$$

Or,

$$N = \frac{350}{32.47} = 11 \qquad \text{sector}$$

Therefore, 54 omni sites are needed for region 4.

The results of Example 11.1 are shown in Table 11.6. A quick review of the table reflects that for regions 1, 2, and 3 sector cell sites were chosen since they met the coverage requirements and capacity with the fewest cell sites. Region 4, however, required omni cell sites since it was primarily driven by coverage requirements and not capacity. The option in this case was to select omni cell sites and deploy them around the regions that had sector cell sites.

TABLE 11.6

Region:	1	2	3	4
Coverage	7	49	43	54
Capacity				
Omni	57	114	86	29
Sector	22	43	33	11
Design	22 sector*	49 sector	43 sector	54 omni

*Because of the dense urban environment, 22 sector sites were chosen for region 1.

For an existing system the issue is often progressing from an omni configuration to a sector configuration at some time. In the example in Fig. 11.8 the omni cells that make up the core of the system have grown to the point where adjacent channel interference is the limiting performance issue. In order to overcome this dilemma it is necessary to convert cell 1 from an omni to a sector cell site. The configuration in Fig. 11.9 illustrates that while the adjacent channel issues may be resolved further expansion of the system will most likely require either further sectorization or more cell sites.

The example shows that sometimes sectorizing a system when there are not that many cells may not be the best option. If possible the addition of a new cell will help resolve this problem by having its footprint reduced in size. The trunking efficiency loss with going from omni to sector is an important consideration to factor into the design.

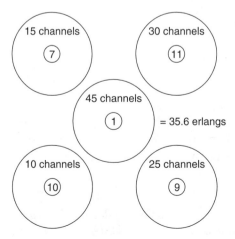

= 35.6 erlangs

Figure 11.8 $N = 12$ pattern.

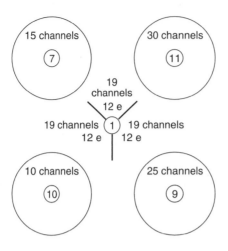

Figure 11.9 Sectorization. ch = channels. e = erlangs. The total requirements are for 57 channels and 36.9 erlangs.

11.3   CDMA

Code division multiple access (CDMA) is a spread-spectrum technology platform which enables multiple users to occupy the same radio channel (frequency spectrum) at the same time. CDMA technology has been championed by many system operators in the United States and Asia. It is based on the principle of direct sequence (DS) and is a wideband spread-spectrum technology. The CDMA channel utilized is reused in every cell of the system and is differentiated by the pseudo-random-number (PN) code that it utilizes. Depending on whether the system will be deployed in an existing AMPS or new PCS band system, the design concepts are fundamentally the same.

Utilizing the IS-95 standard, the same functionality, with the exception of frequency band particulars, applies to both the 800- and 1900-MHz bands. The primary difference is that CDMA has a few nuances that are directly applicable to the channel assignments in an existing cellular band. Beyond that fundamental issue the discussion that follows, with the exception of the examples, applies to both cellular and PCS systems. It should be noted that two different vocoders will also be deployed in CDMA, 8 and 13 kbits/s. The chief difference between the two is the voice clarity. However, since the channel bandwidth did not change, the higher vocoder speed results in a reduction in system capacity, an obvious implication.

In addition to the above, the introduction of CDMA in an existing cellular network is not plug and chug, since there is the issue of immediate capacity reduction, but with a long-term upside. Also for PCS operators, they must relocate existing microwave links to clear the spectrum for their use. The degree of ease or difficulty for implementing CDMA into the PCS market is directly impacted by the ability to clear microwave spectrum.

11.3.1   Spectrum allocation

The spectrum that the IS-95 CDMA technology is designed to operate in is shown in the next few tables. Cellular operation for CDMA is shown in Fig. 11.10 and

Figure 11.10 Cellular CDMA assignment.

Table 11.7. The CDMA channel assignment for cellular is defined as requiring the primary or secondary CDMA channel defined in Table 11.7. The rationale behind this issue lies in the initialization algorithm used for CDMA. Simply put, if the subscriber unit, dual mode, does not find a pilot channel on either the primary or secondary channel, it reverts to an analog mode.

TABLE 11.7   CDMA Channel Designation

	A band	B band
Primary	283	384
Secondary	691	777

A second CDMA carrier could be placed for, say, a B-band operator (Fig. 11.11). Specifically the fact that a preferred channel is used enables the deployment of a second CDMA carrier that is more congenial for the operator. In this case the second channel is planted next to the primary preferred channel and the guard band is now shifted up in frequency.

PCS, on the other hand, has a different set of preferred channels. The initialization algorithm means simply that when the subscriber powers up it will search in its preferred block, *A* through *F*, for a pilot channel using the preferred channel set located in Table 11.8. The pilot channels can, like cellular, also exist in any of the valid ranges shown in Table 11.9. Additionally the comments listed as CV (conditionally valid) are based on the premise that the

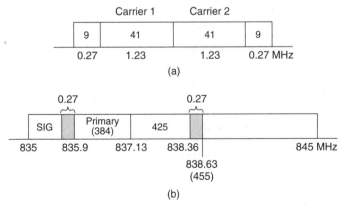

Figure 11.11 Multiple CDMA carriers. (*a*) Primary channel. (*b*) Secondary channel. CDMA channel = 1.23 MHz; AMPS channel = 30 kHz. CDMA guard band = 0.27 MHz × 2 = 0.54 MHz.

TABLE 11.8   PCS CDMA Preferred Pilot Channel

PCS block	CDMA channel no.	Valid CDMA assignment	Preferred set channel numbers
A (15 MHz)	0–24	NV	25, 50, 75, 100, 125, 150, 175, 200, 225, 250, 275
	25–275	V	
	276–299	CV	
D	300–324	CV	325, 350, 375
	325–375	V	
	376–399	CV	
B	400–424	CV	425, 450, 475, 500, 525, 550, 575, 600, 625, 650, 675
	425–675	V	
	676–699	CV	
E	700–724	CV	725, 750, 775
	725–775	V	
	776–799	CV	
F	800–824	CV	825, 850, 875
	825–875	V	
	876–899	CV	
C	900–924	CV	925, 950, 975, 1000, 1025, 1050, 1075, 1100, 1125, 1150, 1175
	925–1175	V	
	1176–1199	NV	

NV = not valid.

V = valid.

CV = conditionally valid.

TABLE 11.9

PCS block	Base station	
	Rx	Tx
A	1850–1865	1930–1945
D	1865–1870	1945–1950
B	1870–1885	1950–1965
E	1885–1890	1965–1970
F	1890–1895	1970–1975
C	1895–1910	1975–1990

$\text{Tx} = (1930.000 + 0.05\,N)$ MHz. $0 \le N \le 1199$
$\text{Rx} = (1850.00 + 0.05\,N)$ MHz.

TABLE 11.10   CDMA Subscriber Power Levels

Station class	EIRP (max), dBw
I	3
II	0
III	−3
IV	−6
V	−9

operator has control of the adjacent block of frequencies, or both of the adjacent blocks like C and F utilize CDMA technology, therefore eliminating the need for a guard band on each side of the allotted spectrum. Table 11.10 represents the different power classes for a CDMA subscriber unit.

11.3.2 Call processing

The channel structure for CDMA is shown in Fig. 11.12, a flow diagram for a mobile originated and terminated call flow.

11.3.3 Standard cell site configurations

Several general types of cell sites are currently usable at this time. The configuration is slightly different for both cellular and PCS owing to colocation issues. However, both cellular and PCS have the commonality of being either an omni or a three-sector cell site.

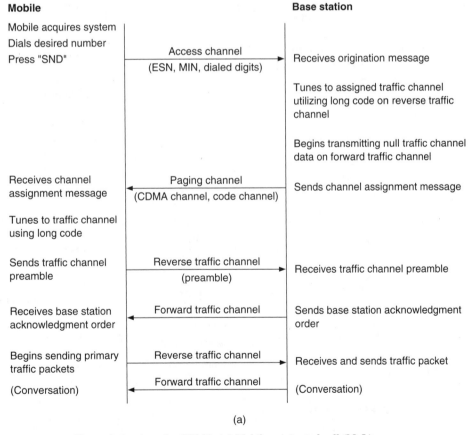

(a)

Figure 11.12 Channel structure for CDMA. (*a*) Mobile originated call (M–L). (*b*) Mobile terminated call (L–M).

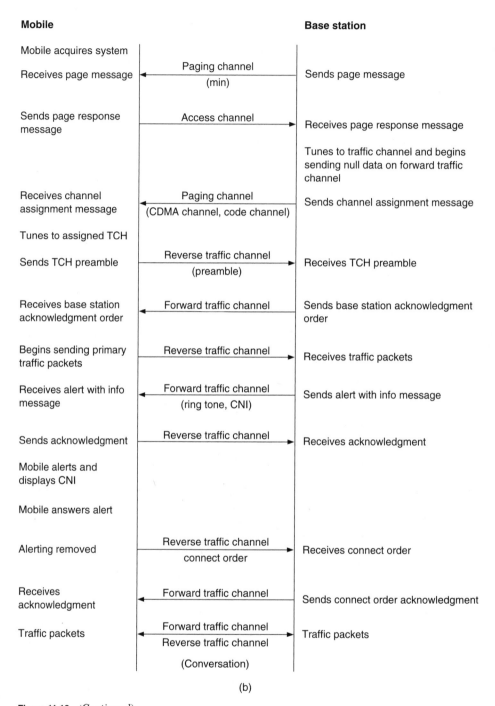

Figure 11.12 (*Continued*)

The radio equipment for both cellular and PCS is fundamentally the same also. The difference between the two is that for PCS the frequency for transmit and receive was up-banded; i.e., an additional mix is taking place. Typically each cell or sector will require a separate transmit antenna per carrier per sector and two receive antennas. The reason for the separate transmit antennas per sector lies in the forward transmit power for the cell in that combing the channels through use of either a cavity or a hybrid results in about a 3-dB loss.

However, for receive antennas masthead electronics is being utilized by one infrastructure vendor as a predominant means of improving the noise figure for their cell site receiver. The configurations in Fig. 11.13 are meant for PCS and cellular CDMA only cells. Figures 11.14 and 11.15 list potential configurations for cellular deployment. Note that, depending on the configuration

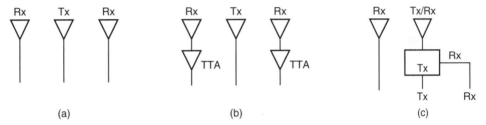

Figure 11.13 PCS and cellular CDMA only installations. (*a*) Optimal sector configuration. (*b*) Utilizing TTAs in receive path. (*c*) Duplex configuration.

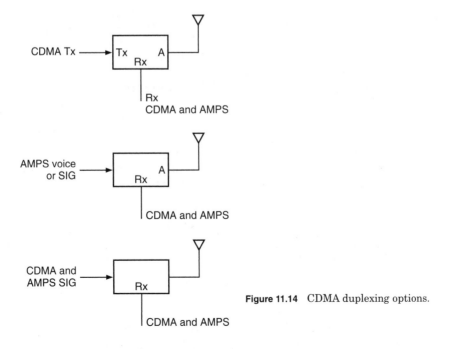

Figure 11.14 CDMA duplexing options.

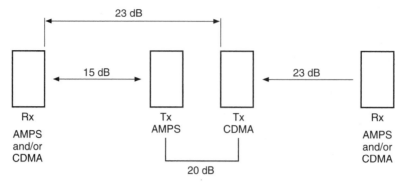

Figure 11.15   AMPS/CDMA colocation antenna isolation.

available, duplexers may be the only viable method available at the site. Therefore, some simple rules apply.

11.3.4   Traffic model

The capacity for a CDMA cell site is driven by several issues. The first and most obvious point for traffic modeling for a CDMA cell site involves how many channel cards the cell site is configured with. A total of 55 possible traffic channels are available for use at a CDMA cell site, but unless the channel cards are installed the full potential is not realizable.

The other factor that fits into the traffic calculations for the site involves system noise. There is a simple relationship between system noise and the capacity of the cell site. Typically the load of the cell site design is somewhere in the vicinity of 40 to 50 percent of the pole capacity, maximum 75 percent.

The third major element in determining the capacity for a CDMA cell is the soft handoff factor. Since CDMA relies on soft handoffs as part of the fundamental design for the network, this must also be factored into the usable capacity at the site. The reason for factoring soft handoffs into capacity is that if 33 percent of the calls are in a soft handoff mode, this will require more channel elements to be installed at the neighboring cell sites to keep the capacity at the desired levels. The traffic model for CDMA is shown in Eq. (11.1). A simple example of impact on capacity when the system noise floor is raised is shown in Table 11.11, and the corresponding relationship between usable traffic channels and the system load is in Fig. 11.16.

$$N = \left(\frac{W/R}{E_b/N_0} \right) \left(\frac{f_u \cdot G}{d} \right) \tag{11.1}$$

where $W = 1.228 \text{ MHz}$
$R = 9.6$
14.4
$E_b/N_0 = 7.5$
$d = \text{duty cycle} = 50\%$
$f_u = \text{reverse link frequency reuse efficiency} = 65\%$
$G = \text{sectorization} - \text{gain} = 2.6$

$$N_{9.6} = \left(\frac{127.9}{7.5}\right) \times \frac{(0.65)(2.6)}{0.5} = 57.64$$

$$N_{14.4} = \left(\frac{85.277}{7.5}\right) \times \frac{0.65\,(2.6)}{0.5} = 38.4$$

As stated earlier, there are a total of 64 Walsh codes. Typically the Walsh codes are allocated as shown in Table 11.12.

TABLE 11.11

Load, %	10	20	30	40	50	60	70	80
Noise floor	0.457	0.97	1.55	2.22	3	3.98	5.23	6.99

Noise rise = $10 \log_{10} (1 \backslash 1 - x)$.

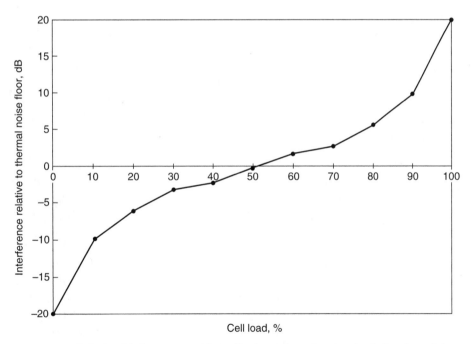

Figure 11.16 Relationship between usable traffic channels and system load, showing points of interference.

TABLE 11.12   Allocation of Walsh Codes

Channel type	No. of Walsh codes
Pilot	1
SYNC	1
Paging	1–7
Traffic channels	55

The pole capacity for CDMA is the theoretical maximum number of simultaneous users that can coexist on a single CDMA carrier (Table 11.13). However, at the pole the system will become unstable, and therefore operating at less than 100 percent of the pole capacity is the desired method of operation.

The effective traffic channels for a CDMA carrier are the number of CDMA traffic channels needed to handle the expected traffic load. However, since soft handoffs are an integral part of CDMA they need to be also included in the calculation for capacity. In addition, for each traffic channel that is assigned for the site a corresponding piece of hardware is needed at the cell site also.

TABLE 11.13   Pole Capacity for CDMA

Pole	25%	50%	60%	75%
Traffic channels	10	20	24	30
Erlang B (2%)	5.08	13.2	16.6	21.9

The actual traffic channels for a cell site are determined using Eq. (11.2).

Actual traffic channels = (effective traffic channels

$$+ \text{ soft handoff channels}) \quad (11.2)$$

The maximum capacity for a CDMA cell site should be 75 percent of the pole.

11.3.5 Handoffs

Several types of handoffs are available with CDMA. They involve soft, softer, and hard. The difference between the types is dependent upon what is to be accomplished. Several user-adjustable parameters help the handoff process take place. The parameters that need to be determined involve the values to add or remove a pilot channel from the active list and the search window sizes. Several values determine when to add or remove a pilot from consideration. In addition the size of the search window cannot be too small nor can it be too large.

Search window.   There are several search windows in CDMA. Each has its own role in the process, and it is not uncommon to have different search window sizes for each of the windows for a particular cell site. Additionally the search window for each site needs to be set based on actual system conditions. However, several system startup values are shown that can be used to get you in the ball park initially.

The search windows to be determined for CDMA involve the active, neighbor, and remaining windows. The search window is defined as an amount of time, in terms of chips, that the CDMA subscriber's receiver will hunt for a pilot channel. There is a slight difference in how the receiver hunts for pilots depending on its type.

If the pilot is an active set the receiver center for the search window will track the pilot itself and adjust the center of the window to correspond to fading conditions. The other search windows are set as defined sizes. Example 11.2 shows the size of a search window. Table 11.14 shows a relationship between the search window and the number of chips.

Example 11.2 The size of a search window is determined as follows:

$$\text{Search window} = 7 = 28 \text{ chips}$$

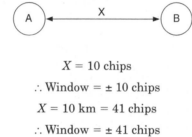

The size of the search window is directly dependent upon the distance between the neighboring cell sites. How to determine what the correct search window is for your situation can be extrapolated using Example 11.3.

Example 11.3 To determine the search window size:

1. Determine the distance between sites in chips.
2. Determine the maximum delay spread in chips.
3. Search window = ± (cell spacing + maximum delay spread).

$$X = 10 \text{ chips}$$
$$\therefore \text{Window} = \pm 10 \text{ chips}$$
$$X = 10 \text{ km} = 41 \text{ chips}$$
$$\therefore \text{Window} = \pm 41 \text{ chips}$$

The search window for the neighbor and remaining sets consists of parameters SRCH_WIN_N, SRCH_WIN_R which represent the search window sizes associated with the neighbor set and remaining set pilots. The subscriber unit

TABLE 11.14   Search Window Sizes

Search window A, N, R	Window size PN chips
0	2
1	4
2	6
3	8
4	10
5	14
6	**20**
7	28
8	40
9	56
10	80
11	114
12	160
13	226
14	320
15	452

centers its search window around the pilot's PN offset and compensates for time variants with its own time reference.

The SRCH_WIN_N should be set so that it will encompass the whole area in which a neighbor pilot can be added to the set. The largest the window should be set is 1.75 $D+3$ chips, where D is the distance between the cells.

SRCH_WIN_A is the value that is used by the subscriber unit to determine the search window size for both the active and candidate sets. The difference between the search window for the active and candidate sets versus the neighbor and remaining sets is that the search window effectively floats with the active and candidate sets based on the first arriving pilot it demodulates.

Soft handoffs. Soft handoffs are an integral part of CDMA. The determination of which pilots will be used in the soft handoff process has a direct impact on the quality of the call and the capacity for the system. Therefore, setting the soft handoff parameters is a key element in the system design for CDMA. The parameters associated with soft handoffs involve the determination of which pilots are in the active, candidate, neighbor, and remaining sets. The list of neighbor pilots is sent to the subscriber unit when it acquires the cell site or is assigned a traffic channel.

A brief description of each type of pilot set follows:

The *active set* is the set of pilots associated with the forward traffic channels assigned to the subscriber unit. The active set can contain more than one pilot since a total of three carriers, each with its own pilot, could be involved in a soft handoff process.

The *candidate set* are the pilots that the subscriber unit has reported as of sufficient signal strength to be used. The subscriber unit also promotes the neighbor set and remaining set pilots that meet the criteria of the candidate set.

The *neighbor set* is a list of the pilots that are not currently on the active or candidate pilot list. The neighbor set is identified by the base station via the neighbor list and neighbor list update messages. The remaining set is the set of all possible pilots in the system that can possibly be used by the subscriber unit. However, the remaining set pilots that the subscriber unit looks for must be a multiple of the Pilot_INC.

Figure 11.17 shows a soft handoff region. The region shown is an area between cells A and B. Naturally as the subscriber unit travels farther away from cell A, cell B, in this example, increases in signal strength for the pilot. When the pilot from cell B reaches a certain threshold it is added to the active pilot list. Example 11.4 describes how a pilot channel moves from the neighbor to candidate to active and then to neighbor sets.

Example 11.4 The depiction of how a pilot channel moves from neighbor to candidate to active and then to neighbor sets is shown in Fig. 11.18.

1. Pilot exceeds T_ADD and subscriber unit sends a PSMM and transfer pilot to candidate set.
2. Base station sends an extended handoff direction message.
3. Subscriber unit transfers pilot to active set and acknowledges this with a handoff completion message.

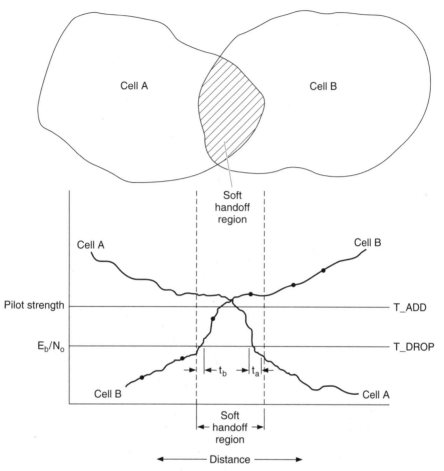

Figure 11.17   Soft handoff region. t_a = area where cell A is below T_DROP but is still in active set. t_b = area where cell B is below T_DROP but still in active set.

4. Pilot strength drops below T_DROP and subscriber unit begins handoff drop time.
5. Pilot strength goes above T_DROP prior to handoff drop time expiring and T_DROP sequences topping.
6. Pilot strength drops below T_DROP and subscriber unit begins the handoff drop times.
7. Handoff drop timer expires and subscriber unit sends a PSMM.
8. Base station sends an extended handoff direction message.
9. Subscriber unit transfers the pilot from the active set to the neighbor set and acknowledges this with a handoff completion message.

Figure 11.19 highlights how T_COMP is factored into the decision matrix for adding and removing pilots from the neighbor, candidate, and active sets. Note that the subscriber unit only reports that a candidate set pilot is stronger than the active set when the candidate set pilot is T_COMP × 0.5 dB greater.

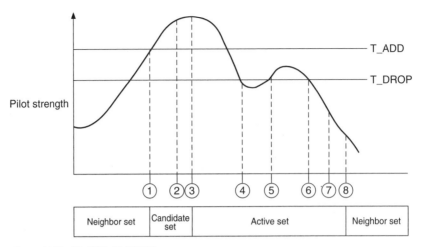

Figure 11.18   T_ADD, T_DROP.

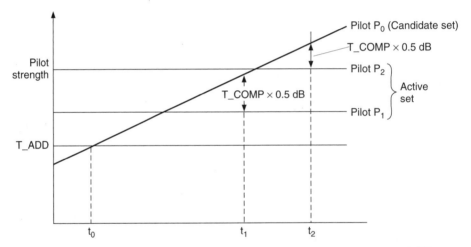

Figure 11.19   Active set. t_0–PSMM sent $P_0 > $ T_ADD (neighbor $\rightarrow$ candidate. t_1– PSMM sent $P_0 > P_1 + $ T_COMP $\times$ 0.5 dB (candidate $\rightarrow$ active). t_2– PSMM sent $P_0 > P_2 + $ T_COMP $\times$ 0.5 dB (active).

Table 11.15 represents an initial set of values that can be used for establishing the parameters associated with T_ADD, T_DROP, T_COMP, and T_TDROP.

Table 11.16 represents the relationship between the T_TDROP and the number of seconds each increment of T_TDROP represents.

11.3.6   Pilot channel

The pilot channel carries no data but it is used by the subscriber unit to acquire the system and assist in the process of soft handoffs, synchronization, and channel estimation. A separate pilot channel is transmitted for each sector of the cell site. The pilot channel is uniquely identified by its PN offset or rather PN short code that is used.

The PN sequence has some 32,768 chips, which when divided by 64 results in a total of 512 possible PN codes available for potential use. The fact that there are 512 potential PN short codes to pick from almost ensures that there will be no problems associated with the assignment of these codes. However, some simple rules must be followed in order to ensure that no problems are encountered with the selection of the PN codes for the cell and its surrounding cell sites. Equations (11.3) show the relationship between chips and distance.

$$\frac{32{,}768}{64} = 512 \text{ possible PN offsets}$$

$$f_{chip} = 1.228 \times 10^6 \text{ chips/s}$$

$$\text{Time} = \frac{1}{f_{chip}} = 0.8144 \ \mu\text{s/chip}$$

$$\text{Distance} = 244 \text{ m/chip}$$

$$(11.3)$$

There are numerous perturbations of how to set the PN codes. However, it is suggested that a reuse pattern be established for allocating them to facilitate

TABLE 11.15 CDMA Parameters

	Suggested	Range
T_ADD	−16 dB	−31.5–0
T_DROP	−20 dB	−31.5–0
T_COMP	3 dB	0–7.5
T_TDROP	5 s	0–15

T_COMP hysteresis=2×T_COMP=6 dB window

TABLE 11.16 Handoff Drop Timer Expiration Values

T_TDROP	Second
0	0
1	1
2	2
3	4
4	6
5	9
6	13
7	19
8	27
9	39
10	55
11	79
12	112
13	159
14	225
15	319

TABLE 11.17   PN Code Reuse Pattern

Sector	PN code
Alpha	$3 \cdot P \cdot N - 2\,P$
Beta	$3 \cdot P \cdot N$
Gamma	$3 \cdot P \cdot N - P$
Omni	$3 \cdot P \cdot N$

N = reuse cell.

P = PN code increment.

the operation of the network for maintenance and growth. Table 11.17 can be used for establishing the PN codes for any cell site in the network. Example 11.5 shows how to determine whether you wish to have a 4, 7, 9, 19, etc., reuse pattern for the PN codes.

Example 11.5

$$C/I = 10 \log_{10} \left(\frac{D\,(P, P_0)}{D\,(P, P_1)} \right)^{-3} \geq a$$

$$M \geq (R + S) \cdot (10^{a/(\alpha)10} - 1)$$

where M = offset
   R = radius in chips
   S = ½ Search window_A
   a = C/I
   α = attenuation factor, propagation exponent

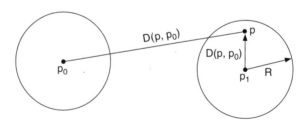

The suggested PN reuse pattern is an $N = 19$ pattern for a new PCS system. If you are overlaying the CDMA system onto a cellular system an $N = 14$ pattern should be used when the analog system utilizes an $N = 7$ voice channel reuse pattern.

Figure 11.20 is an example of an $N = 19$ PN code reuse pattern. Note that not all the codes have been utilized in the $N = 19$ pattern. The remaining codes should be left in reserve for use when a PN code problem arises. In addition a suggested PN_INC value of 6 is also recommended for use. The PN short code used by the pilot is an increment of 64 from the other PN codes, which requires an offset value to be defined. The Pilot_INC is the value that is used

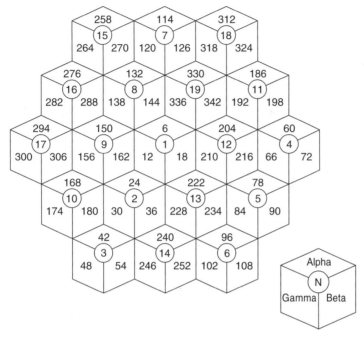

Figure 11.20   PN offset pattern, $N = 19$, Pilot_INC = 6.

to determine the number of chips, or rather phase shift, one pilot has versus another pilot.

The method for calculating the PN offset is by use of the equations in Example 11.4. Pilot_INC is valid from the range of 0 to 15. Pilot_INC is the PN sequence offset index and is a multiple of 64 chips. The subscriber unit uses the Pilot_INC to determine which are the valid pilots to be scanned. Table 11.18 can be used to determine the Pilot_INC as a function of the distance between reusing sites.

TABLE 11.18   Pilot Offsets

R, km	R (chips)	S	C/I	m (chips)	Pilot_INC	No. of offsets
25	103	14	24	622	10	50
20	82	12	24	499	8	64
15	61	12	24	390	6	85
12.5	51	10	24	325	5	102
10	41	10	24	271	4	128
7	29	10	24	207	4	128
5	21	10	24	165	3	170
3	12	10	24	117	2	256
2.5	10	10	24	106	2	256
2	8	10	24	96	2	256

11.3.7 Link Budget

The link budget calculations directly influence the performance of the CDMA system since it is used to determine power setting and capacity limits for the network. Proper selection of the variables which comprise the link budget is a very obvious issue. There are two links, forward and reverse. The forward and reverse links utilize different coding and modulation formats. The first step in the link budget process is to determine the forward, then the reverse link maximum path losses.

Forward link. The forward link maximum path loss is determined using Example 11.6.

Example 11.6

$$\frac{E_c}{I_t} = \frac{\zeta_p\, P_t^{\,c}\, L(R)\, G_c\, G_m}{(N_0 W)_m + I_0 c(R)\, W + I_0 c(R)\, W}$$

where ζ_p = portion of cell Tx power allocated to pilot channel

$I_0 c(R)$ = interference spectral density, other cells

$I_0\,(R)$ = interference spectral density, own cell

$(N_0 W)_m$ = thermal noise of mobile LNA input

$$T(R) = (E_c/I_t)_{\min} - P_t^{\,L} + (N_0 W)_m + 10\log_{10}\left\{\zeta_p - 10^{[(E_c/I_t)\min]/10} \times [1 + 10^{((Ic_0/I_0)/10)}]\right\}$$

$$\therefore \quad If\,(E_c/I_t)_{\min} = -15\ \mathrm{dB}$$

$$P_t^{\,c} = +44\ \mathrm{dBm}$$

$$(NF)_m = 8\ \mathrm{dB}$$

$$I_0 c/I_0 \approx 2.5\ \mathrm{dB}$$

$$\zeta = 15\ (15\%)$$

$$(N_0 W)_m = -174\ \mathrm{dBm/Hz} + 10\log_{10}(BW) + NF_m$$

$$= -174 + 60.89 + 8$$

$$= -105\ \mathrm{dBm}$$

$$y = 10\log_{10}\left\{\zeta_p - 10^{[(E_c/I_t)m/10]} \times [1 + 10^{[(I_0 c/I_0)/10]}]\right\}$$

$$= 10\log_{10}[15 - (0.0316) \times (1 + 1.778)]$$

$$= 10\log_{10}(15 - 0.08778)$$

$$= 11.735$$

$$\therefore \quad T(R) = -15 - 44 - 105 + 11.735 = -152.265\ \mathrm{dB}$$

$$L(R) = T(R) - G_c - G_m = -152.265 - 11 - 0 = -163.265\ \mathrm{dB}$$

Therefore, the maximum path loss is -163.265 dB.

The data gathered show that the maximum path loss sustainable is about -163 dB using the parameters selected. Figure 11.21 is a series of plots showing the effects of altering the E_c/I_t value for the pilot channel. The chart indicates that coverage for the cell can be increased by reducing the pilot's power. Figure 11.22 is an indication of the impact on path loss in the forward direction by altering the antenna gain alone. As suspected, that range of the site is increased owing to increased antenna gain.

Reverse link.   The reverse link calculations are shown in Example 11.7.

Example 11.7[4]

$$\text{CNR}_{(R)} = \frac{P_t^{\,m}(R)\,L(R)\,G_c\,G_m}{(N_0W)_c + (N/F - 1)\,v\,P_t^{\,m}(R)\,L(R)\,G_c\,G_m}$$

where

$P_t^{\,m}(R)$	= mobile Tx
$L(R)$	= reverse link path loss
G_c	= cell antenna system gain
G_m	= mobile antenna system gain
v	= average voice activity factor
F	= frequency reuse efficiency
$(N_0W)_c$	= thermal noise, cell LNA input

Maximum transmission loss:

$$T(R) = \text{CNR}_{\min} + (N_0W)_c - P_t^{\,m} - 10\log_{10}(1-x)$$

where x = system loading

Maximum path loss:

$$L(R) = \text{CNR}_{\min} + (N_0W)_c - P_t^{\,m} - G_c - G_m - 10\log_{10}(1-x)$$

$$\therefore\; P_t^{\,m}(R) = 200\text{ mW} \qquad G_m = 0\text{ dB}$$

$$E_b/N_t = 7\text{ dB} \qquad G_c = 13\text{ dBi} = 11\text{ dBd}$$

$$x = 0.9\ (90\%)$$

$$\text{NF}_{\text{cell receive}} = 5\text{ dB}$$

$$E_b/N_t = \text{CNR} + \text{processing gain}$$

$$\therefore\; \text{CNR} = E_b/N_t - \text{processing gain} \begin{cases} 21\text{ dB} - 8\text{ k} \\ 19\text{ dB} - 13\text{ k} \end{cases}$$

$$\text{CNR (13 k)} = 7 - 19 = -12\text{ dB}$$

$$\begin{aligned} (N_0W)_c &= -174\text{ dBm/Hz} + 10\log(\text{BW}) + \text{NF} \\ &= -174\text{ dBm/Hz} + 60.89 + 5 \\ &= -108\text{ dBm} \end{aligned}$$

$$T(R) = -12\text{ dB} - 108\text{ dBm} - 23\text{ dBm} + 10 = -133.0\text{ dB}$$

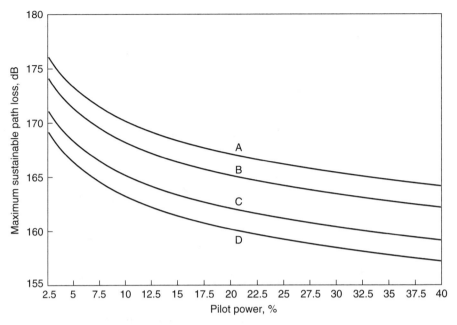

Figure 11.21 Path loss f (pilot and E_b/I_t). Curve A, $E_c/I_t = -20$ dB; B, $E_c/I_t = -18$ dB; C, $E_c/I_t = -15$ dB; D, $E_c/I_t = -13$ dB.

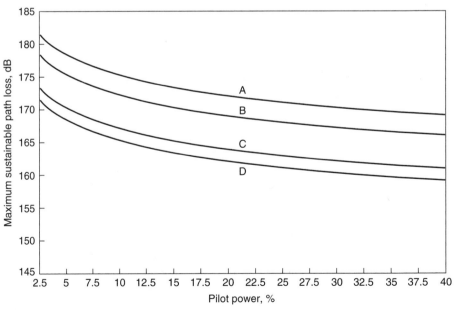

Figure 11.22 Path loss f (antenna gain, pilot, E_b/I_t). Curve A, 11 dB. B, 13 dB; C, 18 dB; D, 21 dB.

Now, calculating for path loss:

$$L(R) = T(R) - G_c - G_m$$
$$= -133.0 - 11\ \text{dB} - 0 = -144.0\ \text{dB}$$

Therefore, -144.0 dB is the maximum allowable path loss.

The graph in Fig. 11.23 indicates the path loss effects for different vocoders as a function of system load. As expected, the higher vocoder experiences more path loss and the range of 40 to 60 percent appears to be the area of system load that is desired for the network.

Figures 11.24 and 11.25 represent plots of path loss for different vocoders as a function of antenna gain and system load. As expected, the graphs show as the cell load increases the size of the cell decreases.

Calculations.   For the link budget taking into account system design issues is shown in Fig. 11.26. The link budget represents most of the variables and parameters that make up a link budget itself.

Referencing the Tx power for the forward link, a 20 percent value was used for determining the power that the pilot would be allocated out of the entire forward link budget. The allocation of power is dependent upon the type of channel. The relationship between the power levels and the type of channel is the SCDMA carrier power allocation:

Pilot power total	20 percent
SYNC	(TCH − 3 dB)
Paging	(TCH + 3 dB)

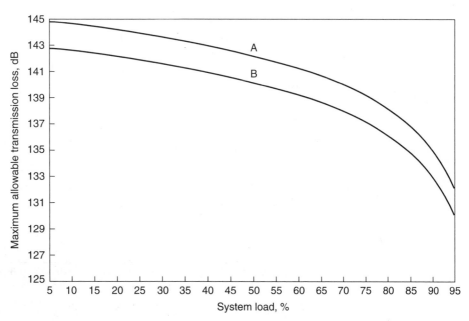

Figure 11.23 Percent of system load vs. transmission loss. Curve A, 8-k vocoder; B, 13-k vocoder.

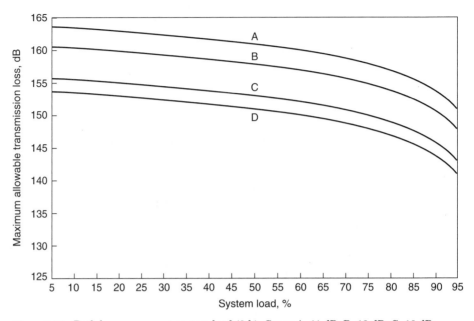

Figure 11.24 Path loss vs. percent system load (8 k). Curve A, 11 dB; B, 13 dB; C, 18 dB; D, 21 dB.

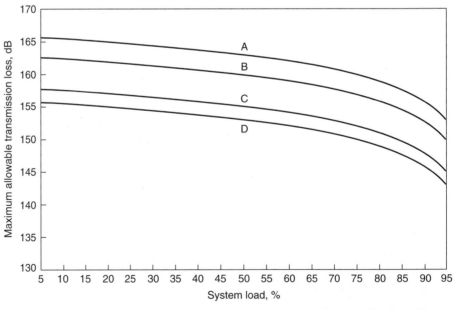

Figure 11.25 Path loss vs. percent system load (13 k). Curve A, 11 dB; B, 13 dB; C, 18 dB; D, 21 dB.

```
System: Example 1                          Date _____
Frequency of operation
     Tx
     Rx


Down-link path

Base station parameters
     Tx PA output power (+ 44dBm 20% pilot)      37    dBm
     Tx combiner loss                            1.5   dB
     Tx duplexer loss                            0     dB
     Jumper and connector loss                   0.25  dB
     Lightning arrester loss                     0.2   dB
     Feedline loss                               1.5   dB
     Jumper and connector loss                   0.25  dB
     Tower top amp gain                          0     dB
     Antenna gain                                18    dBd

     Cell ERP                                    51.3  dBm

Environmental margins
     Tx diversity gain                           0     dB
     Fading margin                               8     dB
     Environmental attenuation                   15    dB
     Cell overlap                                5     dB

     Environmental margin                        28    dB

Subscriber unit parameters
     Antenna gain                                0     dBd
     Rx diversity gain                           0     dB
     Antenna cable loss                          0     dB
     C/I or $E_b/N_0$                            7.5   dB
     Rx sensitivity                             −108.5 dBm

     Subscriber unit effective sensitivity      −101   dBm

Maximum down-link path loss allowed
     Base station ERP                            51.3  dBm
     Subscriber unit effective sensitivity       109.5 dBm

     Maximum down-link path loss                 151.3 dB


Up-link path

Subscriber unit parameters
     Tx PA output                                23    dBm
     Cable and jumper loss                       0     dB
     Antenna gain                                0     dBd

     Subscriber unit ERP                         23    dBm

Environmental margins
     Fading margin                               8     dB
     Environmental attenuation                   15    dB
     Cell overlap                                5     dB

     Environmental margin                        28    dB
```

Figure 11.26 CDMA link budget.

```
Base station parameters
    Rx antenna gain                            18     dBd
    Tower top amp net gain                      0     dB
    Jumper and connector loss                0.25    dB
    Feedline loss                             1.5     dB
    Lightning arrester loss                   0.2     dB
    Jumper and connector loss                0.25    dB
    Duplexer loss                               0     dB
    Receive configuration loss                  0     dB
    Rx diversity gain                           3     dB
    C/I or E_b/N_0                            7.5     dB
    Rx sensitivity                         −108.5    dBm

    Base station effective sensitivity     −119.8    dBm

Maximum up-link path loss allowed
    Subscriber unit ERP                        23     dBm
    Base station effective sensitivity      119.8    dBm

    Maximum up-link path loss               142.8    dB
```

Figure 11.26 *(Continued)*

$$\text{Cell ERP} = \text{traffic} + \text{pilot} + \text{paging} + \text{SYNC}$$

$$\text{TCH} = \frac{\text{traffic ERP}}{\text{No. of active users} + \text{voice activity factor}}$$

Example 11.8 shows how to determine the percentage of power the pilot is to receive, which in this example equates to 10.54 percent. However, considering the number of variables and unknowns when setting up a system or cell site, it is recommended that a value of 20 percent be used for the pilot channel in terms of overall power budget.

Example 11.8

Receive	Forward
PA = 200 MW	Cell PA = 40 dBm
$G_m = 0$	Pilot $E_c/I_t = 15$ dB
$E_d N_t = 7$	$NF_m = 8$ dB
V = 0.4	$I_0 c / I_0 = 2.5$ dB
User = 20 (50% loss)	
NF cell = 5 dB	
$a_c = 6$	

$$T(R) = -14 - 108 - 23 - 10 \log (1 - 0.5) = -142 \text{ dB}$$

$$L(R) = -142 - 6 = -148$$

Pilot power allowed:

$$\frac{E_c/I_t \times (N_0 W) + P_c^{\,t}\, T(R) \times (1 + I_0 c/I_0)}{P_c^{\,t}\, T(R)} = 10.54\%$$

11.3.8 PCS system

The initial system layout for a PCS system first needs to start with a marketing report and follow the steps outlined at the beginning of this chapter. Assuming you have received all the pertinent information, a sample high-level design is shown. In the actual design, actual terrain, realistic building heights, land use issues, and interference analysis need to be factored into the process. The interference analysis requires the use of a computer system, available from several good firms which peddle propagation models. For the system layout refer back to Fig. 11.2. Table 11.19 includes traffic projections for the system. The system is then broken down into regions.

TABLE 11.19

Region	Area, km^2	Traffic, %	Erlangs	Erlangs/km^2
1	100	20	700	7.0
2	1,500	40	1,400	0.933
3	2,500	30	1,050	0.420
4	5,900	10	350	0.0593
Total	10,000	100	3,500	0.35

500 k subscribers at 7 merlangs per subscriber = 3500 erlangs at system busy hour.

Example 11.9 To determine capacity:

Cell erlangs = sector erlangs × sectorization gain

= 13.2 erlangs/sector × 2.64 = 34.848 erlangs/cell

Now to determine the number of cells this equates to for region 1:

$$N_{\text{capacity}} = \frac{\text{capacity region 1}}{\text{capacity cell}} = \frac{700}{34.848} = 20 \text{ cells}$$

Therefore. region 1 is driven by coverage.

Region 1's traffic capacity is

39 cells × 34.848 erlangs = 1359 erlangs

which is almost twice that required if the traffic load is uniformly distributed. If the decision was to use omni cell sites a total of 53 cells would be needed for region 1 alone. However, the antenna system would most likely change.

Region 1:

$$PL = 132 + 42 \log_{10}(d) + \text{margin} + \text{overlap} + \text{soft handoff}$$

$$142.8 = 132 + 42 \log_{10}(d) + 10 + 6 - 4$$

$$-1.2 = 42 \log_{10}(d)$$

$$\therefore d = 0.93632 \text{ km}$$

$$= 0.9019 \text{ km}$$

$$A_{\text{cell}} = \pi r^2 = 2.7542 \text{ km}^2$$

Now to determine the number of cells this equates to for region 1 for coverage:

$$N_{\text{cell}} = \frac{\text{area region 1}}{\text{area cell}} = \frac{100 \text{ km}^2}{2.7542 \text{ km}} = 37 \text{ cells} \quad \text{coverage}$$

Region 2.   Cell radius:

$$PL = 132 + 42 \log_{10}(d) + 5 + 6 - 4 = 142.8$$

$$\therefore d = 1.2316 \text{ km}$$

$$A_{\text{cell region (2)}} = \pi r^2 = 4.7654 \text{ km}^2$$

$$N_{\text{cell}} = \frac{\text{area region 2}}{\text{area cell}} = \frac{1500 \text{ km}^2}{4.7654 \text{ km}^2} = 315 \text{ cells} \quad \text{coverage}$$

$$N_{\text{capacity}} = \frac{\text{capacity region 2}}{\text{capacity cell}} = \frac{1400}{34.848} = 41 \text{ cells} \quad \text{capacity}$$

This would imply that the area is significantly driven by coverage requirements. Therefore, a gain of 5 dB in the uplink path would reduce the cell count in region 2 to 182 cells.

Region 3:

$$\text{Radius}_{\text{region 3}} = 2.7 \text{ km}$$

$$A_{\text{cell region 3}} = \pi r^2 = 22.9 \text{ km}^2$$

$$N_{\text{cell}} = \frac{\text{area region 3}}{\text{area cell}} = \frac{2500}{22.9} = 109 \text{ cells} \quad \text{coverage}$$

$$N_{\text{capacity}} = \frac{\text{capacity region 3}}{\text{capacity cell}} = \frac{1050}{34.848} = 31 \text{ cells} \quad \text{capacity}$$

Region 4:

$$\text{Radius} = 5 \text{ km}$$

$$A_{\text{cell region 4}} = \pi r^2 = 78.54 \text{ km}^2$$

$$N_{\text{cell}} = \frac{\text{area region 4}}{\text{area cell}} = \frac{5900}{78.54} = 75 \text{ cells} \quad \text{coverage}$$

$$N_{\text{capacity}} = \frac{\text{capacity region 4}}{\text{capacity cell}} = \frac{350}{34.848} = 10 \text{ cells} \quad \text{capacity}$$

The number of coverage and capacity cells for Example 11.9 is shown in Table 11.20. The result is a system that will be laid out on an N 5 19 pattern using the parameters in Fig. 11.27.

TABLE 11.20

Region:	1	2	3	4	Total
Coverage	37	182	109	75	403
Capacity	20	41	31	10	102

11.3.9 Cellular system layout

The introduction of CDMA into an existing AMPS system will require the establishment of a guard band and guard zone. They are required for CDMA to ensure that the interference received from the AMPS system does not negatively impact the ability for CDMA to perform well. The specific location that the CDMA channel or channels occupy in a cellular system is dependent upon a multitude of issues. The first issue is how much spectrum will be dedicated to the use of CDMA for the network. The spectrum issue ties into the fact that one CDMA channel occupies 1.77 MHz of spectrum, 1.23 MHz per CDMA channel, and 0.27 MHz of guard band on each side of the CDMA channel.

With a total of 1.77 MHz per CDMA the physical location in the operator's band that CDMA will operate in needs to be defined. Presently for the B-band carrier (wireline operators) two predominant locations are being utilized. The first location in the spectrum is the band next to the control channels, and the other section is in the lower portion of the extended AMPS band. The upper end of the EAMPS band is not as viable owing to the potential of AGT interference since AGT transmit frequencies have no guard band between AMPS receive and AGT transmit. The lower portion of the EAMPS band has the disadvantage of receiving A-band mobile-to-base interference which will limit the size of the CDMA cell site. Sharper filters could be utilized for the lower EAMPS band but group delay with the use of high selective filters then becomes a problem.

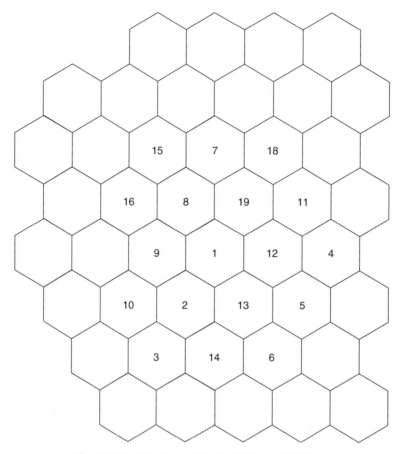

Figure 11.27 $N = 19$. Pilot PN offset = 19; pilot INC = 6; T_ADD = −16; T_DROP = −20; T_COMP = 3; T_TDROP = 5. Search window−active = 6; Search window − neighbor = 11. Search window − remaining = 0. Pilot power = 20 percent.

The other issue with the guard band ties into the actual amount of spectrum unavailable for use by AMPS subscribers in the cellular market. With the expansive growth of cellular, assigning 1.77 MHz of spectrum to CDMA reduces the spectrum available for AMPS usage by 15 percent or 59 radio channels from the channel assignment chart. The reduction in the available number of channels for regular AMPS requires the addition of more cell sites to compensate for the number of radio channels no longer available for use in the AMPS system. Utilizing a linear evaluation the reduction in usable spectrum by 15 percent involves a reduction on traffic-handling capacity by the AMPS system by a maximum of 21 percent at an erlang B 2 percent GOS with a maximum of 16 channels per sector versus 19. The reduction of 21 percent in initial AMPS traffic-handling capacity requires the need to build more analog cell sites to compensate for this reduction. The only way to offset the reduction experienced by partitioning the spectrum is to preload the CDMA subscriber utilizing dual-mode phones or to build more analog cell sites. The guard zone is the physical area outside the CDMA coverage area that can no longer utilize the AMPS channels now occupied by the CDMA system. Figure 11.28 shows a guard zone versus a CDMA system coverage area. The establishment and size of the guard zone is dependent upon the traffic load expected by the CDMA system. The guard zone is usually defined in terms of a signal strength level from which analog cell sites operating with the CDMA channel sets cannot contribute to the overall interference level of the system. The interesting point about the guard zone is when the operator on one system who wishes to utilize CDMA requires the adjacent system operator to reduce their channel utilization in their network to accommodate their neighbor's introduction of this new technology platform.

For the CDMA system deployed there are two distinct methods of implementation to be considered (Table 11.21). The first method is to deploy CDMA in every cell site, for the defined service area, on a 1:1 basis. The other method available is to deploy CDMA on an N:1 basis. Both the 1:1 and N:1 deployment strategies have advantages and disadvantages.

The deployment strategies are shown in Figs. 11.29 and 11.30. A third potential deployment strategy is to combine the one-to-one and the N-to-one methods.

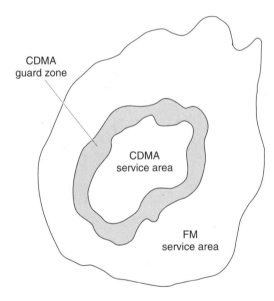

Figure 11.28 Guard zone.

TABLE 11.21

Layout	Advantages	Disadvantages
1:1	Consistent coverage	Cost
	Facilitates gradual growth	Guard zone requirements
	Integrate into existing system	Digital to analog boundary handoff
	Easier to engineer	Slower deployment
	Larger initial capacity gain	
N:1	Lower cost	Harder to engineer properly
	Faster to implement	Lower capacity gain

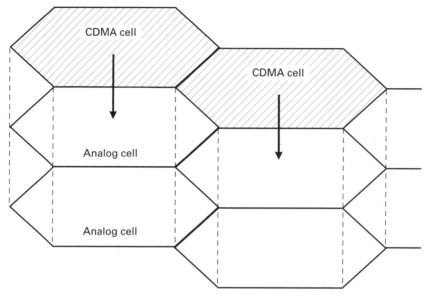

Figure 11.29   1:1 CDMA analog development.

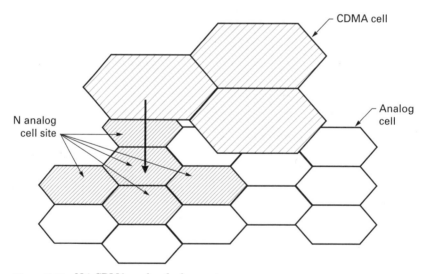

Figure 11.30 N:1 CDMA analog deployment.

Several methods can be used for deploying CDMA into an existing AMPS system. The system design(s) presented here will assume that the primary CDMA channel is selected and that spectrum reallocation will need to take place. No additional services which consume spectrum, like CDPD, will be factored into this design.

Additionally an $N = 7$ three-sector reuse pattern is assumed throughout this discussion. The initial system layout is shown in Fig. 11.31.

The three basic approaches involve

1. *N*:1 layout (Fig. 11.32)
2. 1:1 layout (Fig. 11.33)
3. Core only (Fig. 11.34)

Each of the designs has positives and negatives. Each method requires the establishment of a guard zone and guard band. The parameters to be used for each configuration should be established as follows.

The system offloading and new additions can follow the procedures outlined in Chap. 10 with the exception of the maximum allowable channels. The available channels should be derated accordingly to account for the spectrum reduction due to addition of a CDMA carrier.

Figure 11.32 should involve the following parameters.

Pilot PN offset	14
Pilot_INC	6
T_ADD	-16
T_DROP	-20
T_COMP	3
T_TDROP	5
Search window_Active	8
Search window_Neighbor	11

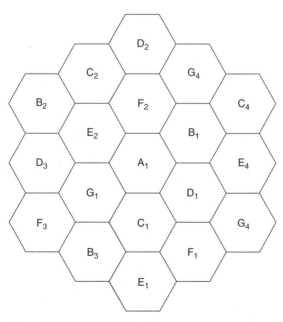

Figure 11.31 Initial $N = 7$ system.

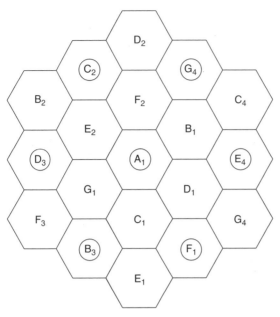

Figure 11.32 N:1 layout. The cells with the circled numbers are CDMA and analog.

Search window_Remaining   0

Pilot power = 20 percent

For Figs. 11.33 and 11.34 the following are the recommended values:

Pilot PN offset   14

Pilot_INC   6

T_ADD   −16

T_DROP   −20

T_COMP   3

T_TDROP   5

Search window_Active   6

Search window_Neighbor   11

Search window_Remaining   0

Pilot power = 20 percent

11.4   IS-136 and IS-54

IS-136 is the NADC standard that is the next evolution to IS-54. IS-136 is currently being deployed in cellular and for new PCS systems throughout the United States. It is similar to IS-54 with a few exceptions noted in Chap. 2 which tie into the digital control channel and short message services, to mention two of the key differentiators.

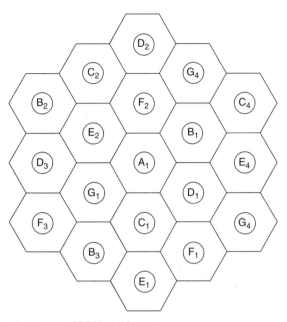

Figure 11.33   CDMA. 1:1 layout.

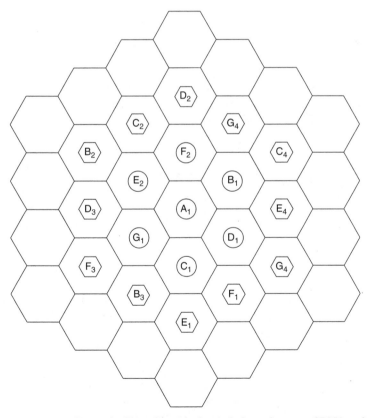

Figure 11.34   Core only. The cells with the circled numbers are CDMA and analog; those with the numbers in hexagons are guard zone; the blank cells are full analog.

The allocation of channels for IS-136 is the same as that of IS-54 in that both use the same modulation format and occupy 30 kHz of spectrum for each physical radio channel. Each of the radio channels is divided into six time slots of which two are used per call, that is, three subscribers per physical radio.

IS-136 brings to the table the digital control channel (DCCH) and it enables the delivery of adjunct features that in cellular were not really possible. The DCCH occupies two of the six time slots, and therefore if a physical radio also has a DCCH assigned to it only two subscribers can use the physical radio for communication purposes.

11.4.1   Spectrum allocation

The spectrum allocation for IS-136 in the cellular and PCS bands is shown in Fig. 11.35. There is a preferred location for the assignment of DCCHs. The DCCHs can be located anywhere in the allocated frequency band; however, there are preferred combinations of channels. The preference is based on the method of the subscriber unit that scans the available spectrum looking for the DCCH.

The preferred channel sets are broken down into 16 relative probability blocks for each frequency band of operation, both cellular and PCS. Relative probability block 1 is the first group of channels the subscriber unit uses to find the DCCH for the system and cell. The subscriber unit then scans through the entire frequency band going through channel sets according to the relative probability blocks until it finds a DCCH.

In the case of cellular if no DCCH is found it reverts to the control channel for a dual-mode phone and then either acquires the system through the control channel or is directed to a specified channel that has the DCCH.

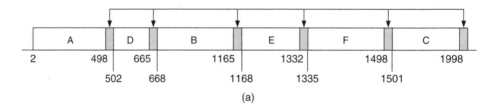

(a)

PCS block	Channels
A	2–498
B	668–1165
C	1501–1998
D	502–665
E	1168–1332
F	1335–1498

(b)

Figure 11.35   (*a*) Preferred DCCH locations. (*b*) Channels.

Table 11.22 is the preferred values or rather channels that the DCCH should reside on for both the A- and B-band cellular operators. The preferred location for the A- and B-band operators is in the extended-spectrum portion of the frequency bands. For the A-band operators to assign, the DCCHs are in the A″ band and for the B-band operators they are in the upper B′ portion of the band. The preferred DCCH allocation for the PCS bands is shown in Table 11.23. The suggested locations for the DCCHs are the upper portion of each of the PCS blocks.

Table 11.24 represents the channel assignments for each PCS block that conforms to an $N = 7$ three-sector channel chart. Glancing at the charts, it is obvious that the D, E, and F blocks do not have enough channels in the preferred

TABLE 11.22   Cellular DCCH Preferred Assignments

Block number	Channel number	Band	Number of channels	Relative probability
		A block		
1	1–26	A	26	4
2	27–52	A	26	5
3	53–78	A	26	6
4	79–104	A	26	7
5	105–130	A	26	8
6	131–156	A	26	9
7	157–182	A	26	10
8	183–208	A	26	11
9	209–234	A	26	12
10	235–260	A	26	13
11	261–286	A	26	14
12	287–312	A	26	15
13	313–333	A	21	16 (lowest)
14	667–691	A′	25	3
15	692–716	A′	25	2
16	991–1023	A″	33	1 (highest)
		B block		
1	334–354	B	21	16 (lowest)
2	355–380	B	26	15
3	381–406	B	26	14
4	407–432	B	26	13
5	433–458	B	26	12
6	459–484	B	26	11
7	485–510	B	26	10
8	511–536	B	26	9
9	537–562	B	26	8
10	563–588	B	26	7
11	589–614	B	26	6
12	615–640	B	26	5
13	641–666	B	26	4
14	717–741	B′	25	3
15	742–766	B′	25	2
16	767–799	B′	33	1 (highest)

TABLE 11.23   DCCH Preferred Assignments

Block number	Channel number	Band	Number of channels	Relative probability
		A block		
1	2–31	A	30	16 (lowest)
2	32–62	A	31	15
3	63–93	A	31	14
4	94–124	A	31	13
5	125–155	A	31	12
6	156–186	A	31	11
7	187–217	A	31	10
8	218–248	A	31	9
9	249–279	A	31	8
10	280–310	A	31	7
11	311–341	A	31	6
12	342–372	A	31	5
13	373–403	A	31	4
14	404–434	A	31	3
15	435–465	A	31	2
16	466–498	A	33	1 (highest)
		B block		
1	668–698	B	31	16 (lowest)
2	699–729	B	31	15
3	730–760	B	31	14
4	761–791	B	31	13
5	792–822	B	31	12
6	823–853	B	31	11
7	854–884	B	31	10
8	885–915	B	31	9
9	916–946	B	31	8
10	947–977	B	31	7
11	978–1008	B	31	6
12	1009–1039	B	31	5
13	1040–1070	B	31	4
14	1071–1101	B	31	3
15	1102–1132	B	31	2
16	1133–1165	B	33	1 (highest)
		C block		
1	1501–1531	C	31	16 (lowest)
2	1532–1562	C	31	15
3	1563–1593	C	31	14
4	1594–1624	C	31	13
5	1625–1655	C	31	12
6	1656–1686	C	31	11
7	1687–1717	C	31	10
8	1718–1748	C	31	9
9	1749–1779	C	31	8
10	1780–1810	C	31	7
11	1811–1841	C	31	6
12	1842–1872	C	31	5
13	1873–1903	C	31	4
14	1904–1934	C	31	3
15	1935–1965	C	31	2
16	1966–1998	C	33	1 (highest)

TABLE 11.23 (*Continued*)

	D block			
1	502–511	D	10	16 (lowest)
2	512–521	D	10	15
3	522–531	D	10	14
4	532–541	D	10	13
5	542–551	D	10	12
6	552–561	D	10	11
7	562–571	D	10	10
8	572–581	D	10	9
9	582–591	D	10	8
10	592–601	D	10	7
11	602–611	D	10	6
12	612–621	D	10	5
13	622–631	D	10	4
14	632–641	D	10	3
15	642–651	D	10	2
16	652–665	D	14	1 (highest)
	E block			
1	1168–1177	E	10	16 (lowest)
2	1178–1187	E	10	15
3	1188–1197	E	10	14
4	1198–1207	E	10	13
5	1208–1217	E	10	12
6	1218–1227	E	10	11
7	1228–1237	E	10	10
8	1238–1247	E	10	9
9	1248–1257	E	10	8
10	1258–1267	E	10	7
11	1268–1277	E	10	6
12	1278–1287	E	10	5
13	1288–1297	E	10	4
14	1298–1307	E	10	3
15	1308–1317	E	10	2
16	1318–1332	E	15	1 (highest)
	F block			
1	1335–1344	F	10	16 (lowest)
2	1345–1354	F	10	15
3	1355–1364	F	10	14
4	1365–1374	F	10	13
5	1375–1384	F	10	12
6	1385–1394	F	10	11
7	1395–1404	F	10	10
8	1405–1414	F	10	9
9	1415–1424	F	10	8
10	1425–1434	F	10	7
11	1435–1444	F	10	6
12	1445–1454	F	10	5
13	1455–1464	F	10	4
14	1465–1474	F	10	3
15	1475–1484	F	10	2
16	1485–1498	F	14	1 (highest)

TABLE 11.24

A block

A1	B1	C1	D1	E1	F1	G1	A2	B2	C2	D2	E2	F2	G2	A3	B3	C3	D3	E3	F3	G3
498	497	496	495	494	493	492	491	490	489	488	487	486	485	484	483	482	481	480	479	478
477	476	475	474	473	472	471	470	469	468	467	466	465	464	463	462	461	460	459	458	457
456	455	454	453	452	451	450	449	448	447	446	445	444	443	442	441	440	439	438	437	436
435	434	433	432	431	430	429	428	427	426	425	424	423	422	421	420	419	418	417	416	415
414	413	412	411	410	409	408	407	406	405	404	403	402	401	400	399	398	397	396	395	394
393	392	391	390	389	388	387	386	385	384	383	382	381	380	379	378	377	376	375	374	373
372	371	370	369	368	367	366	365	364	363	362	361	360	359	358	357	356	355	354	353	352
351	350	349	348	347	346	345	344	343	342	341	340	339	338	337	336	335	334	333	332	331
330	329	328	327	326	325	324	323	322	321	320	319	318	317	316	315	314	313	312	311	310
309	308	307	306	305	304	303	302	301	300	299	298	297	296	295	294	293	292	291	290	289
288	287	286	285	284	283	282	281	280	279	278	277	276	275	274	273	272	271	270	269	268
267	266	265	264	263	262	261	260	259	258	257	256	255	254	253	252	251	250	249	248	247
246	245	244	243	242	241	240	239	238	237	236	235	234	233	232	231	230	229	228	227	226
225	224	223	222	221	220	219	218	217	216	215	214	213	212	211	210	209	208	207	206	205
204	203	202	201	200	199	198	197	196	195	194	193	192	191	190	189	188	187	186	185	184
183	182	181	180	179	178	177	176	175	174	173	172	171	170	169	168	167	166	165	164	163
162	161	160	159	158	157	156	155	154	153	152	151	150	149	148	147	146	145	144	143	142
141	140	139	138	137	136	135	134	133	132	131	130	129	128	127	126	125	124	123	122	121
120	119	118	117	116	115	114	113	112	111	110	109	108	107	106	105	104	103	102	101	100
99	98	97	96	95	94	93	92	91	90	89	88	87	86	85	84	83	82	81	80	79
78	77	76	75	74	73	72	71	70	69	68	67	66	65	64	63	62	61	60	59	58
57	56	55	54	53	52	51	50	49	48	47	46	45	44	43	42	41	40	39	38	37
36	35	34	33	32	31	30	29	28	27	26	25	24	23	22	21	20	19	18	17	16
15	14	13	12	11	10	9	8	7	6	5	4	3	2							

B block

A1	B1	C1	D1	E1	F1	G1	A2	B2	C2	D2	E2	F2	G2	A3	B3	C3	D3	E3	F3	G3
1165	1164	1163	1162	1161	1160	1159	1158	1157	1156	1155	1154	1153	1152	1151	1150	1149	1148	1147	1146	1145
1144	1143	1142	1141	1140	1139	1138	1137	1136	1135	1134	1133	1132	1131	1130	1129	1128	1127	1126	1125	1124
1123	1122	1121	1120	1119	1118	1117	1116	1115	1114	1113	1112	1111	1110	1109	1108	1107	1106	1105	1104	1103
1102	1101	1100	1099	1098	1097	1096	1095	1094	1093	1092	1091	1090	1089	1088	1087	1086	1085	1084	1083	1082

11.56

Left block

1061	1062	1063	1064	1065	1066	1067	1068	1069	1070	1071	1072	1073	1074	1075	1076	1077	1078	1079	1080	1081
1040	1041	1042	1043	1044	1045	1046	1047	1048	1049	1050	1051	1052	1053	1054	1055	1056	1057	1058	1059	1060
1019	1020	1021	1022	1023	1024	1025	1026	1027	1028	1029	1030	1031	1032	1033	1034	1035	1036	1037	1038	1039
998	999	1000	1001	1002	1003	1004	1005	1006	1007	1008	1009	1010	1011	1012	1013	1014	1015	1016	1017	1018
977	978	979	980	981	982	983	984	985	986	-987	988	989	990	991	992	993	994	995	996	997
956	957	958	959	960	961	962	963	964	965	966	967	968	969	970	971	972	973	974	975	976
935	936	937	938	939	940	941	942	943	944	945	946	947	948	949	950	951	952	953	954	955
914	915	916	917	918	919	920	921	922	923	924	925	926	927	928	929	930	931	932	933	934
893	894	895	896	897	898	899	900	901	902	903	904	905	906	907	908	909	910	911	912	913
872	873	874	875	876	877	878	879	880	881	882	883	884	885	886	887	888	889	890	891	892
851	852	853	854	855	856	857	858	859	860	861	862	863	864	865	866	867	868	869	870	871
830	831	832	833	834	835	836	837	838	839	840	841	842	843	844	845	846	847	848	849	850
809	810	811	812	813	814	815	816	817	818	819	820	821	822	823	824	825	826	827	828	829
788	789	790	791	792	793	794	795	796	797	798	799	800	801	802	803	804	805	806	807	808
767	768	769	770	771	772	773	774	775	776	777	778	779	780	781	782	783	784	785	786	787
746	747	748	749	750	751	752	753	754	755	756	757	758	759	760	761	762	763	764	765	766
725	726	727	728	729	730	731	732	733	734	735	736	737	738	739	740	741	742	743	744	745
704	705	706	707	708	709	710	711	712	713	714	715	716	717	718	719	720	721	722	723	724
683	684	685	686	687	688	689	690	691	692	693	694	695	696	697	698	699	700	701	702	703
662	663	664	665	666	667	668	669	670	671	672	673	674	675	676	677	678	679	680	681	682

C block

1978	1979	1980	1981	1982	1983	1984	1985	1986	1987	1988	1989	1990	1991	1992	1993	1994	1995	1996	1997	1998
1957	1958	1959	1960	1961	1962	1963	1964	1965	1966	1967	1968	1969	1970	1971	1972	1973	1974	1975	1976	1977
1936	1937	1938	1939	1940	1941	1942	1943	1944	1945	1946	1947	1948	1949	1950	1951	1952	1953	1954	1955	1956
1915	1916	1917	1918	1919	1920	1921	1922	1923	1924	1925	1926	1927	1928	1929	1930	1931	1932	1933	1934	1935
1894	1895	1896	1897	1898	1899	1900	1901	1902	1903	1904	1905	1906	1907	1908	1909	1910	1911	1912	1913	1914
1873	1874	1875	1876	1877	1878	1879	1880	1881	1882	1883	1884	1885	1886	1887	1888	1889	1890	1891	1892	1893
1852	1853	1854	1855	1856	1857	1858	1859	1860	1861	1862	1863	1864	1865	1866	1867	1868	1869	1870	1871	1872
1831	1832	1833	1834	1835	1836	1837	1838	1839	1840	1841	1842	1843	1844	1845	1846	1847	1848	1849	1850	1851
1810	1811	1812	1813	1814	1815	1816	1817	1818	1819	1820	1821	1822	1823	1824	1825	1826	1827	1828	1829	1830
1789	1790	1791	1792	1793	1794	1795	1796	1797	1798	1799	1800	1801	1802	1803	1804	1805	1806	1807	1808	1809
1768	1769	1770	1771	1772	1773	1774	1775	1776	1777	1778	1779	1780	1781	1782	1783	1784	1785	1786	1787	1788
1747	1748	1749	1750	1751	1752	1753	1754	1755	1756	1757	1758	1759	1760	1761	1762	1763	1764	1765	1766	1767
1726	1727	1728	1729	1730	1731	1732	1733	1734	1735	1736	1737	1738	1739	1740	1741	1742	1743	1744	1745	1746
1705	1706	1707	1708	1709	1710	1711	1712	1713	1714	1715	1716	1717	1718	1719	1720	1721	1722	1723	1724	1725

(Continued)

TABLE 11.24 (*Continued*)

A1	B1	C1	D1	E1	F1	G1	A2	B2	C2	D2	E2	F2	G2	A3	B3	C3	D3	E3	F3	G3
										C block										
1704	1703	1702	1701	1700	1699	1698	1697	1696	1695	1694	1693	1692	1691	1690	1689	1688	1687	1686	1685	1684
1683	1682	1681	1680	1679	1678	1677	1676	1675	1674	1673	1672	1671	1670	1669	1668	1667	1666	1665	1664	1663
1662	1661	1660	1659	1658	1657	1656	1655	1654	1653	1652	1651	1650	1649	1648	1647	1646	1645	1644	1643	1642
1641	1640	1639	1638	1637	1636	1635	1634	1633	1632	1631	1630	1629	1628	1627	1626	1625	1624	1623	1622	1621
1620	1619	1618	1617	1616	1615	1614	1613	1612	1611	1610	1609	1608	1607	1606	1605	1604	1603	1602	1601	1600
1599	1598	1597	1596	1595	1594	1593	1592	1591	1590	1589	1588	1587	1586	1585	1584	1583	1582	1581	1580	1579
1578	1577	1576	1575	1574	1573	1572	1571	1570	1569	1568	1567	1566	1565	1564	1563	1562	1561	1560	1559	1558
1557	1556	1555	1554	1553	1552	1551	1550	1549	1548	1547	1546	1545	1544	1543	1542	1541	1540	1539	1538	1537
1536	1535	1534	1533	1532	1531	1530	1529	1528	1527	1526	1525	1524	1523	1522	1521	1520	1519	1518	1517	1516
1515	1514	1513	1512	1511	1510	1509	1508	1507	1506	1505	1504	1503	1502	1501						
										D block										
665	664	663	662	661	660	659	658	657	656	655	654	653	652	651	650	649	648	647	646	645
644	643	642	641	640	639	638	637	636	635	634	633	632	631	630	629	628	627	626	625	624
623	622	621	620	619	618	617	616	615	614	613	612	611	610	609	608	607	606	605	604	603
602	601	600	599	598	597	596	595	594	593	592	591	590	589	588	587	586	585	584	583	582
581	580	579	578	577	576	575	574	573	572	571	570	569	568	567	566	565	564	563	562	561
560	559	558	557	556	555	554	553	552	551	550	549	548	547	546	545	544	543	542	541	540
539	538	537	536	535	534	533	532	531	530	529	528	527	526	525	524	523	522	521	520	519
518	517	516	515	514	513	512	511	510	509	508	507	506	505	504	503	502				
										E block										
1332	1331	1330	1329	1328	1327	1326	1325	1324	1323	1322	1321	1320	1319	1338	1337	1336	1335	1334	1333	1332
1331	1330	1329	1328	1327	1326	1325	1324	1323	1322	1321	1320	1319	1318	1317	1316	1315	1314	1313	1312	1311
1310	1309	1308	1307	1306	1305	1304	1303	1302	1301	1300	1299	1298	1297	1296	1295	1294	1293	1292	1291	1290
1289	1288	1287	1286	1285	1284	1283	1282	1281	1280	1279	1278	1277	1276	1275	1274	1273	1272	1271	1270	1269
1268	1267	1266	1265	1264	1263	1262	1261	1260	1259	1258	1257	1256	1255	1254	1253	1252	1251	1250	1249	1248
1247	1246	1245	1244	1243	1242	1241	1240	1239	1238	1237	1236	1235	1234	1233	1232	1231	1230	1229	1228	1227

1206	1207	1208	1209	1210	1211	1212	1213	1214	1215	1216	1217	1218	1219	1220	1221	1222	1223	1224	1225	1226
1185	1186	1187	1188	1189	1190	1191	1192	1193	1194	1195	1196	1197	1198	1199	1200	1201	1202	1203	1204	1205
				1168	1169	1170	1171	1172	1173	1174	1175	1176	1177	1178	1179	1180	1181	1182	1183	1184

F block

1478	1479	1480	1481	1482	1483	1484	1485	1486	1487	1488	1489	1490	1491	1492	1493	1494	1495	1496	1497	1498
1457	1458	1459	1460	1461	1462	1463	1464	1465	1466	1467	1468	1469	1470	1471	1472	1473	1474	1475	1476	1477
1436	1437	1438	1439	1440	1441	1442	1443	1444	1445	1446	1447	1448	1449	1450	1451	1452	1453	1454	1455	1456
1415	1416	1417	1418	1419	1420	1421	1422	1423	1424	1425	1426	1427	1428	1429	1430	1431	1432	1433	1434	1435
1394	1395	1396	1397	1398	1399	1400	1401	1402	1403	1404	1405	1406	1407	1408	1409	1410	1411	1412	1413	1414
1373	1374	1375	1376	1377	1378	1379	1380	1381	1382	1383	1384	1385	1386	1387	1388	1389	1390	1391	1392	1393
1352	1353	1354	1355	1356	1357	1358	1359	1360	1361	1362	1363	1364	1365	1366	1367	1368	1369	1370	1371	1372
			1334	1335	1336	1337	1338	1339	1340	1341	1342	1343	1344	1345	1346	1347	1348	1349	1350	1351

locations for these groups. Therefore, in the figures listed for channel assignments the remaining channels that fill out the channels that should make up the DCCH list include those in probability block 2. Note that the channels on the top of all the charts represent the suggested channels that should be used for the initial control channels, DCCHs for the PCS band.

11.4.2 Frequency reuse

The frequency reuse for IS-136 is the same as that of IS-54. The C/I level desired is 17 dB and is the same for DCCH and the digital traffic channel (DTC). This is convenient since in all the cellular systems the majority of the channels are analog and they too require a minimum of 17 dB C/I. The fundamental issue here is that the same D/R ratios can and are used when implementing the radio channel assignments for digital.

The additional parameters associated with IS-136/IS-54 involve SDCC, DCC, and DVCC. The DCC is the digital color code, SDCC is the supplementary digital color code, and DVCC is the digital verification color code.

DCC and SDCC must be assigned to each sector, cell, and control channel of the system that utilizes IS-136/IS-54. The DCC is used by analog and dual-mode phones for accessing the system. The SDCC is used by dual-mode phones only and should be assigned to each control channel along with the DCC.

Parameter	Values
DCC	0,1,2,3
SDCC	0 through 15

The reuse pattern for the CDD and SDCC is shown in Fig. 11.36.

The DVCC is assigned to each digital traffic channel, DTC. There are a total of 255 different DVCC values, range 1 to 255, leaving much room for variations in assignments. However, how they should be assigned needs to be in a pattern. The recommended pattern is in Fig. 11.37.

The same DVCC should be assigned to *all* the channels in a sector in the same fashion that SAT is assigned to a frequency group in analog. However, in the event that more than one frequency group is assigned to a particular sector, it is recommended that the same DVCC be used for all the channels in the sector.

11.4.3 Channel structures

The channel structures for NADC were covered in Chap. 2. However, the variations that are particular to IS-136 are listed below for reference. Figure 11.38 is an illustration of the forward control channel slot structure. Figure 11.39 is the reverse control channel. Figure 11.40 is the DCCH channel structure that represents the DCCH structure in the traffic channels. Figure 11.41 is the IS-136 forward traffic channel slot structure.

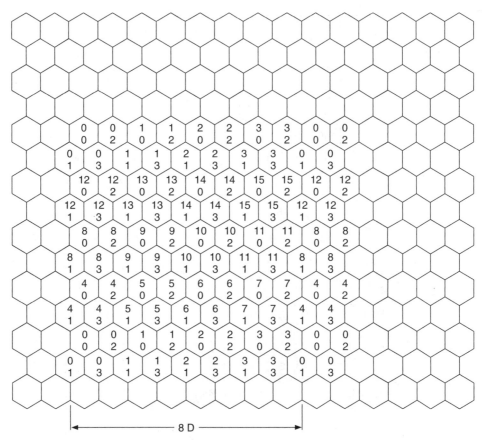

Figure 11.36   SDCC (top number) and DCC (bottom number) reuse pattern.

11.4.4 Bit error rate

The relationship between BER class and bit error rate (BER) is shown in Table 11.25. Note the relationship between BER and voice quality shown below:

BER, %	<1	1–3	>3
Voice quality	Good	Marginal	Bad

11.4.5 Capacity

The capacities for an IS-136 and IS-54 cell site are slightly different, depending on the use of the DCCH. Typically the relationship is that for every digital radio there are three potential traffic channels. For a radio that has a DCCH the capacity is limited to two traffic channels while for a radio that does not have a DCCH a total of three traffic channels are available.

The following tables are meant to help in assigning the traffic capability of the various digital channels when mixed in an analog environment. The same

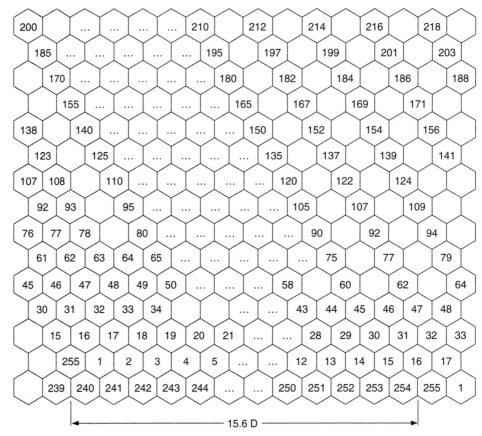

Figure 11.37 DVCC and traffic channel reuse.

S	SCF	DATA	CSFP	DATA	SCF	RSV
No. of bits 28	12	130	12	130	12	2

Figure 11.38 IS-136 forward control channel slot structure (DCC only). S = synchronization word. SCF = shared channel feedback. CSFP = coded superframe. RSV=reserved.

G	R	PREAM	S	DATA	SYNC+	DATA
No. of bits 6	6	16	28	128	24	122

Figure 11.39 IS-136 reverse control channel structure. G = guard time. R = transmitter ramp uptime. PREAM = preamble. SYNCH = second synchronization word.

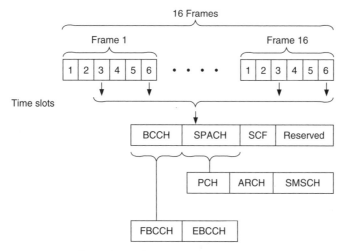

Figure 11.40   IS-136 forward traffic channel slot structure. 1 super-frame = 32 slots = 640 m s. BCCH = broadcast control channel. FBCCH = fast broadcast channel. EBCCH = extend broadcast channel. SPACH = short message, paging, and access response channel. PCH = paging channel. ARCH = access response channel. SMSCH = short message service channel. SCF = shared control feedback.

	S	SACCH	DATA	CDUCC	DATA	RSV	CDL
No. of bits	28	8	130	12	130	1	11

Figure 11.41   IS-136 forward traffic channel slot structure. S = synchronization word. SACCH = slow associated control channel. CDVCC = coded digital voice color code. RSV = reserved. CDL = coded DCC location.

TABLE 11.25

BER class	BER, %
0	<0.1
1	0.01–0.10
2	0.10–0.50
3	0.50–1.0
4	1.0–2.0
5	2.0–4.0
6	4.0–8.0
7	8.0<

configuration of 19 physical radio channels was used for all the examples, and a simple manipulation to your particular situation can be made to Table 11.26. For PCS bands, since there are no analog channels that need to be displaced, with the exception of current microwave users, the channel capacity can be extracted directly from the Erlang B table in Appendix A. Keep in mind that the DCCH has two traffic channels when calculating the capacity.

TABLE 11.26

	Channels			Erlangs, 29% erlang B		
		Digital				
No. of analog	No. of DTC	DCCH		Analog	Digital	Total*
19	0			12.3	0	12.3
18	1	3	0	11.5	0.602	14
	1	2	1		0.223	13.2
17	2	6	0	10.7	2.28	15.8
	2	5	1		1.66	16.6
16	3	9	0	9.83	4.34	17.5
	3	8	1		3.63	16.6
15	4	12	0	9.01	6.61	19.3
	4	11	1		5.84	18.4
14	5	15	0	8.2	9.01	19.3
	5	14	1		8.2	20.2

*Assumes trunking efficiency.

11.4.6 Link budget

The sample link budget for IS-136 is included for reference in Fig. 11.42. Note that an ERP of 100 W was picked and for link path balancing it is possible to decrease the ERP of the cell site significantly following the parameters used.

11.4.7 Number of cells required

The number of cells required for deployment of IS-136 is dependent upon many variables. The first variable is whether the system in which you are deploying this technology is either cellular or PCS. If it is PCS, then the example in Sec. 11.2 can be used for determining the number of cells required for the initial deployment. The chief difference in the calculation for the number of sites lies in the differences in the link budget, to determine cell radius and thus coverage. And the capacity of the cell is also dependent upon the number of channels available and to be utilized.

If the system is being overlaid into an existing cellular system the first issue is the determination of the number of analog channels to be displaced. The determination of channel redistribution can be achieved by following the example used in Chap. 10 for system growth, but putting a hard limit of, say, 17 channels per sector as the top-end channel growth capability before a new cell is required.

11.5 GSM

GSM is a technology platform that has gained worldwide acceptance and deployment. The GSM spectrum occupies several frequency bands. The particular band that GSM operates within is dependent upon the country that it

```
System: Example 1                          Date _____
Frequency of operation
     Tx
     Rx


Down-link path

Base station parameters
     Tx PA output power                    42.7  dBm
     Tx combiner loss                       1.5  dB
     Tx duplexer loss                       0    dB
     Jumper and connector loss             0.25  dB
     Lightning arrester loss               0.2   dB
     Feedline loss                          1.5  dB
     Jumper and connector loss             0.25  dB
     Tower top amp gain                     0    dB
     Antenna gain                          11    dBd

     Cell ERP                              50    dBm

Environmental margins
     Tx diversity gain                      0    dB
     Fading margin                          8    dB
     Environmental attenuation             15    dB
     Cell overlap                           5    dB

     Environmental margin                 −28    dB

Subscriber unit parameters
     Antenna gain                           3    dBd
     Rx diversity gain                      0    dB
     Antenna cable loss                     2    dB
     C/I or Eb/N0                          18    dB
     Rx sensitivity                      −110    dBm

     Subscriber unit effective sensitivity −93   dBm

Maximum down-link path loss allowed
     Base station ERP                      50    dBm
     Subscriber unit effective sensitivity −93   dBm

     Maximum down-link path loss          143    dB


Up-link path

Subscriber unit parameters
     Tx PA output                          28    dBm
     Cable and jumper loss                  0    dB
     Antenna gain                           0    dBd

     Subscriber unit ERP                   28    dBm

Environmental margins
     Fading margin                          8    dB
     Environmental attenuation             15    dB
     Cell overlap                           5    dB

     Environmental margin                  28    dB
```

Figure 11.42 IS-136 link budget.

```
Base station parameters
    Rx antenna gain                          11     dBd
    Tower top amp net gain                    0     dB
    Jumper and connector loss              0.25    dB
    Feedline loss                           1.5    dB
    Lightning arrester loss                 0.2    dB
    Jumper and connector loss              0.25    dB
    Duplexer loss                             0     dB
    Receive configuration loss                0     dB
    Rx diversity gain                         3     dB
    C/I or E_b/N_0                           18     dB
    Rx sensitivity                         -110    dBm

    Base station effective sensitivity     103.8   dBm

Maximum up-link path loss allowed
    Subscriber unit ERP                      28    dBm
    Base station effective sensitivity     103.8   dBm

    Maximum up-link path loss              131.8   dB
```

Figure 11.42 (*Continued*)

is in. For instance, GSM operates in the 900-MHz range, DCS 1800 in the 1800-MHz range, and DCS1900 or PCS1900 in the 1900-MHz range. Each of the names given (GSM, DCS1800, DCS1900, or PCS1900) is a GSM system. The features and functionality of GSM are the same for each of the bands mentioned. The general GSM system is shown in Fig. 11.43.

11.5.1 Channel allocation

The channel allocations associated with GSM are defined in Fig. 11.44 for GSM and Fig. 11.45 for DCS 1900. The channel chart for PCS1900 follows in Table 11.27.

Note that a three-sector cell was assumed using an $N = 3$ reuse pattern. If another reuse pattern is chosen for the system, a reallocation of the channels is easily achieved for PCS blocks A, B, and C. However, for PCS blocks D, E, and F the flexibility to relax the reuse distance does not really exist because of the allotted spectrum.

11.5.2 Call processing

The initialization process for a GSM subscriber unit trying to acquire the GSM system is shown in Fig. 11.46. It should be noted that the BCCHs can be located on any GSM channel in the system. An extremely important concept in GSM is the ability of the subscriber unit to hand off to another cell to maintain the same call quality as it traverses through the network. The flow diagram of the handover process is shown in Fig. 11.47. In Fig. 11.47 the first step taken was to adjust the power level of the up- and down-link, i.e., subscriber unit or base station, prior to the initiation of the handover process algorithm. Once the

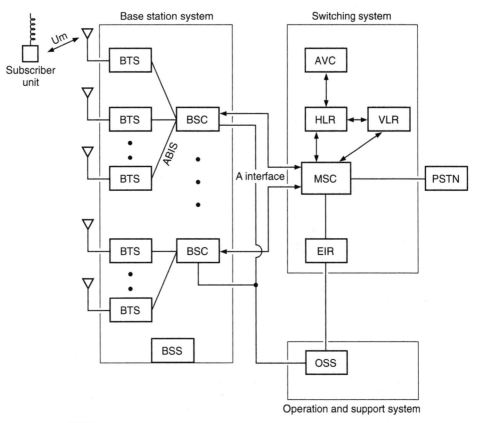

Figure 11.43 GSM system.

Figure 11.44 GSM channel. Base Tx:
[((channel no.) × 0.2) + 935.0] MHz.
Base Rx: [((channel no.) × 0.2) + 890.0] MHz.

A MTA	D BTA	B MTA	E BTA	F BTA	C BTA

512   586 611   686 711 736   810

Figure 11.45 DCS 1900 channels. Base Tx: [((channel no. −
511) × 0.2) + 1930] MHz. Base Rx: [((channel no. − 511)
× 0.2) + 1850] MHz.

TABLE 11.27   Channel chart for PCS 1900

A1	B1	C1	A2	B2	C2	A3	B3	C3
A block								
512	513	514	515	516	517	518	519	520
521	522	523	524	525	526	527	528	529
530	531	532	533	534	535	536	537	538
539	540	541	542	543	544	545	546	547
548	549	550	551	552	553	554	555	556
557	558	559	560	561	562	563	564	565
566	567	568	569	570	571	572	573	574
575	576	577	578	579	580	581	582	583
584	585							
B block								
611	612	613	614	615	616	617	618	619
620	621	622	623	624	625	626	627	628
629	630	631	632	633	634	635	636	637
638	639	640	641	642	643	644	645	646
647	648	649	650	651	652	653	654	655
656	657	658	659	660	661	662	663	664
665	666	667	668	669	670	671	672	673
674	675	676	677	678	679	680	681	682
683	684	685						
C block								
736	737	738	739	740	741	742	743	744
745	746	747	748	749	750	751	752	753
754	755	756	757	758	759	760	761	762
763	764	765	766	767	768	769	770	771
772	773	774	775	776	777	778	779	780
781	782	783	784	785	786	787	788	789
790	791	792	793	794	795	796	797	798
799	800	801	802	803	804	805	806	807
808	809	810						
D block								
586	587	588	589	590	591	592	593	594
595	596	597	598	599	600	601	602	603
604	605	606	607	608	609	610		
E block								
686	687	688	689	690	691	692	693	694
695	696	697	698	699	700	701	702	703
704	705	706	707	708	709	710		
F block								
711	712	713	714	715	716	717	718	719
720	721	722	723	724	725	726	727	728
729	730	731	732	733	734	735		

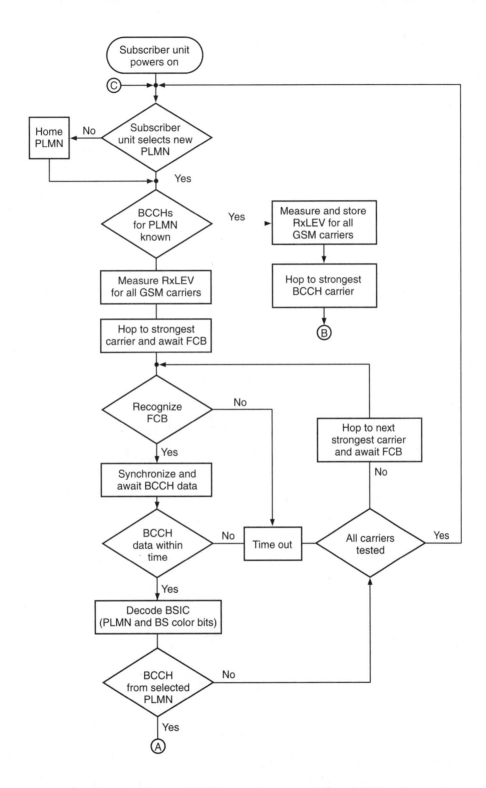

Figure 11.46 GSM initialization. BCCH = broadcast control channel; FCB = frequency control bursts; BSIC = base system ID code.

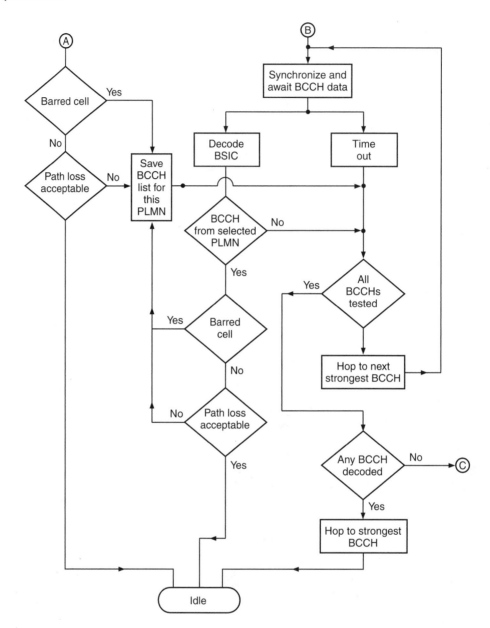

Figure 11.46   (*Continued*)

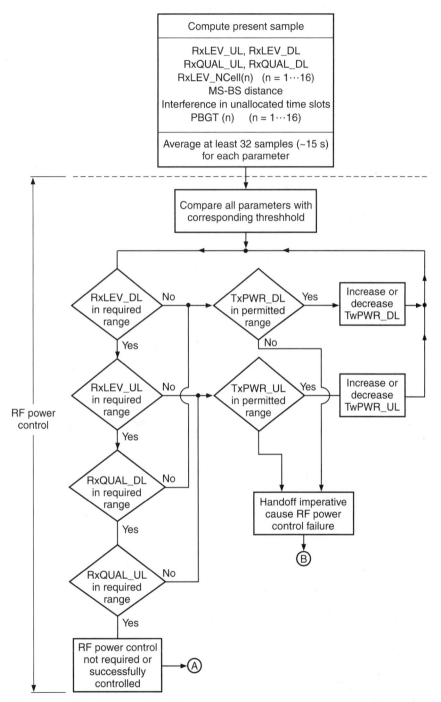

Figure 11.47   Flow diagram of handover process.

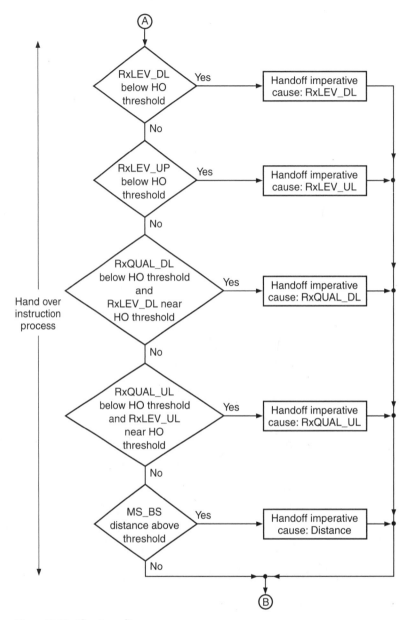

Figure 11.47 (*Continued*)

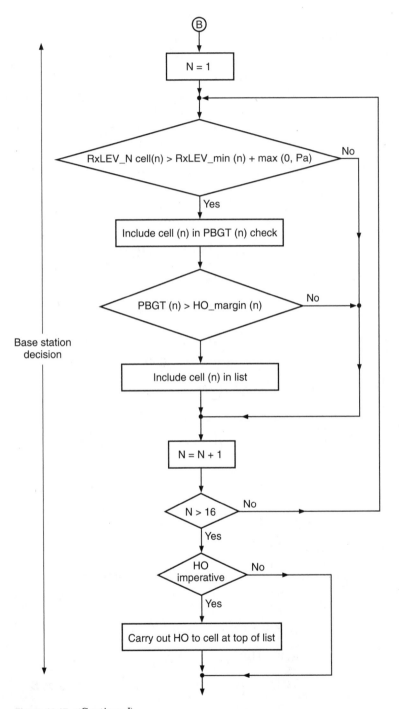

Figure 11.47   *(Continued)*

handover algorithm is completed the next phase of the process is to have the base station make the final decision on what to do with the subscriber.

Equations (11.4)–(11.6) will help define the handover process.

$$\text{PBGT}(n) = [\min(\text{Ms\_TxPWR\_MAX}, P) - \text{RxLEV\_DL}]$$
$$- [\min(\text{Ms\_Tx PWR\_MAX}(R), P) - \text{RxLEV\_N cell}(n)] \quad (11.4)$$

$$\text{Ms\_TxPWR\_MAX} = \text{maximum allowed Ms transmitted power on traffic chart}$$

$$\text{Ms\_TxPWR\_MAX}(n) = \text{Ms\_TxPWR\_MAX in adjacent } (n) \text{ cell}$$

$$P = \text{maximum transmitted power capability of mobile}$$

$$\text{RxLEV\_N cell } (n) > \text{Rx LEV\_MIN}(n) + \text{MAX } (0, \text{Pa})$$

$$\text{when Pa} = [\text{Ms\_TxPWR\_MAX}(n) - P]$$

SDCCH time slots:

$$x = \text{no. of call attempts/s} + \text{no. of location updates/s} \quad (11.5)$$

$$x = 1 \text{ call attempt/s} + 2 \text{ location updates/s}$$

$$x = 3 \text{ activities/s} = 3 \times 3600 = 10{,}800 \text{ activities/h}$$

$$y = \text{average SDCCH occupancy} = 4 \text{ s} = 0.00111 \text{ h/activity}$$

$$\text{SDCCH traffic} = x \cdot y = 10{,}800 \cdot 0.0011 = \underline{12 \text{ erlangs}}$$

$$\text{SDCCH erlang} = 12 \text{ erlangs} = 15 \text{ channels } 10 \text{ percent } \underline{\text{erlang B}}$$

$$\text{SDCCH time slots} = \frac{\text{SDCCH channels}}{8} = 1.875 = \underline{2}$$

Therefore, 2 SDCCH time slots are needed.

Number of radios:

No. of radios = no. of TCHs + SDCCH/8 time slots

$$+ \text{ no. of BCCH/CCCH} \quad (11.6)$$

$$\text{BCCH/CCCH} \quad = 1$$

$$\text{SDCCH/8 slots} \quad = 2$$

$$\text{TCHs (10 erlangs)} = 17 \quad 2 \text{ percent erlang B}$$

$$= 17 + 2 + 1 = 20$$

Therefore, $\frac{20}{8} = \underline{3}$ GSM carriers are needed.

Lap D:

$$\frac{\text{No. of carriers/no. of sectors}}{4} = \frac{3}{4} = 0.75 \approx 1$$

Therefore, 1 Lap D is needed.

Table 11.28 equates RxQUAL to BER percent. RxQUAL is an important parameter in GSM, and its relationship to BER percent is important.

Table 11.28   RxQUAL vs. BER

RxQUAL	BER, %
0	< 0.2
1	0.2 – 0.4
2	0.4 – 0.8
3	0.8 – 1.6
4	1.6 – 3.2
5	3.2 – 6.4
6	6.4 – 12.8
7	12.8 >

11.5.3   Capacity

With GSM systems the traffic load for a particular cell is dependent upon not only the number of traffic channels available for use but also the link capacity between major subsystems in a GSM environment. The following guidelines can be used in determining the dimensioning requirements for a GSM network.

The Um interface is the link that connects the subscriber unit to the base station itself. The Um interface utilizes a LAPDm protocol for signaling. The link between the BSC and BTS is the interface referred to as Abis and is the protocol used between them.

MS→BTS	Air interface (Um)
BTS–BSC	Abis
BSC–MSC	A interface

The Um interface dimensioning depends on the number of traffic channels (TCH), signaling channels (BCCH, SDCCH/8), and control channels (RACH, PCH, and AGCH). Since each radio channel in GSM can support up to eight traffic channels and signaling or control channels, the relationships between them need to be defined. In addition the BTS dimensioning, unlike other wireless platforms, is dependent upon the radio and Abis dimensioning.

The preceding equations show the method that can be used for calculating the required SDCCH time slots, number of physical radios, and the LAPDs needed for the site. Most vendors provide a lookup table that has these values already computed, but in the event that one does not exist, this is how you can figure it out.

```
System: Example 1                        Date _____
Frequency of operation
    Tx
    Rx


Down-link path

Base station parameters
    Tx PA output power                   42.7   dBm
    Tx combiner loss                      1.5   dB
    Tx duplexer loss                      0     dB
    Jumper and connector loss             0.25  dB
    Lightning arrester loss               0.2   dB
    Feedline loss                         1.5   dB
    Jumper and connector loss             0.25  dB
    Tower top amp gain                    0     dB
    Antenna gain                         11     dBd

    Cell ERP                             50     dBm

Environmental margins
    Tx diversity gain                     0     dB
    Fading margin                         8     dB
    Environmental attenuation            15     dB
    Cell overlap                          5     dB

    Environmental margin                -28     dB

Subscriber unit parameters
    Antenna gain                          3     dBd
    Rx diversity gain                     0     dB
    Antenna cable loss                    2     dB
    C/I or $E_b/N_0$                      11    dB
    Rx sensitivity                     -102     dBm

    Subscriber unit effective sensitivity -92   dBm

Maximum down-link path loss allowed
    Base station ERP                     50     dBm
    Subscriber unit effective sensitivity -92   dBm

    Maximum down-link path loss         142     dB


Up-link path

Subscriber unit parameters
    Tx PA output                         28     dBm
    Cable and jumper loss                 0     dB
    Antenna gain                          0     dBd

    Subscriber unit ERP                  28     dBm

Environmental margins
    Fading margin                         8     dB
    Environmental attenuation            15     dB
    Cell overlap                          5     dB

    Environmental margin                 28     dB
```

Figure 11.48 GSM link budget.

Base station parameters		
Rx antenna gain	11	dBd
Tower top amp net gain	0	dB
Jumper and connector loss	0.25	dB
Feedline loss	1.5	dB
Lightning arrester loss	0.2	dB
Jumper and connector loss	0.25	dB
Duplexer loss	0	dB
Receive configuration loss	0	dB
Rx diversity gain	3	dB
C/I or E_b/N_0	11	dB
Rx sensitivity	−107	dBm
Base station effective sensitivity	107.8	dBm
Maximum up-link path loss allowed		
Subscriber unit ERP	28	dBm
Base station effective sensitivity	107.8	dBm
Maximum up-link path loss	135.8	dB

Figure 11.48   (*Continued*)

11.5.4   Link budget

Figure 11.48 is a standard link budget that can be used for GSM designs. The link budget values then can be used to determine the number of sites needed for a given area. The methodology that should be used for a new system is the same that is used in previous sections of this chapter.

If the system exists, the methodology that should be used is that defined in Chap. 10 with adjustments made to the maximum traffic-handling capacity based on the available spectrum for the system and the reuse pattern chosen.

11.6   Final Report

With any system design aspect a document should be generated indicating for review by peers what the design is and the underlying assumptions that drove the creation of the design. The report should take on the same look and feel of that covered in Chap. 10. At the minimum it is necessary to include some key elements in the report. Items to include:

1. What is the objective?

2. Time frames for report

3. Technology-specific issues

4. Design criteria and sources (include link budget report)

5. Design

6. Implementation time frame

7. Budget considerations

References

1. Smith, Clint, and Curt Gervelis, *Cellular System Design and Optimization,* McGraw-Hill, New York, 1996.
2. Mouly, Michel, and Marie-Bernadette Pautet, *The GSM System for Mobile Communications,* 1992.
3. Steele, Raymond, *Mobile Radio Communications,* IEEE, 1992.
4. *CDMA Network Engineering Handbook,* Qualcom, 1995.
5. J-STD-008, TIA.
6. Kaufman, M., and A. H. Seidman, *Handbook of Electronics Calculations,* 2d ed., McGraw-Hill, New York, 1988.
7. Lee, W.C.Y., *Mobile Cellular Telecommunications Systems,* 2d ed., McGraw-Hill, New York, 1996.
8. Qualcom, *An Overview of the Application of Code Division Multiple Access (CDMA) to Digital Cellular Systems and Personal Cellular Networks,* Qualcom, San Diego, CA, May 21, 1992.
9. Simo, "IS-95 Based SS-CDMA: Operational ISSUES," seminar, January 1995.
10. White, Duff, *Electromagnetic Interference and Compatibility,* Interference Control Technologies, Gainesville, GA, 1972.
11. Yarborough, *Electrical Engineering Reference Manual,* 5th ed., Professional Publications, Inc., Belmont, CA, 1990.
12. Qualcom, *CDMA Handbook,* Qualcom, San Diego, CA, 1996.
13. Fink, Donald, and Donald Christiansen, *Electronics Engineers Handbook,* 3d ed., McGraw-Hill, New York, 1989.
14. Fink, Beaty, *Standard Handbook for Electrical Engineers,* 13th ed., McGraw-Hill, New York, 1995.
15. Steele, *Mobile Radio Communications,* IEEE, New York, 1992.
16. Jakes, W.C., *Microwave Mobile Communications,* IEEE, New York, 1974.
17. Rappaport, *Cellular Radio & Personal Communications,* IEEE, New York, 1995.
18. *Reference Data for Radio Engineers,* 6th ed., Sams, 1983.
19. Mouly, Pautet, *The GSM System for Mobile Communications,* Mouly Pautet, 1992.

Erlang B Grade of Service

Erlang B Grade of Service

Channels	1	1.5	2	3	5
1	0.01	0.012	0.0204	0.0309	0.0526
2	0.153	0.19	0.223	0.282	0.381
3	0.455	0.535	0.602	0.715	0.899
4	0.869	0.992	1.09	1.26	1.52
5	1.36	1.52	1.66	1.88	2.22
6	1.91	2.11	2.28	2.54	2.96
7	2.5	2.74	2.94	3.25	3.74
8	3.13	3.4	3.63	3.99	4.54
9	3.78	4.09	4.34	4.75	5.37
10	4.46	4.81	5.08	5.53	6.22
11	5.16	5.54	5.84	6.33	7.08
12	5.88	6.29	6.61	7.14	7.95
13	6.61	7.05	7.4	7.97	8.83
14	7.35	7.82	8.2	8.8	9.73
15	8.11	8.61	9.01	9.65	10.6
16	8.88	9.41	9.83	10.5	11.5
17	9.65	10.2	10.7	11.4	12.5
18	10.4	11	11.5	12.2	13.4
19	11.2	11.8	12.3	13.1	14.3
20	12	12.7	13.2	14	15.2
21	12.8	13.5	14	14.9	16.2
22	13.7	14.3	14.9	15.8	17.1
23	14.5	15.2	15.8	16.7	18.1
24	15.3	16	16.6	17.6	19
25	16.1	16.9	17.5	18.5	20
26	17	17.8	18.4	19.4	20.9
27	17.8	18.6	19.3	20.3	21.9
28	18.6	19.5	20.2	21.2	22.9
29	19.5	20.4	21	22.2	23.8
30	20.3	21.2	21.9	23.1	24.8
31	21.2	22.1	22.8	24	25.8
32	22	23	23.7	24.9	26.7
33	22.9	23.9	24.6	25.8	27.7
34	23.8	24.8	25.5	26.8	28.7
35	24.6	25.6	26.4	27.7	29.7
36	25.5	26.5	27.3	28.6	30.7
37	26.4	27.4	28.3	29.6	31.6
38	27.3	28.3	29.2	30.5	32.6
39	28.1	29.2	30.1	31.5	33.6
40	29	30.1	31	32.4	34.6
41	29.9	31	31.9	33.4	35.6
42	30.8	31.9	32.8	34.3	36.6
43	31.7	32.8	33.8	35.3	37.6
44	32.5	33.7	34.7	36.2	38.6
45	33.4	34.6	35.6	37.2	39.6
46	34.3	35.6	36.5	38.1	40.5
47	35.2	36.5	37.5	39.1	41.5
48	36.1	37.4	38.4	40	42.5
49	37	38.5	39.3	41	43.5
50	37.9	39.2	40.3	41.9	44.5
51	38.8	40.1	41.2	42.9	45.5

Erlang B Grade of Service (*Continued*)

Channels	1	1.5	2	3	5
52	39.7	41	42.1	43.9	46.5
53	40.6	42	43.1	44.8	47.5
54	41.5	42.9	44	45.8	48.5
55	42.4	43.8	44.9	46.7	49.5
56	43.3	44.7	45.9	47.7	50.5
57	44.2	45.7	46.8	48.7	51.5
58	45.1	46.6	47.8	49.6	52.6
59	46	47.5	48.7	50.6	53.6
60	46.9	48.4	49.6	51.6	54.6
61	47.9	49.4	50.6	52.5	55.6
62	48.8	50.3	51.5	53.5	56.6
63	49.7	51.2	52.5	54.5	57.6
64	50.6	52.2	53.4	55.4	58.6
65	51.5	53.1	54.4	56.4	59.6
66	52.4	54	55.3	57.4	60.6
67	53.4	55	56.3	58.4	61.6
68	54.3	55.9	57.2	59.3	62.6
69	55.2	56.9	58.2	60.3	63.7
70	56.1	57.8	59.1	61.3	54.7

Cable and Antenna System
Sweeping Procedure

Outline

The following is a general procedure to be used when sweeping a site for potential problems with the antenna system or Tx/Rx cables. The intent of this procedure is to establish guidelines for a pass-fail criterion for cables, feedlines, and antennas at the site under test. With the number of different configurations available in the cellular industry alone, it is not deemed feasible to have a separate procedure for every configuration available. However, the procedures outlined below can be used for all the types of infrastructure equipment and configurations available.

This procedure can be used for AT&T, Ericsson, and Motorola cell sites and easily adapted to Nortel, Hughes, Samsung, Hyundi, Seimens, LGI, and the like. Note that the calibration procedure outlined must be completed prior to any analysis or data collection. A printout of the test cable and the operating parameters must be included with the test data.

Equipment Required

The following equipment is required for conducting this test:

1. HP network analyzer model 8753a with option 010
2. HP transmission reflections test set model 85044A
3. HP 50-Ω type N accessory kit model 1853A
4. HP RF cable kit model 11841B
5. HP think jet printer with HB-IB interface and cable
6. Two 10-ft superflex cables (0.5-in-diameter with N-type connectors)
7. Paper for printer

8. Ohmmeter

9. Tape measure

A network analyzer with S parameters built into its normal functions would simplify the process involving equipment to carry into the cell site. A wide variety of network analyzer equipment is currently on the market, and many of the units will perform the task adequately.

Cell Site Configuration

AT&T

The primary type of AT&T infrastructure equipment used in cellular is the Series II equipment. It can be made into many different configurations depending upon the operating system requirements. However, there are Series I cell sites, both Mod 1 and Mod 2 vintages, which are still in operation. In addition there are Series II M and MM series cell sites in addition to LMTs and compact base stations. The procedure that follows is geared for a Series II cell site, but with the replacement of the name to other equipment vintages the procedure is functionally the same.

A tick list is included here to use as a guide for anticipating the amount of work required for the site to be tested and also to ensure that the test was carried out to its logical conclusion.

1. Type of site (SII) (SI Mod 1) (S1 Mod 2) (S2 M) (S2 MM) (CB) (LMT)

2. Primary transmit configuration (omni/sector)

3. Secondary transmit configuration (omni/sector)

4. Receive configuration (omni/sector)

5. Number of cell site equipment bays RTU_____ LAC_____ AIF_____

6. Receive filter type (KS model number) (A or B band), bandpass poles, notch

7. Setup configuration: (<u>omni, directional setup, simulcast</u>)

8. Number and type of transmit antennas (by sector/face)

    Setup_____

    Primary_____

    Secondary_____

9. Number and type of receive antennas (by sector/face)

    Setup_____

    Primary_____

    Secondary_____

10. Number and type of hybrid combiners_____

11. Number and type of duplexers_____

12. Attenuators in receive line (location and dB value)_____

Ericsson

Currently several types of Ericsson infrastructure equipment are utilized in cellular. However, for this procedure the RBS 8800 is assumed to be the configuration. The other configurations utilized can be adapted to this procedure. The RBS equipment from Ericsson will be of the general configuration of an omni/omni or sector/sector cell site. As with all the other equipment vendors the sweeping procedures and basic principles for testing the cables and the antenna system will be the same.

A tick list is included here to use as a guide for anticipating the amount of work required for the site to be tested and also to ensure that the test was carried out to its logical conclusion.

1. Type of site (omni/omni) (sector/sector)

2. Number of RBS bays

3. Number and type of transmit antennas

4. Number and type of receive antennas

5. Number and type of hybrid combiners

6. Number and type of duplexers

Motorola

Four general types of cell sites will be used for the procedure, LD mixed, HDII, HD, and HD/HDII. The receive and transmit configuration will naturally be different for omni/omni, omni/sector, and sector/sector. However, the basic principle for testing an O/O, O/S, or S/S site will be functionally the same. The configurations for the Motorola super cell and mini cell are not included in this procedure, but again these configurations are a simple adaptation to this procedure.

A tick list is included here to use as a guide for anticipating the amount of work required for the site to be tested and also to ensure that the test was carried out to its logical conclusion.

1. Type of site (HD) (LD mixed) (HDII) (HD/HDII)

2. Composite cell (Y/N)

3. Primary transmit configuration (omni/sector)

4. Secondary transmit configuration (omni/sector)

5. Receive configuration (omni/sector)

6. Number of sites HD_____ LD mixed_____ HDII_____

7. EAMPS capable (Y/N)

8. Multicoupler (Y/N)

9. Extenders or 7-dB pads_____

10. Number of extenders_____

11. Sig on sector (Y/N)

12. Number and type of transmit antennas

   Sig (primary and secondary)_____

   Primary voice_____

   Secondary_____

13. Number and type of receive antennas_____

14. Number and type of hybrid combiners_____

15. Number and type of duplexers_____

Procedure

Obtain a cell site drawing (current configuration) or make the drawing at the cell site. If a current volume two or as built is available, this can be used. If not, a drawing must be made as detailed as possible (i.e., drawing should contain the same information that the configuration schematic in a volume two has). The drawing is used as a guide for determining what has been tested and what still needs to be tested.

Currently, this procedure does not test the hybrids, duplexers, bandpass filters, multicouplers, or Sig on sector equipment. It is intended only to verify the antenna systems and Tx/Rx cables at a site.

For the test, each cable in the site (transmit and receive) must be checked to verify its integrity. In addition the feedlines and antennas will also be checked after each measurement of the feedlines and antenna. A printout of the screen's display will be done so that a permanent record will be available. The Tx and Rx cables will be passed or failed on site and printouts will be made only for those that fail and are not immediately replaced.

Attention must be paid to labeling the printouts since this is all that the engineer reviewing the test data will see and he or she will need to know what the printout is.

The following part of the test procedure will cover how to test

1. Cables (in the site)

2. Feedline

3. Antennas

Calibration

Prior to any test the equipment needs to be calibrated. The manual for the HP network analyzer is included or should be with the test equipment. It is highly recommended that the following sequence be followed in calibrating the equipment. The procedure to follow is an S11 one-port 50-Ω N-type calibration.

First the network analyzer should be calibrated with the transmission reflection test set only and no superflex cable attached. Once the integrity of the equipment is verified the unit will be recalibrated with the superflex cable and bullet attached. During part of the text sequence the bullet will not be used. In this case the unit will need to be recalibrated.

Enable Printer

1. Connect the printer to the HP network analyzer utilizing the HP-IB cable (with paper in the printer).
2. Press the LOCAL button on the network analyzer.
3. Select the SYSTEM CONTROLLER on the screen.
4. Press COPY button on the network analyzer.
5. Select PRINT on the screen and verify the printer is operating.

Calibration Procedure

1. Let the network analyzer warm up for 30 min.
2. Connect the test setup.
3. Press the RECALL button on the network analyzer.
4. Select the RECALL REG 1 on the screen (the parameters that the network analyzer will use for the test such as the frequency band; output power and other settings have been stored in this register).
5. Press the CAL button on the network analyzer.
6. Select the CALIBRATE MENU on the screen.
7. Select the S11 1-PORT test on the screen.
8. Select the OPENS on the screen.
9. Connect the SHORT FEMALE CONNECTOR to the test port of the transmission reflection test set.
10. Select the SHORTS on the screen.
11. Connect the open FEMALE CONNECTOR to the text port of the transmission reflection test set.
12. Select the SHORT (M) on the screen.
13. Wait for beep and press DONE on the screen.
14. Connect the load FEMALE LOAD to the test port of the transmission reflection test set.
15. Select the LOAD on the screen.

16. Wait for beep, press DONE on the screen, and wait for screen menu to change (takes a few seconds). "COR" should appear on the left-hand side of the screen indicating that the correction has been turned on.

17. Select REG1 on the screen.

18. Press the FORMAT button on the network analyzer.

19. Select the LOG MAG on the screen.

20. With the 50-Ω load still attached to the transmission reflection test set, check if there is a better than −60-dB return loss (i.e., return loss should be at least −60 dB down; −50 dB is unacceptable). If there is, continue or recalibrate the unit. The reference scale should be 10 dB per division.

21. If you cannot achieve better than a −60-dB return loss for the 50-Ω load (i.e., −60 dB or below), something is wrong with the equipment or it has been set up improperly.

    NOW THE CABLE'S INTEGRITY MUST BE CHECKED BEFORE THE ACTUAL MEASUREMENTS ARE CONTINUED.

22. Connect the cable without the bullet to the test port of the transmission reflection test set and connect the male 50-Ω load to the end of the cables.

23. With the network analyzer still in LOGMAG measure the return loss. The return loss should be at least −20 dB down. If the return loss is greater than −20 dB (i.e., −19 dB) then the cable should be replaced with a better cable for the test.

24. Connect the male short to the end of the cable.

25. With the network analyzer still in LOGMAG measure the insertion loss of the cable. It may be necessary to change the scale to 1 dB per division. This is done by pressing SCALE FEF on the network analyzer. Then select SCALE/DIVISION on the screen, and then enter the desired scale (1 in this case). If the units you are looking for are not there press x1. The insertion loss should be no worse than −3 dB (i.e., −3.1 dB is unacceptable). If the insertion loss is working at −3 dB the cable should be replaced with a better one.

26. Connect the bullet to the cable, use the female 50-Ω load now, and repeat step 23. If a problem occurs, the point of failure is probably the bullet, and it will need to be replaced.

27. A hard copy of the return and insertion loss for the cable that will ultimately be used for testing should be generated at this point by pressing the COPY button on the network analyzer and then selecting PRINT on the screen. (Note that only what appears on the screen can be printed so you will have to generate a copy for each of these measurements while it is still on the screen.)

Two separate calibration procedures follow. One is for the cable with the bullet attached and the other is for the cable without the bullet attached.

Cable without the Bullet

1. Press the CAL BUTTON on the network analyzer.
2. Select the CALIBRATE MENU on the screen.
3. Select the S11 1-PORT test on the screen.
4. Select the OPENS on the screen.
5. Connect the open MALE CONNECTOR to the end of the cable.
6. Select the OPEN (F) on the screen.
7. Wait for beep and press DONE on the screen.
8. Select the SHORTS on the screen.
9. Connect the SHORT MALE CONNECTOR to the end of the cable.
10. Select the SHORT (F) on the screen.
11. Wait for beep and press DONE on the screen.
12. Connect the MALE LOAD to the end of the cable.
13. Select the LOAD on the screen.
14. Wait for beep, and press DONE on the screen, and wait for screen menu to change.
15. Select RE-SAVE REG1 on the screen.
16. Press the FORMAT button on the network analyzer.
17. Select the LOG MAG on the screen.
18. With the 50-Ω load still attached to the transmission reflection test set is there a better than -60 dB RETURN LOSS? If so, continue, or else recalibrate the unit.

Cable with Bullet Attached

1. Press the CAL BUTTON on the network analyzer.
2. Select the CALIBRATE MENU on the screen.
3. Select the S11 1-PORT test on the screen.
4. Select the OPENS on the screen.
5. Connect the OPEN FEMALE CONNECTOR to the end of the bullet.
6. Select the OPEN (F) on the screen.
7. Wait for beep and press DONE on the screen.
8. Select the SHORTS on the screen.
9. Connect the SHORT FEMALE CONNECTOR to the end of the bullet.
10. Select the SHORT (F) on the screen.

11. Wait for beep and press DONE on the screen.

12. Connect the FEMALE LOAD to the end of the bullet.

13. Select the LOAD on the screen.

14. Wait for beep, press DONE on the screen, and wait for screen menu to change.

15. Select RE-SAVE REG1 on the screen.

16. Press the FORMAT button on the network analyzer.

17. Select the LOG MAG on the screen.

18. With the 50-Ω load still attached to the transmission reflection test set, is there a better than -60-dB RETURN LOSS? If so, continue, or else recalibrate the unit.

WHEN CONDUCTING THE TEST THE SITE SHOULD BE OOS TO PREVENT SERVICE-AFFECTING PROBLEMS AND ALSO DAMAGE TO THE EQUIPMENT USED FOR THE TEST.

Testing Procedure

1. At this time the average function should be engaged.

2. Press the AVG button on the network analyzer.

3. Select AVG ON/OFF on the screen. Avg should now appear on the left-hand side of the screen, indicating that averaging is active.

4. If at any time during the testing procedure it becomes necessary to recalibrate, make sure that averaging is turned off. Selecting the AVG ON/OFF again will toggle it off.

Cables

PRIOR TO TESTING THE CABLES IT IS HIGHLY RECOMMENDED THAT A SEPARATE TICK LIST BE GENERATED BY THE FIELD TEAM AT THE SITE LISTING ALL AMOUNTS AND LOCATION OF ALL THE TRANSMIT AND THEN RECEIVE CABLES AT THE SITE.

IT MAY BE NECESSARY TO UTILIZE AN OHMMETER TO TRACE OUT THE CABLES.

MARKERS SHOULD BE PLACED AT THE START AND STOP FREQUENCIES FOR THE SERVICE BAND BEING TESTED.

Receive

The receive cables include all the cables between the following items which are dependent upon the infrastructure vendor utilized. Particular variants will need to be made to reflect the current vintage of equipment utilized at the site.

AT&T. Between AIF and the RTU.

Motorola. Multicoupler. Multicoupler extender, and the radio bays.

Ericsson. Bulkhead cable entry and the multicoupler on top of the equipment frame.

Starting with the receive cables, measure the insertion and return loss on each cable utilizing the SUPERFLEX CABLE WITH A BULLET for connecting the transmission reflection test set to the cable under test. Use the MALE SHORT AND MALE 50- Ω LOAD at the other end of the cable to establish a good short for insertion loss measurements and a good termination for return loss measurements at the other end of the cable.

1. Press the FORMAT button on the network analyzer.

2. Select the LOG MAG on the screen.

3. With the short attached, the insertion loss for the cable should not be more than -3 dB (i.e., 3.1 fails).

4. With the load attached, the return loss for the cable should be better than -20 dB (i.e., -19 dB fails).

5. If the cable fails either of these tests it should be replaced; if this is not possible then press the COPY button on the network analyzer and then select PRINT on the screen to obtain a hard copy of both tests (note only what appears on the screen can be printed so you will have to generate a hard copy for each of these measurements while it is still on the screen).

6. Mark on each printout the cable number and reference the cable number to the site configuration drawing acquired before the test or made on site. The length of the cable should also be measured and marked on the printout.

7. Repeat steps 1 through 6 for each receive cable at the site.

Transmit

The transmit cables involve all the cables between the harmonic or bandpass filters and the feedline connection in the cell site. Starting with the transmit cables, measure the insertion and return loss on each cable utilizing the SUPERFLEX CABLE WITH A BULLET for connecting the transmission reflection test set to the cable under test. Use the MALE SHORT AND MALE 50-Ω LOAD at the other end of the cable to establish a good short for insertion loss measurements and a good termination for return loss measurements at the other end of the cable.

1. Press the FORMAT button on the network analyzer.

2. Select the LOG MAG on the screen.

3. With the short attached, the insertion loss for the cable should not be more than -3 dB (i.e., 3.1 fails).

4. With the load attached, the return loss for the cable should be better than −20 dB (i.e., −19 dB fails).

5. If the cable fails either of these tests it should be replaced; if this is not possible then press the COPY button on the network analyzer and then select PRINT on the screen to obtain a hard copy of both tests (note only what appears on the screen can be printed, so you will have to generate a hard copy for each of these measurements while it is still on the screen).

6. Mark on each printout the cable number and reference the cable number to the site configuration drawing acquired before the test or made on site. The length of the cable should also be measured and marked on the print-out.

7. Repeat steps 1 through 6 for each receive cable at the site.

Antenna System

For the following sequence of tests the reference cable will be connected to the feedline inside the cell site. The objective of this test is to measure the response of the feedline and antenna and also the antenna by itself. Printouts will be made of all measurements regardless of their results.

1. Calibrate the CABLE WITHOUT THE BULLET.

2. Connect the cable to the input of the feedline in the cell site.

3. Press the FORMAT button on the network analyzer.

4. Select the LOG MAG on the screen.

5. This is the response of the feedline and antenna, and the return loss should be at least −20 dB (i.e., −15 dB is not good). If the return loss is less than −20 dB there is a potential problem in the antenna system.

6. Press the COPY button on the network analyzer and then select PRINT for the hard copy of the measurement.

7. Mark on each printout the feedline/antenna number and reference the cable number to the site configuration drawing acquired before the test or made on site.

8. Press the FORMAT button on the network analyzer.

9. Select SWR on the screen.

10. Adjust the scale to be 0.1 per division.

11. This is the return loss of the feedline and antenna, and the SWR should be better than 1:1.3 across the band.

12. Press the COPY button on the network analyzer and then select PRINT for the hard copy of the measurement.

13. Mark on each printout the feedline antenna number and reference the cable number to the site configuration drawing acquired before the test or made on site.

14. Press the FORMAT button on the network analyzer.

15. Select LOG MAG on the screen.

16. Set the scale to 10 dB per division.

17. Press the SYSTEM button on the network analyzer.

18. Select the TRANSFORM MENU on the screen.

19. Select the TRANSFORM ON on the screen.

20. Select SPECIFY GATE on the screen.

21. Adjust the gates to encompass the antenna only. Do this by selecting GATE CENTER on the screen and using the wheel to position the "t" over the peak of the last big bump that is visible. This is the response of the antenna. The location of this response can be estimated by performing the following simple calculation (2×1.25 ns $\times$ length of feedline). Select GATE START and position the flag at the beginning of the antenna response. Select GATE STOP and position the flag at the end of the antenna response.

22. Press the COPY button on the network analyzer and then select PRINT for the hard copy of the measurement.

23. Mark on each printout the feedline/antenna number and reference the cable number to the site configuration drawing acquired before the test or made on site.

24. Press SYSTEM on the network analyzer.

25. Select TRANSFORM MENU on the screen.

26. Select SPECIFY GATE on the screen.

27. Select GATE ON/OFF on the screen. "GAT" will appear on the screen after a few seconds to indicate that gating has been activated.

28. Select RETURN on the screen.

29. Select TRANSFORM OFF on the screen.

30. Press the FORMAT button on the network analyzer.

31. Select SWR on the screen.

32. Adjust the scale to be 0.1 per division.

33. This is the return loss of the antenna alone, and the SWR should be better than 1:1.2 across the band.

34. Press the COPY button on the network analyzer and then select PRINT for the hard copy of the measurement.

35. Mark on each printout the feedline/antenna number and reference the cable number to the site configuration drawing acquired before the test or made on site.

36. Press SYSTEM on the network analyzer.

37. Select TRANSFORM MENU on the screen.

38. Select SPECIFY GATE on the screen.

39. Select GATE ON/OFF on the screen. "gat" on the left-hand side of the screen should turn off, indicating that the gating has been deactivated.

40. Repeat the above procedure for each antenna (feedline).

Network Analyzer Parameters

The setup parameters of the analyzer should be printed out to verify that the unit was configured properly for the test.

1. Press COPY on the network analyzer.

2. Select SYSTEM PARAMETERS on the screen.

3. Select PRINT on the screen.

Motorola specific

Bay level offsets.   The bay level offsets for each scan should be dumped and a hard copy generated. List site number on the printout. All hard copies should be attached with a cover sheet on top stating the date, cell number, personnel present, and any other miscellaneous comments.

Trigonometric Identities

$$(\cos u)(\cos v) = \frac{\cos(u - v) + \cos(u + v)}{2}$$

$$(\sin u)(\sin v) = \frac{(\cos u - v) - (\cos u + v)}{2}$$

$$\sin u \cos v \;\;= \frac{\sin(u + v) + \sin (u - v)}{2}$$

$$\cos u \sin v \;\;= \frac{(\sin u + v) - \sin (u - v)}{2}$$

$$\sin (t + s) = \sin t \cos s + \cos s \sin t$$

$$\sin (t - s) = \sin t \cos s - \cos t \sin s$$

$$\cos (t + s) = \cos t \cos s - \sin t \sin s$$

$$\cos (t - s) = \cos t \cos s + \sin t \sin s$$

$$\sin^2(t) + \cos^2(t) = 1$$

Index

ABOUT THE AUTHOR

Clint Smith, P.E. is the Vice President of Communications Consulting Services and the coauthor of *Cellular System Design and Optimization*.